Energy in Construction and Building Materials

Energy in Construction and Building Materials

Editor

Antonio Caggiano

MDPI • Basel • Beijing • Wuhan • Barcelona • Belgrade • Manchester • Tokyo • Cluj • Tianjin

Editor
Antonio Caggiano
Università di Genova
Italy

Editorial Office
MDPI
St. Alban-Anlage 66
4052 Basel, Switzerland

This is a reprint of articles from the Special Issue published online in the open access journal *Materials* (ISSN 1996-1944) (available at: www.mdpi.com/journal/materials/special_issues/Energy_Construction).

For citation purposes, cite each article independently as indicated on the article page online and as indicated below:

LastName, A.A.; LastName, B.B.; LastName, C.C. Article Title. *Journal Name* **Year**, *Volume Number*, Page Range.

ISBN 978-3-0365-6407-4 (Hbk)
ISBN 978-3-0365-6406-7 (PDF)

Contents

Editorial

Energy in Construction and Building Materials

Antonio Caggiano

Dipartimento di Ingegneria Civile, Chimica e Ambientale—DICCA, Università degli Studi di Genova, Via Montallegro 1, 16145 Genova, Italy; antonio.caggiano@unige.it

Energy efficiency in buildings has become a major challenge in both science and industry. It is driven by the urgent need to strongly reduce the anthropogenic emissions of greenhouse gases and to cut back on the inefficient usage of the worldwide primary energy demand [1]. Building stock is, in fact, responsible for over one-third of the global energy consumption and is, additionally, responsible for nearly 40% of total direct and indirect CO_2 emissions, making it the largest European energy consumer [2].

Therefore, a major leap in energy-saving is vital to protect our environment and to boost the EU's green economy. However, the main problem we face is that we still construct our buildings with obsolete technologies and/or materials. We still believe that energy efficiency in buildings means completely insulating from all outer heat fluxes. The current trend is to deal with developing new challenging materials, and concepts, based on dynamic thermal manipulations [3,4], which can provide excellent building envelopment performance, in contrast with most classical solutions which are extremely inefficient because of being based on outdated concepts of insulation [5] and/or the R-value parameter, the latter defined as thermal resistances per unit area [6–8].

In this context, innovations in the construction sector are seeking breakthrough answers by using smart and intelligent components, materials and composites [9], energy saving concepts [10], and cost-effective solutions, in order to ultimately reach technologies with nearly zero CO_2 emissions.

The aim of this Special Issue was to explore the current state of the art, new ideas, and novel developments on the relevant topics that link energy efficiency to construction and building materials. A wide range of research outputs on various topics, which are contributing to enhanced energy efficiency and sustainable materials used for residential and non-residential buildings, was provided.

The emphasis of these works has been on collecting fundamental studies, experimental research, numerical approaches, analysis tools, and design receipts for energy-efficient materials and constructions. It has the ambition to stimulate and spread the latest knowledge on energy and construction and building materials, making the basis for new ideas on various topics for young investigators as well as leading experts in the field of Materials Science and Engineering.

The collection counts fifteen research papers and one review study. Most of the research studies covered the topic of thermal energy storage (TES) in construction and building components: i.e., in wooden façade [11], optimum placement of heating tubes [12], use of iron (III) oxide powders for modifying the mortar thermal conductivity and diffusivity [13], fiber-reinforced geopolymers for sensible TES [14], thermal insulation waste extruded polystyrene [15], highly insulated wall systems with exterior insulation of polyisocyanurate [16], thermal properties of high-strength concrete containing CBA fine aggregates [17], heat conductivity properties of hemp-lime composites [18], insulating glass units subjected to climatic loads [19], conduction mechanisms in graphene nanoplatelets (GNPs)-cement composite [20], and bio-waste thermal insulation panels [21]. The remaining articles directly disseminated research on storing solar and/or environmental latent heat to level-out daily temperature differences through the smart use of Phase Change

Citation: Caggiano, A. Energy in Construction and Building Materials. *Materials* **2023**, *16*, 504. https://doi.org/10.3390/ma16020504

Received: 23 December 2022
Accepted: 28 December 2022
Published: 4 January 2023

Material (PCM) [22–25]. They provided experimental and numerical studies on advanced PCM (latent) composites, consisting of porous cementitious materials, which have the potential to store/release large latent TES energy during phase changes, i.e., from solid to liquid and vice versa. Major research contributions address their physical, TES, and mechanical design and how to achieve stable integrated systems where PCMs are homogeneously distributed among the porous cementitious materials. Finally, a review study discussed the potential of PCMs in building wall constructions [26].

Conflicts of Interest: The authors declare no conflict of interest.

References

1. European Commission, Causes of Climate Change. Available online: https://ec.europa.eu/clima/change/causes_en (accessed on 1 March 2022).
2. IEA, Key World Energy Statistics. Available online: https://www.iea.org/reports/key-world-energy-statistics-2020 (accessed on 1 March 2022).
3. Peralta, I.; Fachinotti, V.D.; Koenders, E.A.; Caggiano, A. Computational design of a Massive Solar-Thermal Collector enhanced with Phase Change Materials. *Energy Build.* **2022**, *274*, 112437. [CrossRef]
4. Sam, M.; Caggiano, A.; Dubyey, L.; Dauvergne, J.L.; Koenders, E. Thermo-physical and mechanical investigation of cementitious composites enhanced with microencapsulated phase change materials for thermal energy storage. *Constr. Build. Mater.* **2022**, *340*, 127585. [CrossRef]
5. Gilka-Bötzow, A.; Folino, P.; Maier, A.; Koenders, E.A.; Caggiano, A. Triaxial Failure Behavior of Highly Porous Cementitious Foams Used as Heat Insulation. *Processes* **2021**, *9*, 1373. [CrossRef]
6. Saber, H.H.; Maref, W.; Hajiah, A.E. Effective R-value of enclosed reflective space for different building applications. *J. Build. Phys.* **2020**, *43*, 398–427. [CrossRef]
7. Khoukhi, M. The combined effect of heat and moisture transfer dependent thermal conductivity of polystyrene insulation material: Impact on building energy performance. *Energy Build.* **2018**, *169*, 228–235. [CrossRef]
8. Atsonios, I.A.; Mandilaras, I.D.; Kontogeorgos, D.A.; Founti, M.A. A comparative assessment of the standardized methods for the in–situ measurement of the thermal resistance of building walls. *Energy Build.* **2017**, *154*, 198–206. [CrossRef]
9. Bre, F.; Caggiano, A.; Koenders, E.A. Multiobjective Optimization of Cement-Based Panels Enhanced with Microencapsulated Phase Change Materials for Building Energy Applications. *Energies* **2022**, *15*, 5192. [CrossRef]
10. Tarpani, E.; Piselli, C.; Fabiani, C.; Pigliautile, I.; Kingma, E.J.; Pioppi, B.; Pisello, A.L. Energy Communities Implementation in the European Union: Case Studies from Pioneer and Laggard Countries. *Sustainability* **2022**, *14*, 12528. [CrossRef]
11. Almusaed, A.; Yitmen, I.; Almsaad, A.; Akiner, İ.; Akiner, M.E. Coherent investigation on a smart kinetic wooden façade based on material passport concepts and environmental profile inquiry. *Materials* **2021**, *14*, 3771. [CrossRef]
12. Ghalambaz, M.; Mohammed, H.I.; Naghizadeh, A.; Islam, M.S.; Younis, O.; Mahdi, J.M.; Chatroudi, I.S.; Talebizadehsardari, P. Optimum placement of heating tubes in a multi-tube latent heat thermal energy storage. *Materials* **2021**, *14*, 1232. [CrossRef]
13. Masdeu, F.; Carmona, C.; Horrach, G.; Muñoz, J. Effect of Iron (III) Oxide Powder on Thermal Conductivity and Diffusivity of Lime Mortar. *Materials* **2021**, *14*, 998. [CrossRef] [PubMed]
14. Frattini, D.; Occhicone, A.; Ferone, C.; Cioffi, R. Fibre-reinforced geopolymer concretes for sensible heat thermal energy storage: Simulations and environmental impact. *Materials* **2021**, *14*, 414. [CrossRef] [PubMed]
15. Li, A.; Zhang, W.; Zhang, J.; Ding, Y.; Zhou, R. Pyrolysis kinetic properties of thermal insulation waste extruded polystyrene by multiple thermal analysis methods. *Materials* **2020**, *13*, 5595. [CrossRef] [PubMed]
16. Iffa, E.; Tariku, F.; Simpson, W.Y. Highly insulated wall systems with exterior insulation of polyisocyanurate under different facer materials: Material characterization and long-term hygrothermal performance assessment. *Materials* **2020**, *13*, 3373. [CrossRef]
17. Yang, I.H.; Park, J. A study on the thermal properties of high-strength concrete containing CBA fine aggregates. *Materials* **2020**, *13*, 1493. [CrossRef]
18. Pochwała, S.; Makiola, D.; Anweiler, S.; Böhm, M. The heat conductivity properties of hemp–lime composite material used in single-family buildings. *Materials* **2020**, *13*, 1011. [CrossRef]
19. Respondek, Z. Heat transfer through insulating glass units subjected to climatic loads. *Materials* **2020**, *13*, 286. [CrossRef]
20. Goracci, G.S.; Dolado, J. Elucidation of conduction mechanism in graphene nanoplatelets (GNPs)/Cement composite using dielectric spectroscopy. *Materials* **2020**, *13*, 275. [CrossRef]
21. Pavelek, M.; Adamová, T. Bio-waste thermal insulation panel for sustainable building construction in steady and unsteady-state conditions. *Materials* **2019**, *12*, 2004. [CrossRef]
22. Guardia, C.; Barluenga, G.; Palomar, I. PCM Cement-Lime Mortars for Enhanced Energy Efficiency of Multilayered Building Enclosures under Different Climatic Conditions. *Materials* **2020**, *13*, 4043. [CrossRef]
23. Fachinotti, V.D.; Bre, F.; Mankel, C.; Koenders, E.A.; Caggiano, A. Optimization of multilayered walls for building envelopes including PCM-based composites. *Materials* **2020**, *13*, 2787. [CrossRef] [PubMed]

24. Sam, M.N.; Caggiano, A.; Mankel, C.; Koenders, E. A comparative study on the thermal energy storage performance of bio-based and paraffin-based PCMs using DSC procedures. *Materials* **2020**, *13*, 1705. [PubMed]
25. Mankel, C.; Caggiano, A.; König, A.; Schicchi, D.S.; Sam, M.N.; Koenders, E. Modelling the thermal energy storage of cementitious mortars made with PCM-recycled brick aggregates. *Materials* **2020**, *13*, 1064. [CrossRef]
26. Kurdi, A.; Almoatham, N.; Mirza, M.; Ballweg, T.; Alkahlan, B. Potential Phase Change Materials in Building Wall Construction—A Review. *Materials* **2021**, *14*, 5328. [CrossRef] [PubMed]

 materials

MDPI

Review

Potential Phase Change Materials in Building Wall Construction—A Review

Abdulaziz Kurdi [1], Nasser Almoatham [1], Mark Mirza [2], Thomas Ballweg [2] and Bandar Alkahlan [1,*]

1 The National Centre for Building and Construction Technology, King Abdulaziz City for Science and Technology, P.O. Box 6086, Riyadh 11442, Saudi Arabia; akurdi@kacst.edu.sa (A.K.); nalmoatham@kacst.edu.sa (N.A.)
2 Fraunhofer Institute for Silicate Research ISC, Neunerplatz 2, 97082 Würzburg, Germany; mark.mirza@isc.fraunhofer.de (M.M.); thomas.ballweg@isc.fraunhofer.de (T.B.)
* Correspondence: alkahlan@kacst.edu.sa

Abstract: Phase change materials (PCMs) are an effective thermal mass and their integration into the structure of a building can reduce the ongoing costs of building operation, such as daily heating/cooling. PCMs as a thermal mass can absorb and retard heat loss to the building interior, maintaining comfort in the building. Although a large number of PCMs have been reported in the literature, only a handful of them, with their respective advantages and disadvantages, are suitable for building wall construction. Based on the information available in the literature, a critical evaluation of PCMs was performed in this paper, focusing on two aspects: (i) PCMs for building wall applications and (ii) the inclusion of PCMs in building wall applications. Four different PCMs, namely paraffin wax, fatty acids, hydrated salts, and butyl stearate, were identified as being the most suitable for building wall applications and these are explained in detail in terms of their physical and thermal properties. Although there are several PCM encapsulation techniques, the direct application of PCM in concrete admixtures is the most economical method to keep costs within manageable limits. However, care should be taken to ensure that PCM does not leak or drip from the building wall.

Keywords: phase change materials (PCMs); paraffin; fatty acid; hydrate salts; butyl stearate; encapsulation

Citation: Kurdi, A.; Almoatham, N.; Mirza, M.; Ballweg, T.; Alkahlan, B. Potential Phase Change Materials in Building Wall Construction—A Review. *Materials* **2021**, *14*, 5328. https://doi.org/10.3390/ma14185328

Academic Editor: Antonio Caggiano

Received: 3 August 2021
Accepted: 10 September 2021
Published: 15 September 2021

1. Introduction

The storage of thermal energy for later use is a growing trend and is known as thermal energy storage (TES) in the literature. In TES systems, energy can be stored in a medium in the form of either "heat" or "cold" and is thus available for later use. It bridges the gap between the supply and demand of energy, mainly electricity. Ever-increasing energy consumption in various industrial sectors is causing global warming, and researchers are increasingly seeking newer ways to convert energy with a limited impact in terms of greenhouse gas (GHG) production [1]. According to the California Energy Commission, the potential deployment of TES systems could result in a reduction in NO_x levels of 560 tons and a reduction in CO_2 emissions of 260,000 tons from the building sector across the state [2]. If we consider all the associated costs in the building industry, such as the ongoing costs of heating and cooling, the costs of fossil fuel electricity generation, and the associated greenhouse gas emissions (CO_2, SO_2, NO_x, CFCs), TES, along with other conventional and unconventional energy storage systems, is very promising for the sustainable development of the building industry [3]. Research on the efficient use of solar thermal energy has been established worldwide, because it is one of the cleanest energy sources and is abundant in nature. However, limiting factors include its absence at night and on cloudy days, posing a major challenge for a continuous energy supply. One answer to this challenge is to develop a medium that is sensitive enough to store thermal energy in the absence of sunlight. In this regard, phase change materials (PCMs) have found new applications in the construction industry, especially in green buildings. Green buildings are referred to as energy-efficient buildings, using much less traditional energy (e.g., electricity) while maintaining the same level of comfort. In short, PCMs tend to be characterized by a phase transition that occurs at a more or less constant temperature. Here, the phase transition is not limited to melting, but can also be evaporative, for example. It is interesting to note that PCMs are neither new nor exotic materials, as is often claimed in the literature [4]. PCMs have low density and form a viscous/semi-viscous mass when melted (compared to free-flowing water), thus avoiding the problem of the use of heavy materials in building design and construction.

Several studies on the development of PCMs, as well as their incorporation into the building envelope, have been reported in the literature. Nevertheless, it is difficult to obtain a good understanding because most of the reported work is not scientifically linked to previous work and is disorganized. There are also a number of reviews in the literature, but they are disjointed. The main reason for this is that researchers have tried to combine all the available information into one paper, ignoring the complexity and vastness of the topic. Therefore, it is necessary to address each aspect of PCMs separately. With respect to phase change materials, two different aspects are considered in this review. They are:

- Phase change materials themselves;
- How to use them in building construction.

When incorporating PCMs into building envelopes, each building envelope requires PCMs of a specific nature, shape and size. Considering this, the aim of this article is to provide a comprehensive overview of PCMs applied in the building envelope. Therefore, they will be referred to as "smart walls" in the following. In this study, all the possible PCMs available for this purpose have been critically analyzed, and the related information, such as the development of PCMs, their physical and thermal properties, and their integration into the building envelope, has been presented. In this way, industry partners and researchers can benefit from this comprehensive overview and overcome the associated limitations and drawbacks to meet the future challenge of sustainable development. In addition, the associated challenges in the use of PCMs, in terms of materials and methodology, are also discussed. The new feature of this study is to assess the advantageous usage of PCMs in the Middle East region in general and Saudi Arabia in particular, in terms of lower energy consumption, energy efficiency, and low-cost home insulation systems. The current study will enable engineers to select the particular PCMs for a given application, as well as helping researchers to carry out more innovative research work in this area.

2. Impact of PCMs on Building Construction

As mentioned earlier, latent heat storage provides a higher density of energy storage than sensible heat storage. In addition, the temperature variation is small for the former. The scenario can be modeled on that of a thermal switch. Once the switch is turned on, i.e., the PCMs reach their melting point, the temperature is kept constant until the material is completely melted. This melting process allows a large amount of heat to be absorbed while keeping the system temperature constant at the melting temperature of the respective PCMs [5]. Since PCMs store heat or cold for later use, they can thus regulate or dampen temperature fluctuations within a building, automatically leading to energy savings. In other words, PCMs provide a virtual building mass. If the room temperature exceeds the melting temperature of the PCM, it melts and absorbs heat. Later, when the outside temperature rises, the interior of the room will not easily reach a high temperature. In colder weather, the PCMs release the heat. A brief justification for the potential use of PCMs is shown in Figure 1, in which the thickness of the thermal mass is compared to traditional building materials such as gypsum, wood, concrete, sandstone and brick. Thus, the use of PCMs not only reduces the thermal mass, but also reduces the footprint of the building and provides significantly more usable space inside the buildings.

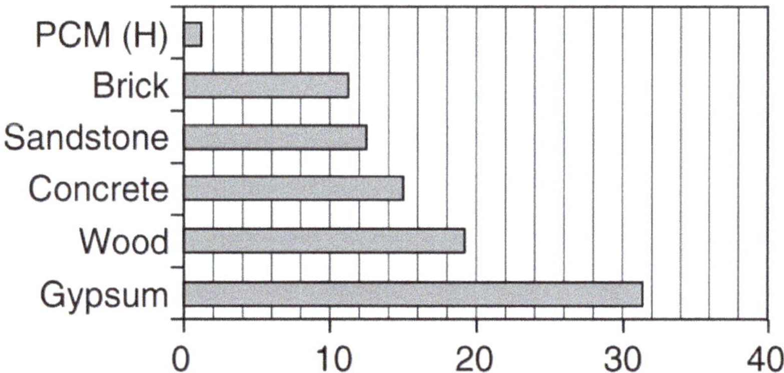

Figure 1. The thickness of PCMs required, compared to conventional thermal masses (e.g., gypsum, wood, concrete, sandstone and brick), for building envelope applications. This figure was redrawn based on [3].

As recently reported by Lagou et al. [4] in their numerical and simulation work, in addition to the PCM itself, the optimal positioning of the PCM in a given building envelope also plays a significant role. According to the authors [4–6], the optimal location to incorporate PCMs is the interior edge of the building and also depends on the geographical location. In addition, the energy payback time for PCM incorporated building elements is less than 7 days/m^2. Such promising data on energy payback time creates a motivation to further incorporate PCMs into building wall construction. The following section will present the classification and general properties of PCMs for building wall applications.

3. Classification and Properties of PCMs and Their Application in Building Walls

PCM is not a single material, but a group of materials that have a number of specific properties. Broadly speaking, PCMs are divided into the following categories:

1. Organic;
2. Inorganic; and
3. Eutectic, as explained in subsequent sections.

3.1. Organic PCMs

Several factors should be considered when developing PCMs. However, the most important of these are their cost, thermal conductivity, latent heat content and freezing/melting range. Examples of potential organic PCMs are waxes, oils, fatty acids and polyglycols. These types of organic PCMs are in the form of a long molecular chain with a carbon backbone. The melting point of such a material is determined by the length of the carbon molecular backbone. Generally, the longer the chain, the higher the melting point. Since these materials originated as single-chain molecules, this makes their melting point more specific, along with an increase in their latent heat content. Examples include pure linear hydrocarbons, which require complex processing and are therefore expensive. On the other hand, naturally occurring fats such as vegetable oils and animal fats are economically attractive compared to synthetic fats. However, these naturally occurring fats consist of a wide range of molecules with different chain lengths. Consequently, instead of a sharp melting/freezing point, these materials exhibit a temperature range where complete melting/freezing occurs. Since PCMs must operate in a narrow temperature band for efficient TES operation, these naturally occurring fatty materials are not suitable for this purpose. In addition, due to their highly oxidizing behavior over time, these types of fats are fire hazards and can only be used with the use of flame-retardants.

On the positive side, organic PCMs do not suffer from subcooling problems and thus avoid phase separation over time. In addition, they are chemically stable, non-corrosive and non-toxic, but have the disadvantage of relatively low thermal conductivity and potential flammability. Organic PCMs can be further classified as

- Paraffin and
- Non-paraffin types.

3.2. Inorganic PCMs

Compared to organic PCMs, inorganic PCMs are non-reactive (fire-resistant) and have higher latent heat content and high thermal conductivity. Their disadvantages are their corrosive nature and the fact that they suffer from supercooling and phase separation over time [7]. As reported by Mehling and Cabeza [8], additional measures are required to counteract these drawbacks, through the introduction of additives, additional nucleating agents, the dispersion of highly thermally conductive material such as fused perlite or metal dusts, and a general microencapsulation technique [9]. Inorganic PCMs are mostly water-based hydrated salts with a freezing/melting temperature above 0 °C [10]. It is common practice to use a mixture of salts to fine-tune the phase change temperature, i.e., the freezing/melting temperature. It is also possible to use the same salt in different concentrations (in water). Usually, the water chemically combines with the salts to form a crystalline structure commonly known as a "hydrated salt". Interestingly, some hydrated salts can contain up to 50% water, although their physical appearance is in solid crystal form and usually has a distinct color, depending on their water content. When hydrated salts are heated, the water portion of them is separated and the salt dissolves in this water. During this process, the system absorbs the heat flux in the form of latent heat. Exactly the opposite takes place during freezing, as heat escapes from the system. The practical application of this theory produces PCMs that can melt/freeze at up to 117 °C. In other words, these PCMs are frozen at temperatures higher than the boiling point of water. Congruent hydrated salts are transparent when melted and the chemical composition of the melt phase is the same as that of the solid phase before melting. Others are semi-congruent, i.e., they form a different hydrated salt with a lower melting point upon melting. Recently, Guo et al. [5] has proposed inorganic VO_2 as a "smart" phase change material for building applications. This material changes its phase upon external thermal excitation at around 67 °C, which is useful in external walls. This material, used as a coating, is highly manufacturable and scalable with high-throughput.

3.3. Eutectic PCMs

Eutectic PCMs are a mixture of the above-mentioned organic and/or inorganic PCMs and can be sub-divided into different groups. The groups include:

- Organic–organic,
- Inorganic–inorganic and
- Inorganic–organic.

Eutectic PCMs are congruent in nature and occur in crystal forms [11]. In its simplest form, the theory behind eutectic PCMs is as follows. When salts of any kind are added to an aqueous medium, the freezing point of the water is lowered. This is the reason why gritting materials are spread on icy/snowy roads in the winter to keep the roads open. As salts are added to the solution, the freezing temperature drops further. At a certain composition, the mixture takes the form of a slush. However, under certain circumstances, such as a mixture of certain salts at a certain concentration, it melts and freezes translucent at a certain temperature. During the melting/freezing process, the system releases or stores heat in the form of latent heat. The composition of the mixture at this point is called the eutectic composition and this particular temperature is called the eutectic temperature. A brief comparison of PCMs in terms of their classification and common properties is given in Table 1.

Table 1. Classification of PCMs, together with their common properties [10,12–23].

PCM Properties	PCM Type		
	Organic	Inorganic	Eutectics
Examples of PCMs	Paraffin, non-paraffin compounds, fatty acids, acetamide, butyl stearate	$CaCl_2.6H_2O$, $Na_2SO_4.10H_2O$, $Na_2CO_3.10H_2O$	Octadecane + heneicosane, Octadecane + docosane, 34% $C_{14}H_{28}O_2$ + 66% $C_{10}H_{20}O_2$
Melting temperature (°C)	19–32	29–36	25–27
Thermal conductivity (w/m.k)	0.15–0.21	0.54–1.09	-
Heat of fusion (J/kg K)	140–236	105–192	136–203
Density (kg/m^3)	756–815	1600–1800	-

In this regard, the most studied PCMs are hydrated salts, paraffin/non-paraffin waxes, animal/plant-based fatty acids, and eutectic PCMs combining any of the above materials. The details and properties of these materials are described below.

3.3.1. Paraffin Waxes

Paraffin wax (PAR) is a by-product of petroleum refining and can be used as a value-added material as a PCM. It is a mixture of several linear alkyl hydrocarbons. The melting point of PAR is comparable to that of salt hydrate, with reasonable latent heat and without the problem of supercooling associated with salt hydrate. A major disadvantage of PAR is its high flammability; therefore, its use is only recommended in combination with fire-retardant fillers. The chemical formula of paraffin wax is C_nH_{2n+2}, that is, an aliphatic chain (e.g., C_3H_8, C_5H_{12}, C_7H_{16}) as shown in Figure 2 [24]. Commercial paraffin waxes are mixtures of different waxes with a wide range of melting temperatures and are usually cheap, with acceptable heat storage densities in the range of ~200 kJ/kg or 150 MJ/m^3. However, the main disadvantage is that they have a low thermal conductivity coefficient (~0.2 W/m °C). To overcome this, the addition of fillers is necessary.

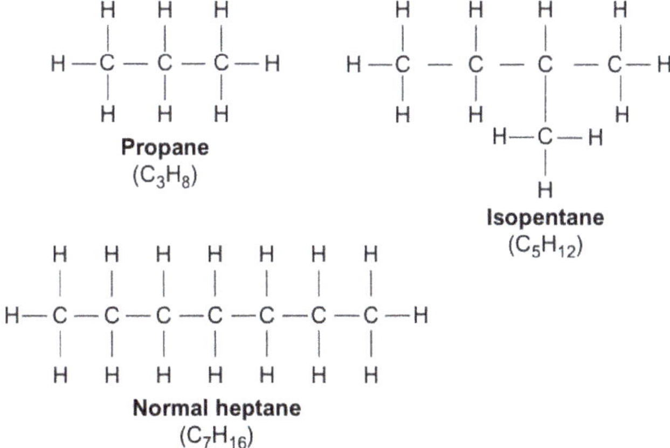

Figure 2. Examples of paraffins. This figure was redrawn based on [24].

The fillers not only increase the thermal conductivity but can also retain a larger volume of PCMs therein. Common fillers include metal dust/particles, molten dolomite/perlite or metal inserts such as finned tubes and aluminum chips [13]. Commercial paraffin, as opposed to pure paraffin, is most commonly used in PCMs due to its high cost, with a melting temperature of about 55 °C [25,26]. Farid et al. [27] studied commercial paraffins with melting points of 44, 53 and 64 °C, which have latent heat densities of 167, 200 and 210 kJ/kg, respectively. The first result is promising in terms of their ability to maintain the temperature level in the comfort zone. Detailed technical information on this can be found in the literature by Himran et al. [21], Faith [28] and Hasnain [13]. A list of the thermo-physical properties of the most common paraffin waxes used in the construction of building walls is given in Table 2.

Table 2. Thermo-physical properties of different paraffin waxes.

PCMs	T_m (°C)	K (W/m.K)	H (kJ/kg)	ρ (kg/m^3)	Ref.
Paraffin RT-18	15.0–19.0	0.20	134.0	0.756	[29]
Paraffin RT-27	28.0	0.20	179.0	0.800	[30]
n-Octadecane	28.0	0.20	179.0	0.750	[31]
	28.0	0.15	200.0	-	[32]
n-Heptadecane	19.0	0.21	240.0	760.0	[32]
	22.0	-	244.0	780.0	[33]
Hexadecane	18.0	0.17–0.26	236.0	780.0	[33]
Polyethylene glycol	21.0–25.0	-	148.0	1128.0	[34]
Paraffin C_{13}–C_{24}	22.0–24.0	0.21	189.0	0.760	[32,35]
n-Octadecane + n-Heneicosane	26.0	-	173.9	-	[36]
Paraffin C_{16}–C_{18}	20.0–22.0	-	152.0	-	[11,35,37]
Paraffin C17	21.70	-	213.0	0.817	[38]
Paraffin C_{18}	28.0	0.15	244.0	0.774	[11,35–38]

Legend: H: heat of fusion. k: thermal conductivity. Tm: melting temperature. ρ: density.

3.3.2. Fatty Acids

Fatty acids are attractive candidates for TES systems such as PCMs for latent heat energy storage in space heating applications. As reported by Feldman et al. [39], the physical and thermal properties of fatty acids (capric, lauric, palmitic and stearic acids) and their binary mixtures meet the requirements for use as PCMs. The melting point of the fatty acid group PCMs ranges from 30 °C to 65 °C and the latent heat content varies from 153 to 182 kJ/kg. In a parametric study of palmitic acid PCMs by Hasan et al. [40], the behavior of such PCMs including the phase transition, solid/liquid interphase in the mixture, freezing/melting temperature and the heat flux rate in a TES system consisting of circular tubes was described in detail. Among the various fatty acids, capric and lauric fatty acids are mainly used in low-temperature TES systems, as reported by Dimaano et al. [41], with a freezing/melting point of about 14 °C. Depending on the composition, the latent heat content is about 113–133 kJ/kg. As mentioned earlier, although a number of materials in this category have been studied at the laboratory scale, only a handful of them have been explored to their full potential, so there are vast opportunities for further research in this area. Examples of such laboratory PCMs include dimethyl sulfoxide, with a melting temperature of 16.5 °C and a latent heat content of 86 kJ/kg [16]; maleic anhydride, with a melting temperature of 52 °C and a latent heat content of 145 kJ/kg [42], etc. Like PAR, fatty acid also has a long chain of molecules, as shown in Figure 3. Depending on the location of the double bonds in fatty acids, they can be further classified as:

- Saturated and
- Unsaturated fatty acids.

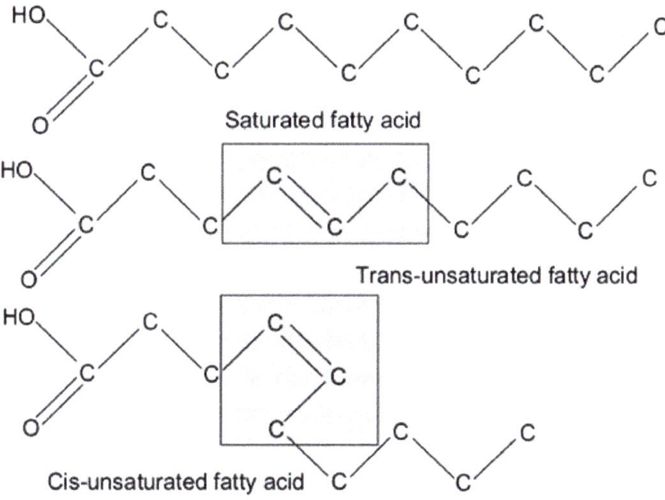

Figure 3. Molecular structure of different types of fatty acids. This figure was redrawn based on [43].

A list of common fatty acid PCMs with their respective thermo-physical properties is given in Table 3.

Table 3. Thermo-physical properties of different fatty acids.

PCMs	T_m(°C)	K (W/m.K)	H (kJ/kg)	ρ (kg/m^3)	Ref.
Lactic acid	26.0	-	184.0	-	[38]
Capric acid	30.2	0.20	142.7	-	[44]
	30.0–30.2	0.2	143.0	815.0	[44,45]
Stearic acid	52.0	0.29	169.0	965.0	[46]
Palmitic acid	62.4	-	183.2	-	[47]
Myristic acid	52.7	-	198.4	-	[47]
Capric acid + Lauric acid	21.0	-	143.0	-	[11]
	19.0	-	132.0	-	[36]
	20.39	-	144.2	-	[48]
Capric acid + Stearic acid	26.8	-	160.0	-	[36]
Capric acid + Palmitic acid	22.1	-	153.0	-	[36]
	26.2	2.20	177.0	784.0	[49]
Myristic acid + Capric acid	21.4	-	152.0	-	[36]
Capric acid + 1-dodecanol	26.5	-	-	-	[44]
	27.0	-	126.9	817.0	[44]
	26.5	0.2	126.9	817.0	[50]
Methyl Palmitate + Methyl Stearate	23.0–26.5	-	180.0	817.0	[51]
Dodecanoic acid	42.5	0.148	182.0	873.0	[52]
Glycerin	18.0	-	199.0	-	[29]
Capric acid (75.2%) + Palmitic acid (24.8%)	22.1	-	153.0	-	[53]

Table 3. *Cont.*

PCMs	T_m(°C)	K (W/m.K)	H (kJ/kg)	ρ (kg/m^3)	Ref.
Capric acid (75%) + Palmitic acid (25%)	17.7–22.8	-	189.0–191.0	-	[54]
Capric acid (86.6%) + Stearic acid (13.4%)	26.8	-	160.0	-	[53]
Capric acid (61.5%) + Lauric acid (38.5%)	19.1	-	132.0	-	[53]
Lauric acid (55.8%) + Myristic acid (32.8%) + Stearic acid (11.4%)	29.29	-	28.38	-	[55]
Expanded Graphite (Lauric acid + Myristic acid + Stearic acid)	29.05	-	137.0	-	[55]
Capric acid + Palmitic acid + Stearic acid	19.93	-	129.5	-	[56]
Myristic acid + Palmitic acid + Stearic acid	41.72	-	163.5	-	[56]
Expanded Graphite (Myristic acid + Palmitic acid + Stearic acid)	41.64	-	153.5	-	[56]
Expanded Perlite (Capric acid + myristic acid)	21.7	-	85.4	-	[57]
Activated Carbon (Lauric acid)	44.1	-	65.14	-	[58]
Expanded Graphite (Stearic acid)	53.12	-	155.5	-	[59]
Diatomite (Capric acid + Lauric acid)	16.7	-	66.8	-	[60]
Activated Montmorillonite (Stearic acid)	59.9	-	84.4	-	[61]
Expanded Graphite (Palmitic acid)	60.9	-	148.4	-	[62]

Legend: H: heat of fusion. k: thermal conductivity. Tm: melting temperature. ρ: density.

3.3.3. Hydrated Salts

Hydrated salts are attractive materials for use as PCMs because they have high thermal energy storage density (~350 MJ/m^3), along with a high thermal conductivity coefficient (~0.5 W/m °C) and a modest cost. Hydrated salts are mainly crystalline salts that contain a certain amount of water in crystalline form and therefore have different colors depending on the water content. For example, Glauber's salt (Na_2SO_4.H2O) contains 44% Na_2SO_4, whereas the water content is higher than the salt content at 56%, although they appear in a solid form at room temperature and pressure. Glauber's salt was one of the first studied PCMs, reported in the 1950s [12,17], with a melting temperature of about 32.4 °C and a latent heat content of 254 kJ/kg (377 MJ/m^3). The physical states of some common hydrated salts at room temperature are discussed below. Despite their high thermal conductivity, problems related to supercooling and phase separation are a major challenge for their wide application. To deal with this problem, the "extra water principle" is the most commonly used solution to prevent the formation of heavy anhydrous salts, as mentioned by Biswas et al. [63]. However, the use of extra water extends the melting/freezing temperature range, which negatively affects the thermal energy storage density of the material. Some researchers have proposed the use of bentonite clay as a thickening agent, but this introduces another challenge, namely, the reduction of the crystallization rate, as well as the heat transfer rate, due to the low thermal conductivity coefficient of the system. Borax can also be used as a nucleating agent to prevent supercooling [12]. In general, hydrated salts suffer from the problems of supercooling, nucleation and phase segregation and therefore require the use of encapsulation, along with thickening and nucleating agents. A list of hydrated salts with their respective thermos-physical properties suitable for wall construction can be found in Table 4.

Table 4. Thermo-physical properties of different hydrated salts.

PCMs	T_m (°C)	K (W/m.K)	H (kJ/kg)	ρ (kg/m^3)	Ref.
Hydrated salts	29.0	1.0	175.0	1490.0	[64]
	31.4	-	150.0	-	[65]
	25.0–34.0	-	140.0	-	[66]
	26.0	0.60	180.0	1380	[30]
Eutectic salt	32.0	-	216.0	-	[67]
Sodium Sulfate Decahydrate	32.50	0.60	180.0	1600	[68]
$FeBr_3 \cdot 6H_2O$	21.0	-	105.0	-	[38]
$Mn(NO_3) \cdot 6H_2O$	25.5	-	126.0	1738	[11,35,38]
	25.8	-	125.9	1728	[46]
$Mn(NO_3)_2 \cdot 6H_2O + MnC_{12} \cdot 4H_2O$	27.0	0.60	125.9	1700	[69]
$CaC_{12} \cdot 6H_2O$	29.0	0.54	187.49	560	[70]
	29.9	0.53	187.0	1710	[71]
Sodium thiosulfate pentahydrate	40.0–48.0	-	210.0	-	[72]
Sodium acetate trihydrate	58.0	1.10	264.0	1280	[73]
$Na_2SO_4 \cdot 10H_2O$-$Na_2CO_3 \cdot 10H_2O$	32.34	-	-	-	[74]
Hexahydrate ($CaCl_2.6H_2O$)	30.0	1.10	170.0	1560.0	[75]
Decahydrate ($Na_2SO_4.10H_2O$)	37.7	-	-	131.7	[76]
KF_4H_2O	18.50	-	231.0	1447.0	[77]
$Na_2SO_4 \cdot 10H_2O$	21.0	0.55	198.0	1480.0	[78]
Calcium chloride	29.8	0.56	191.0	1710	[79]

Legend: H: heat of fusion. k: thermal conductivity. Tm: melting temperature. ρ: density.

3.3.4. Butyl Stearate (BS)

Butyl stearate (BS) is another organic PCM that has been described in the literature as one of the best-known PCMs. It has a melting point of 17 °C–22 °C [80] and the linear chemical formula is $CH_3(CH_2)16COO(CH_2)3CH_3$. BS can be used both through the microencapsulation technique and directly as an admixture in concrete. Liang et al. [81] reported the microencapsulation of butyl stearate in poly-urea microcapsules using the interfacial poly-condensation method with a particle diameter of about 20–35 μm with toluene 2,4-diisocyanate (TDI) and ethylene diamine (EDA) as backbone monomers. This microencapsulated butyl stearate shows a melting temperature of about 29 °C with a latent heat of fusion of about 80 J/g [81]. Examples of such microencapsulated BS in the form of optical images are shown in Figure 4.

Microencapsulated butyl stearate can be used as an additive for interior or exterior coatings or can be included in insulating materials such as foams. Melamine formaldehyde resin is widely used in industry. It has high mechanical strength and desirable thermal resistance properties. The thermal decomposition temperature of melamine formaldehyde is above 300 °C, which is suitable for use as a wall-building material [82]. A number of different polymers and monomers can be used to encapsulate PCMs, namely, amine resins, poly-urea, polyurethane, polyester, polyamide, etc. These polymers form a shell-like container, which then contains PCMs inside it, and the process is referred to as poly-condensation in the literature [81]. For example, Liang et al. [81] used poly-urea as the shell wall material. The particle diameter of the microcapsules in their study was about 20–35 μm [82]. In another work by Hua et al. [83], butyl stearate was microencapsulated over poly-urea polymer. They prepared microcapsules with a size of 4.5–10.2 μm [82]. Bendic and Amza fabricated poly-methyl meta-acrylate (PMMA) microcapsules with butyl stearate as the core with an average diameter of 95 µm [21]. Oliver et al. [84] reported on the role of pH in the composition and thermal stability of melamine microcapsules with butyl stearate on their fabricated microcapsules with a diameter of 20–50 µm [22]. On the other hand, Cellat et al. [80] reported the direct incorporation of BS as an admixture in concrete to improve the thermal performance of building walls. A list of butyl stearate PCMs with their respective thermo-physical properties suitable for wall construction is given in Table 5.

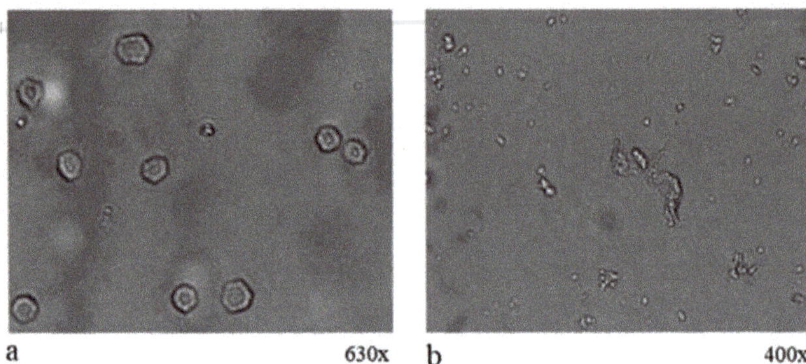

Figure 4. Optical micrographs exhibiting BS micro-encapsulation in polymers with magnification ×630 (**a**), and magnification ×400 (**b**) [81].

Table 5. Thermo-physical properties of different butyl stearate PCMs.

PCMs	T_m (°C)	K (W/m.K)	H (kJ/kg)	ρ (kg/m^3)	Ref.
Butyl stearate	16.0–20.8	0.21	700.0	900	[75]
	18.0	-	30.0	-	[75]
	19.0	-	140.0	760	[53]
	18.0–23.0	0.21	123.0–200.0	-	[53]
BS/MMT	25.30	-	41.81	-	[85]
Butyl Stearate & Butyl Palmitate (49/48)	17.0–20.0	-	137.8	-	[86]
Butyl Stearate & Butyl Palmitate (50/48)	15.0–25.0	-	101.0	-	[87]
$CH_3(CH_2)_{16}COO(CH_2)_3CH_3$	19.0	-	140.0	-	[88]
Butyl stearate (50%) and Butyl palmitate (48%)	16.0–21.0	-	-	-	[52,89]
Butyl stearate (48%) and Butyl palmitate (49%)	17.0–19.3	-	-	-	[90]

Legend: H: heat of fusion. k: thermal conductivity. Tm: melting temperature. ρ: density.

As can be seen from above-mentioned data and properties of PCMs, there are a number of PCMs that can be a candidate for building wall applications. However, each PCM is unique in nature and comes with respective benefits and challenges, as described in the next section.

4. Challenges Associated with the Use of PCMs

For the successful application of PCMs, some of their inherent limitations should be considered and properly addressed. The main limitations arise in terms of the thermal conductivity coefficient, changes in heat storage density and the maintenance of the initial system's efficiency over time, as well as phase segregation and subcooling, as described below.

4.1. Phase Segregation and Supercooling

As mentioned earlier, when hydrated salt PCMs melt, supercooling and phase segregation occur before the next freezing cycle begins. Supercooling in PCMs is defined as a metastable state in which the PCMs remain in a liquid state (phase) even though the temperature is below their respective melting point. This supercooling and phase segregation suppress the efficiency of the PCMs over time and, in the worst case, can prevent the PCMs from solidifying at all. The initial heat storage density deteriorates over time and thus the efficiency decreases at an alarming rate. This is because hydrated salts are a congruent melting material, so melting is accompanied by reduced hydrate formation. This process is irreversible and since lower salt hydrates have a lower thermal energy storage capacity, the efficiency of the system decreases. Subcooling is also blamed for the decrease in efficiency of the hydrated salts. One of the solutions to this is the use of direct contact with the conducting environment, such as an immiscible heat transfer fluid [14,16,18,91,92]. The presence of additional heat transfer fluid causes agitation to occur in the system, which not only minimizes subcooling but also prevents possible phase segregation. Some researchers have used the 'extra water principle,' as described briefly earlier (Section 3.3), to avoid segregation and subsequent plugging. When developing additives/stabilizers to stimulate nucleation, the physical and chemical properties of the salts in question should be considered [19]. Ryu et al. [18] reported extensive research on the development and use of suitable stabilizing/nucleating agents that can be used in a range of PCM systems.

4.2. Stability of PCMs over Time

One of the hurdles to the widespread use of PCMs is their ability to maintain their physical, thermal and chemical properties over the time of use, i.e., over the thermal cycles to which they are exposed. Here, the effect of corrosion by PCMs in the system is also relevant, especially when they are macro encapsulated in a container [10]. PCM containers, both microencapsulation and macro encapsulation, must have sufficient physical and thermal stability as PCMs are subjected to repeated heating/cooling cycles. Kimura et al. [93] reported the use of NaCl in $CaCl_2.6H_2O$ with additional water content, which can withstand up to 1000 heating/cooling cycles. On the other hand, as reported by Gibbs et al. [20], paraffin-based PCMs usually do not suffer because both thermal cycling and the use of the container do not affect their thermal properties. Unfortunately, no information on the corrosion of paraffin is available in the literature, and there are few reports on the corrosion of PCMs, and these lack details about this subject [8,94]. Despite all the above facts, hydrated salts are denser than the paraffin and thus the effective heat capacity per unit volume is high. As a rule of thumb, a 10% volume change in thermal cycling can be considered a minor problem [22]. Hawlader et al. [95] reported on the thermal stability of microencapsulated paraffin and confirmed its stability up to 1000 cycles, as shown in Figure 5.

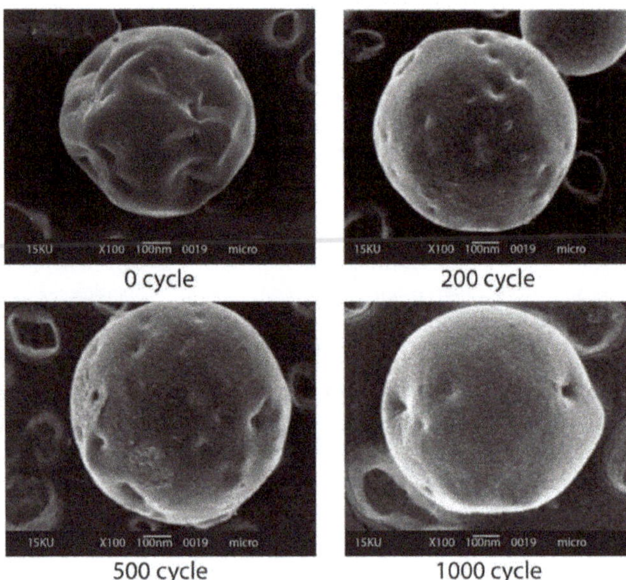

Figure 5. SEM images of paraffin particles after micro-encapsulation and subjected to thermal cycles [74].

It is important to note that there are few studies in the literature that address the thermal stability of PCMs over multiple cycles, and most of these papers are related to accelerated laboratory-scale thermal tests. Accelerated laboratory-scale tests do not always necessarily reflect the real-world scenarios of practical applications. A lack of detailed results on the thermal stability behavior of various PCMs is clearly visible in the literature, and thus further attention by researchers is required.

Cellat et al. [80] reported the thermal stability of microencapsulated BS PCMs subjected to a total of 1000 melting/freezing cycles via differential scanning calorimetry (DSC), as shown in Figure 6 [80]. After 800 cycles, the melting temperature and latent heat storage capacity changed from 21.0 °C to 20.8 °C and from 135.5 J/g to 105.1 J/g, respectively [80]. Although there was no significant change in melting temperature, the change in latent heat (22%) was significant, leading to the need for further investigation, as the current literature is not available.

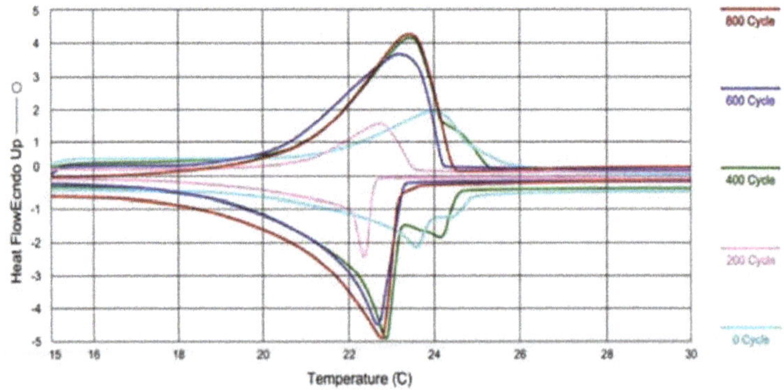

Figure 6. Comparison of the DSC curves of micro-encapsulated BS PCMs subjected to 0, 200, 400, 600 and 800 thermal cycles [75].

4.3. Thermal Conductivity

In general, most PCMs suffer from a low thermal conductivity coefficient. To improve this, measures to increase their heat transfer capability are required to increase their thermal conductivity [96,97]. To increase the surface area of PCMs, which in turn increases the thermal conductivity of the TES system, a common practice is to impregnate a porous matrix with PCMs and then form a composite. These porous materials can be diatomaceous earth, silica, perlite, etc., as reported in the literature [98–100]. An example is shown in Figure 7, which demonstrates the inclusion of PCMs in the diatomaceous earth structure. The diatomaceous earth obtained (Figure 7a) contains impurities and therefore needs to be calcined in order to get rid of these impurities and open the pores of the structure (Figure 7b). After calcination, the organic impurities in the diatomaceous earth have been volatilized and thus the specific surface area has increased. After mixing with paraffin PCMs, the diatomaceous earth absorbs the paraffin in its pores, becomes coated with paraffin (Figure 7c,d), and looks like spheres from the outside.

Figure 7. SEM micrographs of diatomite/paraffin composite: (**a**) diatomite in as-received condition, (**b**) diatomite after the calcination process, (**c**) composite structure of paraffin/diatomite and (**d**) higher magnification view of single composite sphere [76].

Thus, the composite structure is similar to the core-shell structure, where the core is diatomaceous earth, and the shell is paraffin. The size of the composite spheres varies (5–20 µm) depending on the size of the diatomaceous earth, soaking time, temperature, the density of the PCM used and other related process parameters. In general, the composite exhibits a uniform structure due to the homogeneous absorption of kerosene into the diatomaceous earth and has sufficient mechanical strength for handling and standing in the application [100]. So far, different kinds of PCMs and their associated properties has been discussed. However, identifying an optimum PCM does not necessarily mean that their application in building walls will offer the best performance, as the integration of such PCMs is a challenge itself. The next two sections (Sections 5 and 6) summarize various ways of integrating PCMs into building wall applications.

5. PCM Integration into the Building Envelop

Although PCMs can be integrated into various building envelopes, as reported in the literature [19,101], in this paper we will focus mainly on the integration of PCMs into building walls. In buildings with multiple floors, the roof space becomes tight due to the space required for solar panels and air conditioning units. Therefore, to make the best use of the available space, building walls are the second-best place to install PCMs and are referred to here as "smart walls". During the phase transition, PCMs melt into viscous/semi-viscous forms. Therefore, to avoid leakage, they must be properly enclosed in protective containers in a process called encapsulation. To achieve the best possible performance, PCMs must be bonded to the inner surface of the walls whenever possible. This thermally couples the PCMs to prevent the loss of thermal conductivity. Thus, the traditional technique of maintaining an air gap between the inner and outer layers of the walls is not required. Lane et al. [102,103] identified over 200 potential PCMs, covering a wide operating temperature range (10 °C to 90 °C) which can be used for encapsulation. Two types of encapsulation are widely used in commercial applications, namely:

- Micro-encapsulation and
- Macro-encapsulation, as discussed below in detail.

5.1. Micro-Encapsulation

Microencapsulation of PCMs refers to the incorporation of PCMs in microscopic shells, where the shell consists of polymers/monomers and one or more PCMs in colloidal form constitutes the core substance [81]. Depending on the application, the shell can be polymeric or inorganic in nature. The microencapsulation of hydrated salt PCMs (e.g., $CaCl_2.6H_2O$) in a polyester-resin micro-container has been promising, and the application of these PCMs to building walls has been reported in the literature [104]. The final product can take the form of extruded films, thanks to the solvent exchange technique. This process can achieve microencapsulated PCMs with about 40% PCM retention. These microencapsulated PCMs films show good mechanical and thermal constancy under cyclic freeze-melting conditions. As reported by Royon et al. [104], PCMs with water phase (i.e., hydrated salts) can be contained in polyacrylamide polymers. This polymer forms a three-dimensional network of macroscopic polymer chains and acts like a net to trap PCMs within it. The gap between the chains is small enough to prevent the leakage of PCMs due to absorption, as shown schematically in Figure 8.

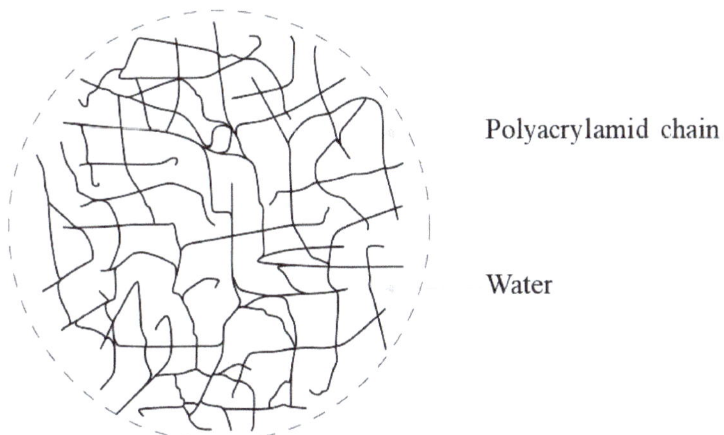

Figure 8. Micro-encapsulation of hydrated salt PCMs in polyacrylamide film [104].

Paraffin wax can also be encapsulated by forming an inorganic shell around it, e.g., a shell of calcium carbonate (CaCO3), using the self-assembly method shown schematically in Figure 9. First, oily paraffin is diffused in an aqueous solution, together with nonionic surfactants, to form a stable emulsion. Then, the chains of surfactants, which are hydrophobic by nature, are oriented to the oil droplets. At the same time, the hydroxyl groups of the surfactants, which are hydrophilic by nature, combine with the water molecules and stay away from the oil droplets. The layer of surfactants covers the surfaces of the oil droplets, forming kerosene micelles. Subsequently, Ca^{2+} ions are bound by the hydroxyl groups of the surfactants when droplets of $CaCl_2$ are added to the same solution. This process is called the complexation process. Finally, $CaCO_3$ is formed because of a precipitation reaction by introducing Na_2CO_3 into the same emulsion system. This $CaCO_3$ is not soluble in the system and acts as a shell encapsulating the oily phase (PCMs) of the emulsion. Since this $CaCO_3$ forms directly on the surface of the paraffin micelles, the process is referred to in the literature as the self-assembly process.

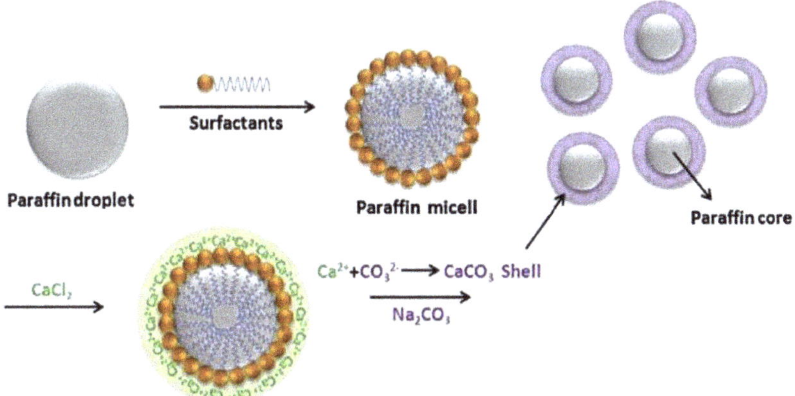

Figure 9. Schematic of paraffin micro-encapsulation in $CaCO_3$ shell [78].

Details of the structure and appearance of microencapsulated paraffin in the initial state are shown in Figure 10 [105]. As can be seen from Figure 10, the morphology of the encapsulated PCMs varies from spherical with some coagulation to shell-like shapes and the diameter is about 1–5 μm. Besides the different paraffin loading, the processing temperature, the presence of additives and the presence of surfactants play a crucial role. As can be seen in the TEM images (Figure 10), a representative core-shell formation of the microcapsules with a shell thickness of about 0.14–0.5 μm is confirmed [105].

Figure 10. Morphology of micro-encapsulated paraffin in $CaCO_3$ shell: SEM images (**a**,**c**,**e**) and TEM images (**b**,**d**,**f**) [78].

As with the encapsulation of paraffin in a $CaCO_3$ shell using the self-assembly process mentioned above, paraffin or other PCMs can also be encapsulated by a polymer shell. As an example, the process of microencapsulation of n-octadecane PCMs with poly-urea shells is shown in Figure 11, and the corresponding SEM images of the microencapsulated PCMs are presented in Figure 12.

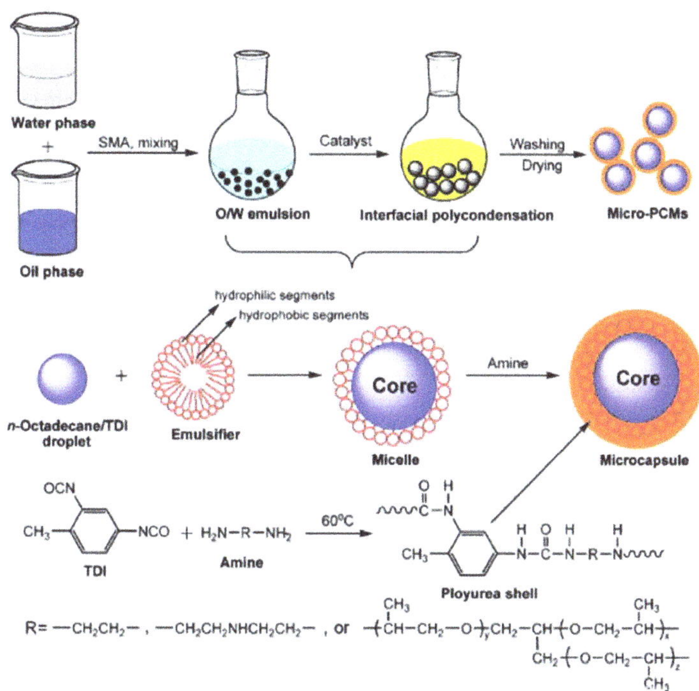

Figure 11. Schematic of microencapsulation process of *n*-octadecane with poly-urea shells [106].

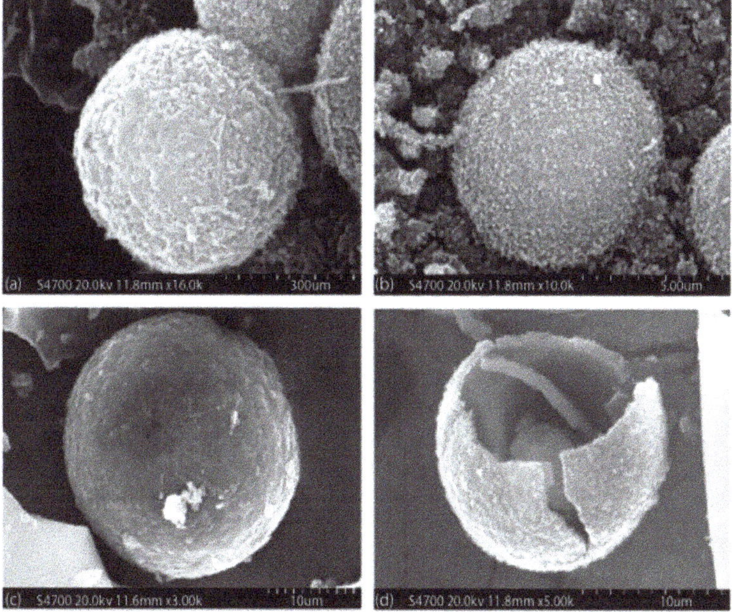

Figure 12. SEM images of micro-encapsulated PCMs using different monomers: (**a**) ethylene diamine (EDA), (**b**) toluene-2,4-diisocyanate (TDI), (**c**) Jeffammine and (**d**) a cracked microcapsule [106].

All this evidence confirms the possibility of various microencapsulation processes, which provide easy and versatile methods for the use of PCMs. The advantages of the microencapsulation of PCMs in various media include increased surface area, resulting in improved heat transfer surface, prevention of the contact of PCMs with the environment, which helps prevent fouling and oxidation of PCMs, and overall maintenance of the volume of the storage materials, which ensures the maximum achievable efficiency of the system over the lifetime of the component [81]. Different kinds of novel and efficient micro-encapsulation techniques are still under investigation and are being reported by researchers. For example, Ballweg et al. [80] has reported ultra-violet (UV)-based micro-encapsulation techniques for hydrated salts and paraffin waxes, which represent a relatively quick and efficient process. However, further investigation on the long-term stability of such micro-encapsulated PCMs are yet to be conducted.

5.2. Macro-Encapsulation

In contrast to microencapsulation, in macro-encapsulation, the PCMs are encapsulated in lifed-size containers that come in different shapes and materials, as shown in Figure 13.

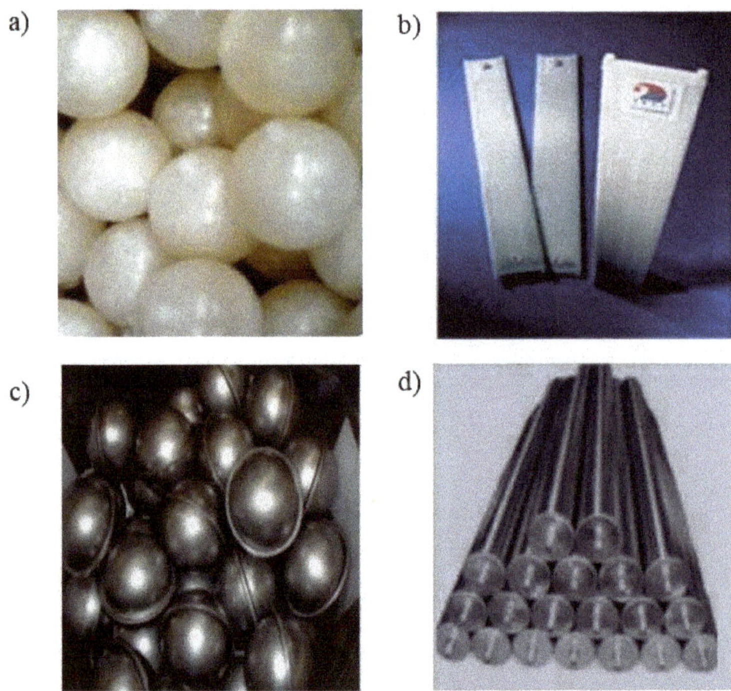

Figure 13. Commercially manufactured macro-capsules to store PCMs: (**a**) Polyolefin spherical balls, (**b**) Polypropylene flat panel, (**c**) stainless ball capsules, and (**d**) modules beam [36].

The most widely used commercial microencapsulation techniques include Ø 2–3 mm spherical capsules, flat plates, metallic (stainless steel) spherical capsules and cylindrical bars filled with PCMs (Figure 13d). PCMs can also be encapsulated in bags made of conductive metallic materials such as thick aluminum foils/plates, as shown in Figure 14.

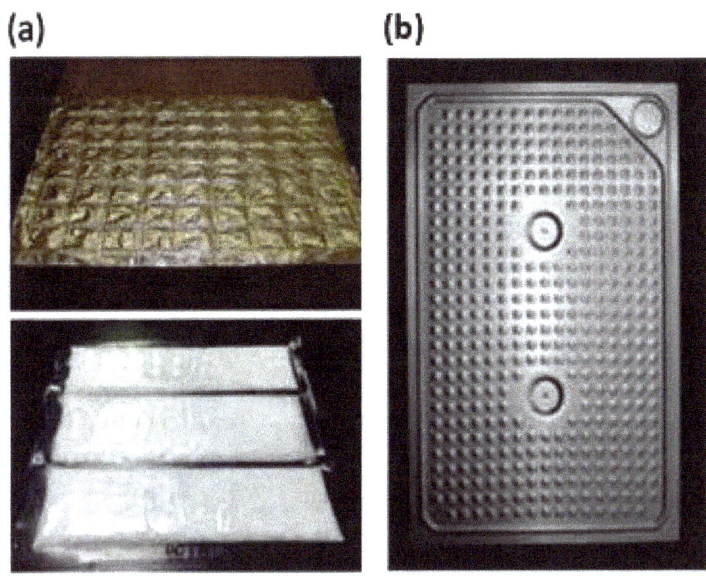

Figure 14. Macro-encapsulation of PCMs in (**a**) pouches and (**b**) flat panels [107].

A list of PCMs that are suitable for micro-/macro-encapsulation, with their respective thermo-physical properties appropriate for wall construction, is given in Table 6.

Table 6. Thermo-physical properties of micro-/macro-encapsulations suitable for wall construction applications available in the literature.

PCMs Core	Shell Materials	Encapsulation Efficiency (%)	Size of Particles (μm)	Maximum Latent Heat (J/g)	Melting Temperature (°C)	Ref.
N–octadecane	Polyurethane	93.4–94.7	5.0–10.0	110.4	28.0	[108]
	Poly(methyl methacrylate-co-methacrylic acid) copolymer	12.0–21.0	1.60–1.68	93.0	29.0–32.9	[109]
	poly(n-butyl methacrylate) & poly(n-butyl acrylate)	47.7–55.6	2.0–5.0	112.0	29.10	[110]
	Melamine Formaldehyde co-polymer	-	34.0	183.2	28.14	[111]
	Silk fibroin	22.6–46.7	4.0–5.0	-	24.99	[112]
	SiO_2/PMMA	19.9–66.4	5.0–15.0	-	21.5–26.3	[113]
	TiO_2/PMMA	26.8–82.8	3.0–16.0	100.0	28.0–31.0	[114]
	Resorcinol-modified melamine	44.0–69.0	5.0–20.0	146.5	26.5–28.4	[115]
	SiO_2/TiC(PMMA)	78.0	-	-	17.2–19.4	[116]
	Poly(MPS-VTMS)	58.7–76.0	-	166.7	17.4–18.2	[117]
	Octadecylamine-grafted	<88.0	-	202.5	27.4–27.5	[118]
	SiO_2	33.6	8.0	210.0	23.3–28.4	[119]
	Calcium Carbonate ($CaCO_3$)	22.4–40.4	5.0	-	28.1–29.2	[120]
	TiO_2	74.3–81.0	2.0–5.0	42.6	25.6–26.1	[112]

Table 6. *Cont.*

PCMs Core	Shell Materials	Encapsulation Efficiency (%)	Size of Particles (μm)	Maximum Latent Heat (J/g)	Melting Temperature (°C)	Ref.
N-nonadecane	Poly(methyl methacrylate)	60.30	0.1–35.0	139.2	31.2	[121]
	$CaCO_3$	40.04	5.0	84.40	29.2	[120]
N-heptadecane	Poly(styrene)	63.3	1.0–20.0	136.9	21.5	[121]
	Starch	49.0–78.3	30.0–175.0	187.3	23.1–24.2	[122]
N-eicosane	Polysiloxane	-	5.0–22.0	240.0	35.0–39.0	[123]
	Crystalline TiO_2	49.9–77.8	1.5–2.0	97.60–195.6	41.5–43.88	[124]
N-octadecane, N-eicosane, and N-nonadecane	Melamine-Urea-Formaldehyde	-	0.30–6.40	165	36.40	[125]
N-octadecane (paraffin wax)	Melamine formalde hyderesin	92.0	2.0–5.0	214.6	28.41	[115]
Paraffin (MPCM24D)	melamine-formaldehyde polymer	-	10.0–30.0	154.0	21.9	[125]
Paraffin wax	Polystyrene	75.60	-	153.50	-	[126]
	Amphiphilic polymer (PE-EVA-PCM)	-	-	98.1	28.4	[127]
	Hydrophobic polymer (St-DVB-PCM)	-	-	96.1	24.2	[127]
Paraffin eutectic	poly(methyl methacrylate)	50.20–65.40	0.01–100	276.41	36.17	[128]
Butyl stearate	Polyurethane	-	10.0–35.0	81.20	22.30	[129]
Paraffin	Melamine-formaldehyde	80.0	5.8–339.0	147.9	21.0–24.0	[130]
	Melamine-formaldehyde	80	10.4–55.2	147.9	22.5	[131]
	Poly-methyl-methacrylate	-	50.0–300.0	100.0	23.0	[66]
	Ethylvinylacetate and polyethylene	-	3.0–10.0	100.0	27.0	[132]
	polymethylmethacrylate	-	6	135.0	23.0	[133]
Graphite-modified MPCM	Polycarboxylate		3.0–3000.0			[133]

6. PCMs for 'Smart Wall' Applications

As mentioned above, PCMs generally have a low coefficient of thermal conductivity and therefore require the use of an additional heat transfer medium, such as metal inserts, etc. This is a major disadvantage in their application and metal inserts are expensive. These disadvantages can be eliminated through the direct application of PCMs in wall surfaces. Thus, the use of wall panels containing PCMs in the building wall can provide smooth temperature variations. The large surface area of the wall allows for a higher thermal conductivity rate with the wall/room [22]. Another popular application is the use of wall panels that contain PCMs within themselves. Wallboards are widely available and economical, and the integration of PCMs into them is very promising. Kedl et al. [134] and Salyer et al. [135] presented the concept of a wallboard impregnated with paraffin wax by simply dipping the wallboard in paraffin wax. This process of PCM integration into the wallboard by simply dipping it in PCMs involves minimal cost, and can be scaled up to any size depending on the size of the wallboard [136]. Neeper et al. [137,138] studied the dynamics of the thermal behavior of gypsum wallboard impregnated with fatty acids and paraffin PCMs. According to these authors, the thermal storage capacity of the PCM-filled wallboards is sufficient to retain the thermal energy for the TES system. Stetiu et al. [139] used a computer-aided simulation technique to investigate the thermal performance of PCM-impregnated wall panels using finite element analysis (FEA) and reported a number of parameters that should be considered, such as the wetting of the PCMs on the wallboards, etc. Zhang et al. [106] investigated fatty acid impregnation and reported direct energy savings of 5–20% [53]. Athienitis et al. [75] reported the results obtained from both experimental and simulation approaches for a life-sized structure made of PCM-impregnated wallboards. The tests were conducted outdoors with rooms made of PCM gypsum boards containing about 25-wt% butyl stearate (BS) as the inner lining of the wall. According to their simulation work, which was supported by experimental results, the use of PCM in gypsum wallboard can reduce the room temperature by 4 °C during the day compared to pure gypsum board. For the direct applications of PCMs in building walls mentioned above, PCMs can be compressed into a sheet form and the surface wrapped and sealed with foil material to prevent leakage. One such commercially researched application is shown in Figure 15, developed by Energain®, which has been achieved by the Dupont de Nemours Society, UK [140].

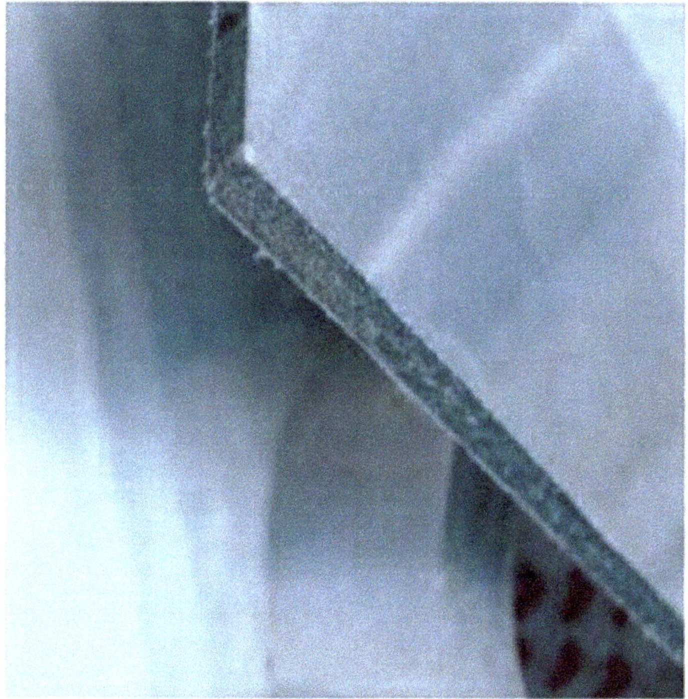

Figure 15. Energain® PCM panel wrapped and sealed in foil tape [141].

Microencapsulated PCMs (e.g., paraffin) can be mixed directly with gypsum and used in the form of a panel, as shown in Figure 16 from Micronal® produced by the BASF (Ludwigshafen, Germany). In this way, the wallboard can be used as a normal wallboard, which is much more energy-efficient than one without PCMs in it [140].

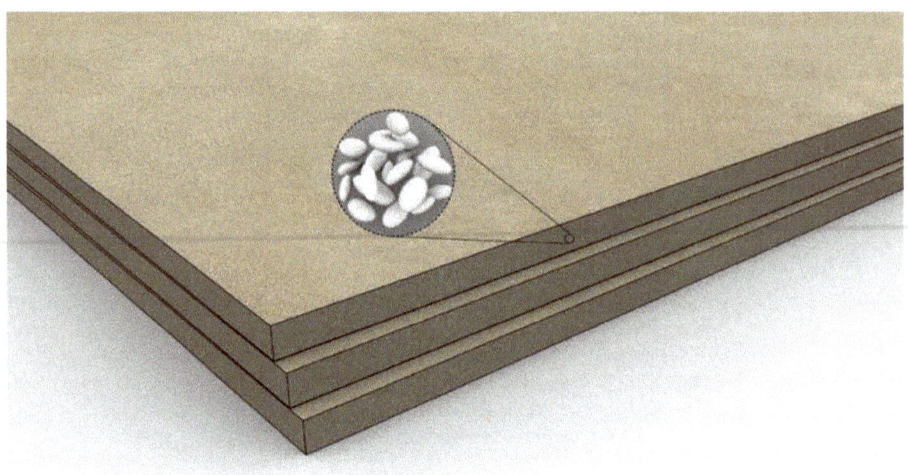

Figure 16. Micronal® PCM-integrated gypsum wallboard. This figure was redrawn based on [140].

Since bricks are one of the most commonly used construction materials to form building walls, incorporating PCMs into bricks is an attractive way to use them with builders and end users. PCMs can be mixed with cement paste or used as a stand-alone PCM paste to fill the cavities of hollow bricks/blocks, as well as the interior plaster of the wall. Figure 17 shows different types of hollow bricks and the construction of a building wall with such bricks, together with the use of PCM paste to bond them. As recently reported by Gao et al. [110], PCM-filled hollow bricks improve the thermal behavior of building walls significantly. According to the authors [110], PCM-filled hollow bricks can reduce the attenuation rate from 13.07% to 0.92–1.93%, together with increasing the delay time from 3.83 h to 8.83–9.83 h. In other words, hollow bricks filled with PCM can reduce the peak heat flux from 45.26 W/m^2 to 19.19–21.4 W/m^2. In addition, inner cavities were the better choice for PCM and there was an extra phase-change extent of close to 90% in favor of the different outdoor thermal environment.

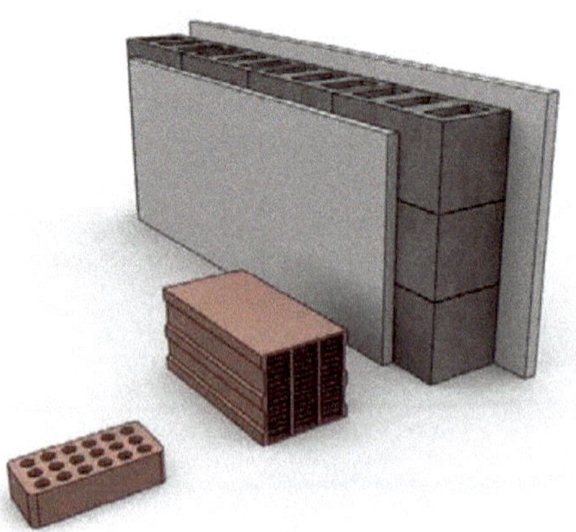

Figure 17. Hollow brick, alveolar brick and wall structure by using PCMs.

An example of the use of microencapsulated PCMs in the interior of plasterboard is shown in Figure 18.

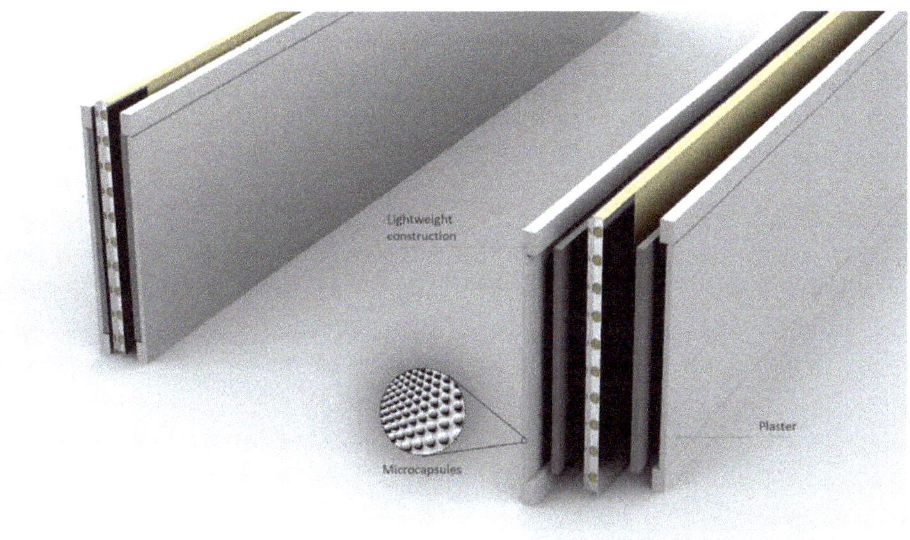

Figure 18. PCM usage as interior plaster on hollow building blocks.

A comparison of common wall-building materials in terms of their respective heat storage capacities is shown in Figure 19. The heat storage capacity of PCMs far exceeds that of all other building materials and is therefore promising as an efficient building material in the present and future.

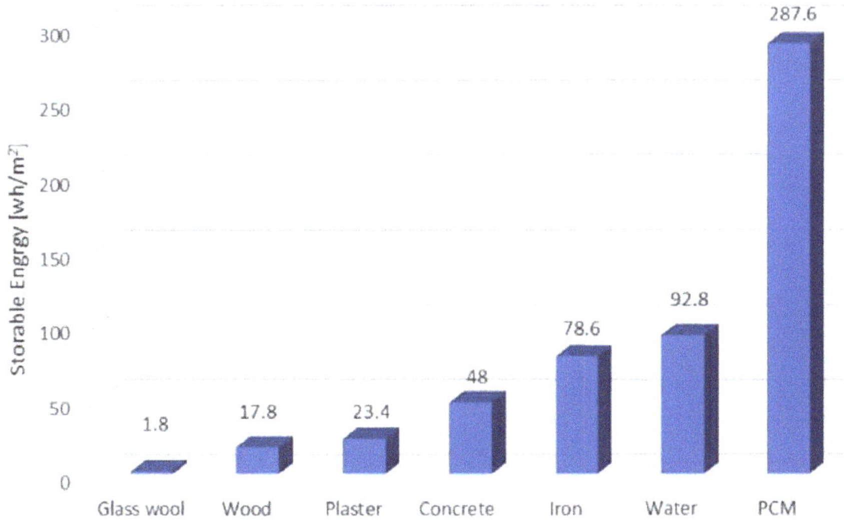

Figure 19. Heat storage capacity of different building materials between 18 °C and 26 °C for a period of 24 h. This figure was redrawn based on [111].

6.1. Case Studies

According to Isaac et al. [142], based on their calculations using computer simulations, GHG emissions for the residential sector only will increase from 0.8 Gt C in 2000 to about 1 Gt C in 2020 and then more than double (2.2 Gt) by 2100 [103]. Much of this can be limited through the use of PCMs, which reduce the reliance on building heating/cooling, which in turn reduces fossil fuel-related electricity generation. Looking at the consumption of electrical energy, there is a staggering from 27 EJ (exajoule) of electricity consumption in 2000 to about 80 EJ (exajoule) in 2100 [103]. According to the International Institute of Refrigeration (IIR), electricity consumption by refrigeration and air conditioning systems in buildings accounts for about 15% of total electricity consumption [142–144]. Due to global warming and climate change, as well as increasing demands in developing countries, more and more buildings are being equipped with conventional air conditioning systems to provide the necessary comfort. In this sense, the widespread application of PCMs in the building sector could lead to savings, even if it can only cover a certain percentage of energy consumption. According to Jeon et al. [6], this could account for up to 55% of the total electricity consumption of buildings in cold climates such as in Korea [104], and the use of PCMs has great potential to reduce dependence on the electricity grid. As reported by Li et al. [14], heating, ventilation and air conditioning (HVAC) systems account for about 48% of building-related energy consumption in the United States [15], and a large portion of this can be managed by using PCMs in building wall construction. A summary of various experimental studies on the use of PCMs as wall construction materials in the last five years, as reported in the literature, is shown in Tables 7–10.

Table 7. Summary of different experimental studies in 2020 from the literature on wall applications.

PCMs	Application in Building Walls	Improvements	Ref.
Paraffin wax	Two-wall models were tested; walls were insulated with wood to achieve one-dimensional heat transfer through the walls.	The PCM layer provides a more uniform temperature rise near the heat source of the building wall, reducing the heat flow through the downstream parts	[106]
OM37 PCM	The experimental setup consists of a PCM-free reference concrete cubicle and a PCM-macro-encapsulated experimental concrete cubicle. The cubicles were both installed in an open area and are directly exposed to solar energy	All four walls of the cubicle show more or less the same temperature profile over 24 h. The test walls of the cubicle interior surface temperature remain below the daytime temperature compared to the reference cubicle walls	[107]
OM35 and Eicosane	Brick	For the double PCM layer within the brick, there is a heat loss of approx. 9.5 °C and for a single-layer PCM brick a temperature reduction of 6 °C. The reduction of heat gain for double-layer PCM bricks is observed up to 60% during the day, and about 40% for single-layer PCM bricks. However, this is not the case when the heat is denied during the night by PCMs. The use of small particles within the PCM encapsulation has a detrimental effect as the heat transfer increases sharply during the day. In these bricks, the internal brick temperature increases more than normal bricks during the night and the PCM was not properly dissipated. Although heat dissipation during the night is ensured by some secondary means other than using the fin configuration, it may not be a viable choice to increase the PCM thickness for cooling a building space	[108]

Table 8. Summary of different experimental studies in 2019 from the literature on wall applications.

PCMs	Application in Building Walls	Improvements	Ref.
Paraffin wax	Blocks made from foamed cement with paraffin	(1) Additional paraffin can make the foamed cement block more effective in storing thermal energy. (2) The method of preparation and absorption is a physical process. During the process shifts in the thermal storage of molded cement, the thermal storage properties and the chemical properties of the paraffin are obtained. (3) The thermal conductivity of the pure cement block is similar to that of the foamed cement block with 20 percent to 25 percent composite PCM. Thirty percent is the optimum mass ratio and displays the best thermal inertness for foamed cement blocks with composite PCM. They effectively slow down the temperature rise and minimize temperature fluctuations, but higher temperatures occur after process changes.	[113]
Paraffin wax (SSPCM)	SSPCM boards with EPS board	(1) It was found that the PCM wallboard keeps more heat out of the room air during the day and releases it at night through the heat flux of the wall. (2) In summer, the room temperature in the PCM area was 1.9 °C below the maximum temperature and 0.6 °C below the average temperature in the reference room. (3) The inside environment temperature in the PCM room in the summer was 1.9 °C below the maximum temperature and 0.6 °C below the reference testing room average temperature. (4) In winter, the indoor air temperature in the PCM space was 1.3 °C lower at altitude and 0.1 °C higher on average than in the comparison room. (5) It was found that the PCM wallboard stores more heat from the indoor air during the day and releases it from the heat flux of the wall surface at night, resulting in less heat being released to the interior during the day. In winter, the case is reversed and it was found that the PCM wallboard stores more heat from the outside air during the day and releases it at night, resulting in more heat being transferred to the inside of the wall at night	[109]
Fatty acids	G/C board consisting of inorganic materials as well as gypsum and cement.	They verified that the heat storage G/C board applied with PCM was effective in reducing the energy inside buildings.	[110]
PH-31 Paraffin wax	Gypsum mortar	The thermal conductivity of conventional gypsum wallboard increases gradually with an increase in temperature. The integration of the micro-PCM into the gypsum would minimize the thermal conductivity on the MPCM-containing wallboard by reducing its density and lowering its conductivity below the density of the gypsum wallboard. The composite wall panel has a much higher apparent specific heat capacity than a gypsum wallboard (2.71 times in the 26 °C to 32 °C temperature range).	[116]
The microcapsules with a paraffin core and a melamine-formaldehyde polymer shell	Geo-polymer concrete walls	The annual energy savings from using walls with 15 cm GPC-MPCM (5.2 wt.%), a 5-cm PCM layer, and 5 cm insulation was approximately 28% compared to the comparison by maintaining the indoor temperature of 19 °C–21 °C. The PCM layer performed better when positioned closer to the outdoor climate.	[139]

Table 9. Summary of different experimental studies in 2018 from the literature on wall applications.

PCMs	Application in Building Walls	Improvements	Ref.
Paraffin wax	Brick wall	(1) PCMs in structures have been used as thermal insulating products and increased thermal comfort. (2) PCM decreases the temperature of the indoor environment and lowers the cooling charge.	[117]
Calcium chloride hexahydrate ($CaCl_2 \cdot 6H_2O$)	Concrete walls contain glass windows with different positions (inside, center, outside)	The use of the PCM in the walls greatly reduced the heat transfer rate and the average internal surface temperature during the working hours. Liquid fraction analysis was performed to determine the effective fraction of the total PCM that could be used to build the PCM layer in the most efficient manner. The heat reduction ratio was used as a measure of the performance of the PCM substrate, varying the shape, thickness, and PCM locations	[118]
(PCM) used is a mixture of ethylene-based polymer (40%) and paraffin wax PCM (60%).	Aluminum sheet	A considerable reduction of thermal losses around the walls. These losses are decreased by 50 percent on average. When placed on a ceiling the PCM is more effective.	[119]
	Wallboard	PCM wall panel has been called low convection heat wallboard, low relative thermal conductivity or low heat ratio wallboard.	[122]
BASF Micronal® DS 5038X	Precast concrete	Not only because of the direct effect of replacing the heavier mixture fraction with the lighter one, but also because the compressed air increases with the implementation of increasing PCM volumes, a statistically important reduction in density is produced by the concentrations of microencapsulated PCM.	[123]

Table 10. Summary of different experimental studies in 2017 from the literature on wall applications.

PCMs	Application in Building Walls	Improvements	Ref.
Paraffin wax	Cement mortar	Only because of the small and fixed-phase temperature range, the composite PCM wall can work efficiently. The advantages of the PCM insert should be optimized by selecting a suitable melting point under appropriate test conditions.	[124]
Paraffin wax	Brick holes	The following was observed: the n-octadecane PCM decreases the cooling rate for the first day in the evaluation by more than 50% compared to the other PCM blocks, with n-eicosane causing a reduction of about 40% and the P116 causing a reduction of about 30%.	[126]

6.2. Cost-Benefit Analysis of the Use of PCM

The design of buildings, whether commercial or residential, depends on a number of factors, and in the end, it is the consumer's decision whether to choose PCMs over conventional insulation or a combination of both together with an efficient HVAC system. This decision also depends on local conditions and specific requirements. In Australia, for example, the cost of integrating PCMs is about AUD 55–110 per square meter, compared with about AUD 10/square meter of conventional insulation. If the building size is large enough, the cost of installing PCMs can break the budget. A number of commercial suppliers (e.g., BioPCM) offer readymade PCM products, in the form of sealed pouches/pallets. The weight of these pouches/pallets varies from 2.7 kg/square meter with a transition temperature in the range of 18 °C to 28 °C from solid to molten state. Typical recommended consumption by suppliers is 1–2 kg of PCM per cubic meter of room volume with a heat storage capacity in the range of 1.1 MJ/m^2 (0.3 kWh/m^2) [62]. The other potential benefits of using PCM as building materials are as follows:

1. The use of PCMs may be advantageous in terms of comparable pricing to very expensive double/triple glazing.
2. The use of PCMs has helped improve the energy rating of buildings from 7.8 to 9 stars through the use of PCMs together with conventional insulation [62].
3. Since PCMs operate as passive heating/cooling materials, no power consumption is required compared to active heating/cooling (e.g., air conditioning). Thus, in the event of a power outage (both natural and system-related), the comfort level in the rooms can be maintained within the desired temperature range, at least for a certain number of hours. This makes it possible to keep the demand for electrical energy low while the repair of the system is carried out.

The use of PCMs also helps to reclaim more floor space, as the footprint is smaller (in the range of 50 mm) than conventional masonry, which can be up to 150 mm high [13]. PCMs are one of the available options, along with conventional insulation, efficient glazing, etc. [62]. A list of commercially available PCMs from different manufacturers with their respective thermal properties is given in Table 11.

Table 11. Thermal properties of commercial PCMs used in different wall components, as discussed in the literature.

PCMs	Location	T_m (°C)	K (W mK^{-1})	H (kJ kg^{-1})	ρ (kg m^{-3})	Type of Wall	Method	PCM Supplier	Ref.
RT 10	Guimarães, Portuga	10.0	-	150.0	880.0	Plaster mortar for exterior wall	Experimental and numerical	Rubitherm GmbH	[136]
RT 18	Shanghai, China	17.0–19.0	0.20	225.0	770.0	Wallboard	Experimental and numerical	Rubitherm GmbH	[136]
	Various Portuguese cities	15.0–19.0	0.20	134.0	0.756	Masonry brick	Experimental study	Rubitherm GmbH	[29]
RT 20	-	18.0–22.0	0.20	172.0	810.0	Gypsum wallboards	Computer simulation	Rubitherm GmbH	[136]
	China	23.2	-	134.1	-	Gypsum wallboards	Gypsum wallboards	Rubitherm GmbH	[85]
RT 20/MMT	China	24.2	-	53.60	-	Gypsum wallboards	Gypsum wallboards	Rubitherm GmbH	[85]
RT 21	Puigverd de Lleida, Spain	21.0	0.20	134.0	0.770	Prefabricated Slab Concrete	Experimental and numerical	Rubitherm GmbH	[133]
	Puigverd de Lleida, Spain	21.0–22.0	-	134.0	-	Prefabricated Slab Concrete	Experimental study	Rubitherm GmbH	[136]
	Sydney, Australia	21.0	0.20	-	880.0	Trombe walls	Experimental study	Rubitherm GmbH	[133]
RT 25	Raebareli Uttar Pradesh & Bhopal, India	26.6	0.18	232.0	749.0	Building bricks	Numerical Study	Rubitherm GmbH	[144]
RT 27	Spain	27.0	0.12	100.0	-	Geopolymer concrete & Cement concrete	Experimental study	Rubitherm GmbH	[136]
	Lawrence, Kansas, USA	27.0	0.20	179.0	760.0	Gypsum wallboard	Experimental and numerical	Rubitherm GmbH	[144]
	Sydney, Australia	25.0–28.0	0.20	-	880.0	Trombe walls	Experimental study	Rubitherm GmbH	[136]
RT 31	Sydney, Australia	27.0–31.0	0.20	-	880.0	Trombe walls	Experimental study	Rubitherm GmbH	[133]
RT 42	-	38.0–43.0	0.20	174.0	760.0	Solar chimney	Experimental study	Rubitherm GmbH	[144]
	Sydney, Australia	38.0–43.0	0.20	-	880.0	Trombe walls	Experimental study	Rubitherm GmbH	[136]

Table 11. *Cont.*

PCMs	Location	T_m (°C)	K (W mK^{-1})	H (kJ kg^{-1})	ρ (kg m^{-3})	Type of Wall	Method	PCM Supplier	Ref.
RT 22 HC	Ljubljana, Slovenia	21.0–22.0	0.18	134.0	677.0	Wallboard	Experimental and numerical	Rubitherm GmbH	[133]
RT 28 HC	Coimbra, Portugal	28.0	0.20	245.0	756.0	Trombe wall	Experimental study	Rubitherm GmbH	[144]
	Coimbra, Portugal	27.55	0.20	258.1	-	Trombe wall	Experimental study	Rubitherm GmbH	[133]
GR 35	Erzurum, Turkey	13.0–41.0	-	41.0	-	Trombe wall	Experimental study	Rubitherm GmbH	[144]
GR 41	Erzurum, Turkey	13.0–51.0	-	55.0	-	Trombe wall	Experimental study	Rubitherm GmbH	[144]
SP29	Shanghai, China	28.0–30.0	0.60	190.0	1520.0	Wallboard	Experimental and numerical	Rubitherm GmbH	[136]
PEG-600	France	21.0–25.0	-	148.0	1128.0	PVC panel	Experimental and numerical	-	[34]
	Slovenia	18.0–27.0	0.20	330.0	-	Wallboard	Simulation	Rubitherm GmbH	[144]
Micronal® PCM	Puigverd de Lleida, Spain	26.0	-	110.0	-	Concrete walls	Experimental study	BASF	[144]
ENERGAIN	Lyon, France	23.5	0.22	107.5	900	Wallboard	Experimental study	Dupont de Nemours	[85]
Methyl Palmitate (Emery 2216)	Canada	24.0–28.0	-	192.0	-	Wallboard	Experimental study	Henkel	[51]
Methyl Stearate (Emery 2218)	Canada	33.0–36.0	-	196.0	-	Wallboard	Experimental study	Henkel	[51]
MC-24	Guimarães, Portuga	24.0	-	162.40	-	Plaster mortar for exterior wall	Experimental and numerical	DEVAN (MC series)	[136]
MC-28	Guimarães, Portuga	28.0	-	170.0	350	Plaster mortar for exterior wall	Experimental and numerical	DEVAN (MC series)	[136]
BSF26	Guimarães, Portuga	26.0	-	110.0	-	Plaster mortar for exterior wall	Experimental and numerical	BASF	[136]
GH 20	Shanghai, China	25.0–25.4	0.82	33.25	1150.0	Shape-stabilized mortar bricks	Experimental study	-	[133]
Energain®	Coimbra, Portugal	18.0–26.0	855	70.0	2500.0	Drywalls	Experimental and numerical	DuPont™	[133]
	Bayern, Germany	-	0.18–0.23	-	855.0	Insulation in lightweight walls	Experimental and numerical		[144]
Q25-BioPCM™	Melbourne Australia	28.2	0.20	242.0	235.0	Internal and external building walls	Experimental and simulation	Phase Change Energy Solutions, Inc.	[133]
Micronal® DS (5001-X)	Coimbra, Portugal	25.67	0.15	111.3	-	Trombe wall	Experimental study	BASF	[144]
	Spain	26.0	-	179.0	-	Gypsum and Portland cement	Experimental study	BASF	[133]
PEG-E600	Madurai, India	25.0–31.0	-	180.0	1126	Inside hollow-brick façades	Experimental study	BASF	[35, 36]
L-30	Würzburg, Germany	30.0	1.02	270.0	-	Building walls	Experimental study	Rubitherm GmbH	[133]
S-27	Würzburg, Germany	27.0	0.48	190.0	-	Building walls	Experimental study	Rubitherm GmbH	[133]
PCM-HDPE	USA	16.6–26.5	-	116.7	505.3	Insulation in wall cavities	Experimental and numerical	PCM Products Ltd.	[63]
GKB®	Athens, Greece	16.0–26.0	0.27	-	787.0	Plasterboards	Experimental study	Knauf	[133]
Micronal-T23®	Southern Italy	19.0–25.5	0.18–0.22	-	545.0	Wallboard	Experimental study	BASF	[136]
Micronal®	Chile	25.0	0.23	-	800	Gypsum board	Simulation study	Knauf	[144]
PT-20	Auckland, New Zealand	20.0	-	180.0	-	Gypsum board	Experimental study	PureTem	[133]

Table 11. *Cont.*

PCMs	Location	T_m (°C)	K (W mK^{-1})	H (kJ kg^{-1})	ρ (kg m^{-3})	Type of Wall	Method	PCM Supplier	Ref.
Emerest 2325	Montreal, Canada	17.0–20.0	-	137.8	-	wallboard	Experimental study	Henkel	[86]
Emerest 2326	Montreal, Canada	15.0–25.0	-	101.0	-	Autoclaved block	Experimental study	Henkel	[87]
MF/PCM24	Oslo, Norway	21.8	0.02–0.10	154.0	-	Insulation material & Geopolymer concrete	Experimental study	Microtek	[139]
Micronal DS-5008X	Lisbon & Porto, Portugal	23.0	0.30	100.0–110.0	300.0	Plastering Mortar	Numerical study	BASF	[88]
Micronal DS-5040X	Australia	23.0	-	100.0	250.0–350.0	Cement-based materials	Experimental study	BASF	[133]
Micronal DS-5008X	Portugal	23.0	-	135.0		Cement, lime, and gypsum mortars	Experimental study	BASF	[133]
Micronal™ Thermal-CORE	United States	23.0	0.20	24.2	800.0	Drywall	Experimental study	ThermalCore Inc	[133]
M-51	United States	23.0	0.15	230.0	860.0	Plastic foil inside wallboard	Experimental study	Bio-PCM	[133]
InfiniteR™	United States	23.0	0.54	200.0	1810.0	PE foil bags inside wallboard	Experimental study	Infinite Business Solutions	[133]
M182Q25	Canada	25.0	0.15–2.5	210.0–250.0	-	Building walls and concrete slab	Experimental study	Bio-PCM	[144]
M51Q25	Canada	25.0	0.15–2.5	210.0–250.0	-	Building walls and concrete slab	Experimental study	Bio-PCM	[136]
PureTemp-20	Los Angeles, USA	10.0–28.0	0.21	100.0–400.0	860.0	Concrete composites walls	Experimental study	Entropy Solution Inc.	[144]
SP 25-A8	Spain	26.0	0.60	180.0	1380	Hollow bricks	Experimental study	Rubitherm GmbH	[30]

7. Conclusions and Future Perspectives

In TES systems, PCMs play an important role and research on them is in the development stage, attracting interest and funding worldwide. The use of PCMs as building materials is a growing trend aiming to limit the energy consumption of buildings, and the direct use of PCMs in building walls is the most favorable. The most common PCMs fall into the category of organic, inorganic or eutectic PCMs with their respective advantages and disadvantages. Inorganic PCMs must be coupled with suitable binders (nucleation and thickening) to avoid phase segregation and sedimentation due to subcooling. For this purpose, considerable efforts are being made both in laboratories and in commercial companies to find versatile and efficient binders. Direct applications of PCMs are more desirable than that of micro/macro encapsulation, not only from an economic point of view, but also in terms of performance. There is still a long way to go to discover new types of PCMs and make them compatible with other common building materials in terms of cost. In the meantime, existing PCMs require more attention in the following aspects:

(a) Due to a huge number of mega projects and a strong demand in the Saudi housing and entertainment industries, the growth of the construction industry in the next years will become more significant. The adoption of new creative technologies and processes, such as integrating PCMs into building walls in the building sector, has therefore become an essential need to increase highly efficient structural operations.

(b) Novel and efficient encapsulation techniques for PCMs, such as UV-based encapsulation process of hydrated salts and paraffin waxes require further investment in terms of their stability over time in real-life applications.

(c) More information on the thermal and physical properties of PCMs is expected, especially in the area of their thermal stability during prolonged freezing/cooling cycles.

The dispersion of nanoparticles can have a positive effect on thermal stability, which has been underestimated in previous research.

(d) The corrosive properties of PCMs is another area where more research is planned. If PCMs lose their anti-corrosive properties with time, this could prove disastrous and appropriate measures should be taken beforehand.

(e) Finally, yet importantly, PCMs are not a substitute for conventional insulation materials; therefore, possibilities to combine the use of PCMs with conventional insulation materials should be investigated. One solution for this may be the direct integration of PCMs into insulation materials.

Funding: This research was funded by King Abdulaziz City for Science and Technology, Project number (20-0003) from the National Center for Building and Construction Technology.

Data Availability Statement: Data will be available on suitable demand.

Acknowledgments: The authors would like to express their deep and sincere gratitude to King Abdulaziz City for Sciences and Technology (KACST) funding and supporting this research project. Project number (20-0003) from the National Center for Building and Construction Technology.

Conflicts of Interest: The authors declare no conflict of interest.

Glossary

Acronyms		Symbols	
TES	Thermal energy storage	CO_2	Carbon-dioxide
GHG	Greenhouse gases	Sn	Tin
PCM	Phase change material	α	Alpha
PAR	Paraffin wax	β	Beta
BS	Butyl stearate	$CaCl_2.6H_2O$	Calcium chloride
TDI	Toluene-2,4-diisocyanate	$CaCO_3$	Calcium carbonate
EDA	Ethylenediamine		
PMMA	Poly-methyl-meta-acrylate		
SEM	Scanning electron microscope		
DSC	Differential scanning calorimetry		
HDPE	High-density polyethylene		
HVAC	Heating, ventilation, air conditioning		

References

1. Anisur, M.; Mahfuz, M.; Kibria, M.; Saidur, R.; Metselaar, H.S.C.; Mahlia, T.M.I. Curbing global warming with phase change materials for energy storage. *Renew. Sustain. Energy Rev.* **2013**, *18*, 23–30. [CrossRef]
2. Dincer, I.; Rosen, M.A. *Energetic, Exergetic, Environmental and Sustainability Aspects of Thermal Energy Storage Systems*; Springer: Dordrecht, The Netherland, 2007.
3. Lizana, J.; Chacartegui, R.; Barrios-Padura, A.; Ortiz, C. Advanced low-carbon energy measures based on thermal energy storage in buildings: A review. *Renew. Sustain. Energy Rev.* **2018**, *82*, 3705–3749. [CrossRef]
4. Lagoua, A.; Kylilia, A.; Šadauskienėb, J.; Fokaides, P.A. Numerical investigation of phase change materials (PCM) optimal melting properties and position in building ele-ments under diverse conditions. *Constr. Build. Mater.* **2019**, *225*, 452–464. [CrossRef]
5. Guo, P.; Biegler, Z.; Back, T.; Sarangan, A. Vanadium dioxide phase change thin films produced by thermal oxidation of metallic vanadium. *Thin Solid Films* **2020**, *707*, 138117. [CrossRef]
6. Jeon, J.; Lee, J.-H.; Seo, J.; Jeong, S.-G.; Kim, S. Application of PCM thermal energy storage system to reduce building energy consumption. *J. Therm. Anal. Calorim.* **2013**, *111*, 279–288. [CrossRef]
7. Zalba, B.; Marín, J.M.; Cabeza, L.F.; Mehling, H. Review on thermal energy storage with phase change: Materials, heat transfer analysis and applications. *Appl. Therm. Eng.* **2003**, *23*, 251–283. [CrossRef]
8. Hasnain, S. Review on sustainable thermal energy storage technologies, Part I: Heat storage materials and techniques. *Energy Convers. Manag.* **1998**, *39*, 1127–1138. [CrossRef]
9. Costello, V.A.; Melsheimer, S.S.; Edie, D.D. Heat transfer and calorimetric studies of a direct contact-latent heat energy storage system. In *Thermal Storage and Heat Transfer in Solar Energy Systems, Proceedings of the Winter Annual Meeting, San Francisco, CA, USA, 10–15 December 1978*; American Society of Mechanical Engineers: New York, NY, USA, 1978; pp. 51–60.
10. Fouda, A.; Despault, G.; Taylor, J.; Capes, C. Solar storage systems using salt hydrate latent heat and direct contact heat exchange—II Characteristics of pilot system operating with sodium sulphate solution. *Sol. Energy* **1984**, *32*, 57–65. [CrossRef]

11. Farid, M.; Yacoub, K. Performance of direct contact latent heat storage unit. *Sol. Energy* **1989**, *43*, 237–251. [CrossRef]
12. Biswas, D.R. Thermal energy storage using sodium sulfate decahydrate and water. *Sol. Energy* **1977**, *19*, 99–100. [CrossRef]
13. Ryu, H.W.; Woo, S.W.; Shin, B.C.; Kim, S.D. Prevention of supercooling and stabilization of inorganic salt hydrates as latent heat storage materials. *Sol. Energy Mater. Sol. Cells* **1992**, *27*, 161–172. [CrossRef]
14. Li, N.; Calis, G.; Becerik-Gerber, B. Measuring and monitoring occupancy with an RFID based system for demand-driven HVAC operations. *Autom. Constr.* **2012**, *24*, 89–99. [CrossRef]
15. Hong, Y.; Ge, X. Preparation of polyethylene–paraffin compound as a form-stable solid-liquid phase change material. *Sol. Energy Mater. Sol. Cells* **2000**, *64*, 37–44. [CrossRef]
16. Himran, S.; Suwono, A.; Mansoori, G.A. Characterization of Alkanes and Paraffin Waxes for Application as Phase Change Energy Storage Medium. *Energy Sources* **1994**, *16*, 117–128. [CrossRef]
17. Dincer, I.; Rosen, M.A. *Thermal Energy Storage: Systems and Applications*; Wiley Online Library: Hoboken, New Jersey, USA, 2021.
18. Abhat, A. Low temperature latent heat thermal energy storage: Heat storage materials. *Sol. Energy* **1983**, *30*, 313–332. [CrossRef]
19. Sharma, A.; Tyagi, V.; Chen, C.; Buddhi, D. Review on thermal energy storage with phase change materials and applications. *Renew. Sustain. Energy Rev.* **2009**, *13*, 318–345. [CrossRef]
20. Mehling, H.; Cabeza, L.F. *Heat and Cold Storage with PCM*; Springer Nature: Cham, Switzerland, 2008. [CrossRef]
21. Mathi, M.; Sundarraja, M.C. Experimental study of passive cooling of building facade using phase change materials to increase thermal comfort in buildings in hot humid areas. *Int. J. Energy Environ.* **2012**, *3*, 739–748.
22. Tyagi, V.V.; Buddhi, D. PCM thermal storage in buildings: A state of art. *Renew. Sustain. Energy Rev.* **2007**, *11*, 1146–1166. [CrossRef]
23. Mehling, H.; Cabeza, L.F. Phase change materials and their basic properties. In *Thermal Energy Storage for Sustainable Energy Consumption*; Springer: Adana, Turkey, 2007; pp. 257–277.
24. Sadeghbeigi, R. Chapter 3—FCC Feed Characterization. In *Fluid Catalytic Cracking Handbook*, 3rd ed.; Butterworth-Heinemann: Oxford, UK, 2012; pp. 51–86.
25. Kenisarin, M.M.; Kenisarina, K.M. Form-stable phase change materials for thermal energy storage. *Renew. Sustain. Energy Rev.* **2012**, *16*, 1999–2040. [CrossRef]
26. Shilei, L.; Neng, Z.; Guohui, F. Impact of phase change wall room on indoor thermal environment in winter. *Energy Build.* **2006**, *38*, 18–24. [CrossRef]
27. Sayyar, M.; Weerasiri, R.R.; Soroushian, P.; Lu, J. Experimental and numerical study of shape-stable phase-change nanocomposite toward energy-efficient building constructions. *Energy Build.* **2014**, *75*, 249–255. [CrossRef]
28. Kong, X.; Lu, S.; Li, Y.; Huang, J.; Liu, S. Numerical study on the thermal performance of building wall and roof incorporating phase change material panel for passive cooling application. *Energy Build.* **2014**, *81*, 404–415. [CrossRef]
29. Koschenz, M.; Lehmann, B. Development of a thermally activated ceiling panel with PCM for application in lightweight and ret-rofitted buildings. *Energy Build.* **2004**, *36*, 567–578. [CrossRef]
30. Ahmad, M.; Bontemps, A.; Sallée, H.; Quenard, D. Thermal testing and numerical simulation of a prototype cell using light wallboards coupling vacuum isolation panels and phase change material. *Energy Build.* **2006**, *38*, 673–681. [CrossRef]
31. Liu, C.; Yuan, Y.; Zhang, N.; Cao, X.; Yang, X. A novel PCM of lauric–myristic–stearic acid/expanded graphite composite for thermal energy storage. *Mater. Lett.* **2014**, *120*, 43–46. [CrossRef]
32. Castell, A.; Martorell, I.; Medrano, M.; Pérez, G.; Cabeza, L.F. Experimental study of using PCM in brick constructive solutions for passive cooling. *Energy Build.* **2010**, *42*, 534–540. [CrossRef]
33. Konuklu, Y.; Unal, M.; Paksoy, H.O. Microencapsulation of caprylic acid with different wall materials as phase change material for thermal energy storage. *Sol. Energy Mater. Sol. Cells* **2014**, *120*, 536–542. [CrossRef]
34. Hussein, H.; El-Ghetany, H.; Nada, S. Experimental investigation of novel indirect solar cooker with indoor PCM thermal storage and cooking unit. *Energy Convers. Manag.* **2008**, *49*, 2237–2246. [CrossRef]
35. Vicente, R.; Silva, T. Brick masonry walls with PCM macrocapsules: An experimental approach. *Appl. Therm. Eng.* **2014**, *67*, 24–34. [CrossRef]
36. Kenisarin, M.; Mahkamov, K. Solar energy storage using phase change materials. *Renew. Sustain. Energy Rev.* **2007**, *11*, 1913–1965. [CrossRef]
37. Jin, X.; Medina, M.A.; Zhang, X. On the placement of a phase change material thermal shield within the cavity of buildings walls for heat transfer rate reduction. *Energy* **2014**, *73*, 780–786. [CrossRef]
38. Kong, X.; Lu, S.; Huang, J.; Cai, Z.; Wei, S. Experimental research on the use of phase change materials in perforated brick rooms for cooling storage. *Energy Build.* **2013**, *62*, 597–604. [CrossRef]
39. Feldman, D.; Banu, D.; Hawes, D. Development and application of organic phase change mixtures in thermal storage gypsum wallboard. *Sol. Energy Mater. Sol. Cells* **1995**, *36*, 147–157. [CrossRef]
40. Murray, R.; Groulx, D. Experimental study of the phase change and energy characteristics inside a cylindrical latent heat energy storage system: Part 1 consecutive charging and discharging. *Renew. Energy* **2014**, *62*, 571–581. [CrossRef]
41. Cabeza, L.; Castell, A.; Barreneche, C.; de Gracia, A.; Fernandez, A.I. Materials used as PCM in thermal energy storage in buildings: A review. *Renew. Sustain. Energy Rev.* **2011**, *15*, 1675–1695. [CrossRef]
42. Browne, M.C.; Norton, B.; McCormack, S. Heat retention of a photovoltaic/thermal collector with PCM. *Sol. Energy* **2016**, *133*, 533–548. [CrossRef]

43. Benatar, J. Trans fatty acids and coronary artery disease. *Open Access J. Clin. Trials* **2010**, *2*, 9–13. [CrossRef]
44. Yuan, Y.; Li, T.; Zhang, N.; Cao, X.; Yang, X. Investigation on thermal properties of capric–palmitic–stearic acid/activated carbon composite phase change mate-rials for high-temperature cooling application. *J. Therm. Anal. Calorim.* **2016**, *124*, 881–888. [CrossRef]
45. Karaipekli, A.; Sari, A. Capric–myristic acid/expanded perlite composite as form-stable phase change material for latent heat thermal energy storage. *Renew. Energy* **2008**, *33*, 2599–2605. [CrossRef]
46. Chen, Z.; Shan, F.; Cao, L.; Fang, G. Synthesis and thermal properties of shape-stabilized lauric acid/activated carbon composites as phase change materials for thermal energy storage. *Sol. Energy Mater. Sol. Cells* **2012**, *102*, 131–136. [CrossRef]
47. Fang, G.; Li, H.; Chen, Z.; Liu, X. Preparation and characterization of stearic acid/expanded graphite composites as thermal energy storage materials. *Energy* **2010**, *35*, 4622–4626. [CrossRef]
48. Li, M.; Wu, Z.; Kao, H. Study on preparation and thermal properties of binary fatty acid/diatomite shape-stabilized phase change materials. *Sol. Energy Mater. Sol. Cells* **2011**, *95*, 2412–2416. [CrossRef]
49. Wang, Y.; Zheng, H.; Feng, H.X.; Zhang, D.Y. Effect of preparation methods on the structure and thermal properties of stearic acid/activated montmorillonite phase change materials. *Energy Build.* **2012**, *47*, 467–473. [CrossRef]
50. Sarı, A.; Karaipekli, A. Preparation, thermal properties and thermal reliability of palmitic acid/expanded graphite composite as form-stable PCM for thermal energy storage. *Sol. Energy Mater. Sol. Cells* **2009**, *93*, 571–576. [CrossRef]
51. Evers, A.C.; Medina, M.A.; Fang, Y. Evaluation of the thermal performance of frame walls enhanced with paraffin and hydrated salt phase change materials using a dynamic wall simulator. *Build. Environ.* **2010**, *45*, 1762–1768. [CrossRef]
52. Lee, K.O.; Medina, M.A.; Raith, E.; Sun, X. Assessing the integration of a thin phase change material (PCM) layer in a residential building wall for heat transfer reduction and management. *Appl. Energy* **2015**, *137*, 699–706. [CrossRef]
53. Jin, X.; Zhang, S.; Xu, X.; Zhang, X. Effects of PCM state on its phase change performance and the thermal performance of building walls. *Build. Environ.* **2014**, *81*, 334–339. [CrossRef]
54. Carbonari, A.; De Grassi, M.; Di Perna, C.; Principi, P. Numerical and experimental analyses of PCM containing sandwich panels for prefabricated walls. *Energy Build.* **2006**, *38*, 472–483. [CrossRef]
55. Principi, P.; Fioretti, R. Thermal analysis of the application of pcm and low emissivity coating in hollow bricks. *Energy Build.* **2012**, *51*, 131–142. [CrossRef]
56. Zhang, C.; Chen, Y.; Wu, L.; Shi, M. Thermal response of brick wall filled with phase change materials (PCM) under fluctuating outdoor temperatures. *Energy Build.* **2011**, *43*, 3514–3520. [CrossRef]
57. Berroug, F.; Lakhal, E.; El Omari, M.; Faraji, M.; El Qarnia, H. Thermal performance of a greenhouse with a phase change material north wall. *Energy Build.* **2011**, *43*, 3027–3035. [CrossRef]
58. Hichem, N.; Noureddine, S.; Nadia, S.; Djamila, D. Experimental and Numerical Study of a Usual Brick Filled with PCM to Improve the Thermal Inertia of Buildings. *Energy Procedia* **2013**, *36*, 766–775. [CrossRef]
59. Hadjieva, M.; Stoykov, R.; Filipova, T. Composite salt-hydrate concrete system for building energy storage. *Renew. Energy* **2000**, *19*, 111–115. [CrossRef]
60. Ma, Z.; Bao, H.; Roskilly, A.P. Study on solidification process of sodium acetate trihydrate for seasonal solar thermal energy storage. *Sol. Energy Mater. Sol. Cells* **2017**, *172*, 99–107. [CrossRef]
61. Xie, N.; Luo, J.; Li, Z.; Huang, Z.; Gao, X.; Fang, Y.; Zhang, Z. Salt hydrate/expanded vermiculite composite as a form-stable phase change material for building energy storage. *Sol. Energy Mater. Sol. Cells* **2018**, *189*, 33–42. [CrossRef]
62. Athienitis, A.; Liu, C.; Hawes, D.; Banu, D.; Feldman, D. Investigation of the thermal performance of a passive solar test-room with wall latent heat storage. *Build. Environ.* **1997**, *32*, 405–410. [CrossRef]
63. Khalifa, A.J.; Abbas, E.F. A comparative performance study of some thermal storage materials used for solar space heating. *Energy Build.* **2009**, *41*, 407–415. [CrossRef]
64. Alkan, C.; Döğüşcü, D.K.; Gottschalk, A.; Ramamoorthi, U.; Kumar, A.; Yadav, S.K.; Yadav, A.S.; Adıgüzel, E.; Altıntaş, A.; Damlıoğlu, Y.; et al. Polyvinyl Alcohol-salt Hydrate Mixtures as Passive Thermal Energy Storage Systems. *Energy Procedia* **2016**, *91*, 1012–1017. [CrossRef]
65. Oró, E.; de Gracia, A.; Castell, A.; Farid, M.; Cabeza, L. Review on phase change materials (PCMs) for cold thermal energy storage applications. *Appl. Energy* **2012**, *99*, 513–533. [CrossRef]
66. Turnpenny, J.; Etheridge, D.; Reay, D. Novel ventilation cooling system for reducing air conditioning in buildings.: Part I: Testing and theoretical modelling. *Appl. Therm. Eng.* **2000**, *20*, 1019–1037. [CrossRef]
67. Rathore, P.K.S.; Shukla, S.K. An experimental evaluation of thermal behavior of the building envelope using macroencapsulated PCM for energy savings. *Renew. Energy* **2019**, *149*, 1300–1313. [CrossRef]
68. Fang, X.; Zhang, Z.; Chen, Z. Study on preparation of montmorillonite-based composite phase change materials and their appli-cations in thermal storage building materials. *Energy Convers. Manag.* **2008**, *49*, 718–723. [CrossRef]
69. Banu, D.; Feldman, D.; Hawes, D. Evaluation of thermal storage as latent heat in phase change material wallboard by differential scanning calorimetry and large scale thermal testing. *Thermochim. Acta* **1998**, *317*, 39–45. [CrossRef]
70. Lee, T.; Hawes, D.; Banu, D.; Feldman, D. Control aspects of latent heat storage and recovery in concrete. *Sol. Energy Mater. Sol. Cells* **2000**, *62*, 217–237. [CrossRef]
71. Sá, A.V.; Almeida, R.; Sousa, H.; Delgado, J.M.P.Q. Numerical Analysis of the Energy Improvement of Plastering Mortars with Phase Change Materials. *Adv. Mater. Sci. Eng.* **2014**, *2014*, 1–12. [CrossRef]

72. Scalat, S.; Banu, D.; Hawes, D.; Parish, J.; Haghighata, F.; Feldman, D. Full scale thermal testing of latent heat storage in wallboard. *Sol. Energy Mater. Sol. Cells* **1996**, *44*, 49–61. [CrossRef]
73. Feldman, D.; Banu, D. DSC analysis for the evaluation of an energy storing wallboard. *Thermochim. Acta* **1996**, *272*, 243–251. [CrossRef]
74. Khudhair, A.M.; Farid, M.M. A review on energy conservation in building applications with thermal storage by latent heat using phase change materials. *Energy Convers. Manag.* **2004**, *45*, 263–275. [CrossRef]
75. Cellat, K.; Beyhan, B.; Kazanci, B.; Konuklu, Y.; Paksoy, H. Direct Incorporation of Butyl Stearate as Phase Change Material into Concrete for Energy Saving in Buildings. *J. Clean Energy Technol.* **2017**, *5*, 64–68. [CrossRef]
76. Sun, Z.; Zhang, Y.; Zheng, S.; Park, Y.; Frost, R.L. Preparation and thermal energy storage properties of paraffin/calcined diatomite composites as form-stable phase change materials. *Thermochim. Acta* **2013**, *558*, 16–21. [CrossRef]
77. da Cunha, S.R.L.; Aguiar, J.; Tadeu, A. Thermal performance and cost analysis of mortars made with PCM and different binders. *Constr. Build. Mater.* **2016**, *122*, 637–648. [CrossRef]
78. Wang, T.; Wang, S.; Luo, R.; Zhu, C.; Akiyama, T.; Zhang, Z. Microencapsulation of phase change materials with binary cores and calcium carbonate shell for thermal energy storage. *Appl. Energy* **2016**, *171*, 113–119. [CrossRef]
79. Khan, R.J. Bhuiyan, M.Z.H.; Ahmed, D.H. Investigation of heat transfer of a building wall in the presence of phase change ma-terial (PCM). *Energy Built Environ.* **2020**, *1*, 199–206. [CrossRef]
80. Ballweg, T.; Von Daake, H.; Hanselmann, D.; Stephan, D.; Mandel, K.; Sextl, G. Versatile triggered substance release systems via a highly flexible high throughput encapsulation technique. *Appl. Mater. Today* **2018**, *11*, 231–237. [CrossRef]
81. Liang, C.; Lingling, X.; Hongbo, S.; Zhibin, Z. Microencapsulation of butyl stearate as a phase change material by interfacial polycondensation in a polyurea system. *Energy Convers. Manag.* **2009**, *50*, 723–729. [CrossRef]
82. Su, J.-F.; Wang, L.-X.; Ren, L. Synthesis of polyurethane microPCMs containing n-octadecane by interfacial polycondensation: Influence of styrene-maleic anhydride as a surfactant. *Colloids Surf. A Physicochem. Eng. Asp.* **2007**, *299*, 268–275. [CrossRef]
83. Tang, X.; Li, W.; Zhang, X.; Shi, H. Fabrication and characterization of microencapsulated phase change material with low supercooling for thermal energy storage. *Energy* **2014**, *68*, 160–166. [CrossRef]
84. Qiu, X.; Song, G.; Chu, X.; Li, X.; Tang, G. Preparation, thermal properties and thermal reliabilities of microencapsulated n-octadecane with acrylic-based polymer shells for thermal energy storage. *Thermochim. Acta* **2013**, *551*, 136–144. [CrossRef]
85. Hu, Q.; Chen, Y.; Hong, J.; Jin, S.; Zou, G.; Chen, L.; Chen, D.-Z. A Smart Epoxy Composite Based on Phase Change Microcapsules: Preparation, Microstructure, Thermal and Dynamic Mechanical Performances. *Molecules* **2019**, *24*, 916. [CrossRef]
86. Zhao, L.; Wang, H.; Luo, J.; Liu, Y.; Song, G.; Tang, G. Fabrication and properties of microencapsulated n-octadecane with TiO 2 shell as thermal energy storage materials. *Sol. Energy* **2016**, *127*, 28–35. [CrossRef]
87. Wang, H.; Zhao, L.; Chen, L.; Song, G.; Tang, G. Facile and low energy consumption synthesis of microencapsulated phase change materials with hybrid shell for thermal energy storage. *J. Phys. Chem. Solids* **2017**, *111*, 207–213. [CrossRef]
88. Li, C.; Yu, H.; Song, Y.; Liang, H.; Yan, X. Preparation and characterization of PMMA/TiO2 hybrid shell microencapsulated PCMs for thermal energy storage. *Energy* **2018**, *167*, 1031–1039. [CrossRef]
89. Zhang, H.; Wang, X. Fabrication and performances of microencapsulated phase change materials based on n-octadecane core and resorcinol-modified melamine–formaldehyde shell. *Colloids Surf. A Physicochem. Eng. Asp.* **2009**, *332*, 129–138. [CrossRef]
90. Wang, H.; Zhao, L.; Song, G.; Tang, G. Organic-inorganic hybrid shell microencapsulated phase change materials prepared from SiO2/TiC-stabilized pick-ering emulsion polymerization. *Sol. Energy Mater. Sol. Cells* **2018**, *175*, 102–110. [CrossRef]
91. Li, W.; Song, G.; Li, S.; Yao, Y.; Tang, G. Preparation and characterization of novel MicroPCMs (microencapsulated phase-change materials) with hybrid shells via the polymerization of two alkoxy silanes. *Energy* **2014**, *70*, 298–306. [CrossRef]
92. Chen, D.-Z.; Qin, S.-Y.; Tsui, C.P.; Tang, C.Y.; Ouyang, X.; Liu, J.-H.; Tang, J.-N.; Zuo, J.-D. Fabrication, morphology and thermal properties of octadecylamine-grafted graphene oxide-modified phase-change microcapsules for thermal energy storage. *Compos. Part B Eng.* **2018**, *157*, 239–247. [CrossRef]
93. He, F.; Wang, X.; Wu, D. New approach for sol–gel synthesis of microencapsulated n-octadecane phase change material with silica wall using sodium silicate precursor. *Energy* **2014**, *67*, 223–233. [CrossRef]
94. Yu, S.; Wang, X.; Wu, D. Microencapsulation of n-octadecane phase change material with calcium carbonate shell for en-hancement of thermal conductivity and serving durability: Synthesis, microstructure, and performance evaluation. *Appl. Energy* **2014**, *114*, 632–643. [CrossRef]
95. Sari, A.; Alkan, C.; Doguscu, D.K.; Biçer, A. Micro/nano-encapsulated n-heptadecane with polystyrene shell for latent heat thermal energy storage. *Sol. Energy Mater. Sol. Cells* **2014**, *126*, 42–50. [CrossRef]
96. Irani, F.; Ranjbar, Z.; Moradian, S.; Jannesari, A. Microencapsulation of n-heptadecane phase change material with starch shell. *Prog. Org. Coatings* **2017**, *113*, 31–38. [CrossRef]
97. Fortuniak, W.; Slomkowski, S.; Chojnowski, J.; Kurjata, J.; Tracz, A.; Mizerska, U. Synthesis of a paraffin phase change material microencapsulated in a siloxane polymer. *Colloid Polym. Sci.* **2012**, *291*, 725–733. [CrossRef] [PubMed]
98. Chai, L.; Wang, X.; Wu, D. Development of bifunctional microencapsulated phase change materials with crystalline titanium dioxide shell for latent-heat storage and photocatalytic effectiveness. *Appl. Energy* **2015**, *138*, 661–674. [CrossRef]
99. Zhang, X.-X.; Fan, Y.-F.; Tao, X.; Yick, K.L. Crystallization and prevention of supercooling of microencapsulated n-alkanes. *J. Colloid Interface Sci.* **2005**, *281*, 299–306. [CrossRef] [PubMed]

100. Sánchez, L.; Sánchez, P.; Carmona, M.; De Lucas, A.; Rodríguez, J.F. Influence of operation conditions on the microencapsulation of PCMs by means of suspension-like polymerization. *Colloid Polym. Sci.* **2008**, *286*, 1019–1027. [CrossRef]
101. Pilehvar, S.; Sanfelix, S.G.; Szczotok, A.M.; Rodríguez, J.F.; Valentini, L.; Lanzón, M.; Pamies, R.; Kjøniksen, A.-L. Effect of temperature on geopolymer and Portland cement composites modified with Micro-encapsulated Phase Change materials. *Constr. Build. Mater.* **2020**, *252*, 119055. [CrossRef]
102. Borreguero, A.M.; Carmona, M.; Sanchez-Silva, L.; Valverde, J.L.; Rodriguez, J.F. Improvement of the thermal behaviour of gypsum blocks by the incorporation of microcapsules containing PCMS obtained by suspension polymerization with an optimal core/coating mass ratio. *Appl. Therm. Eng.* **2010**, *30*, 1164–1169. [CrossRef]
103. Sarı, A.; Alkan, C.; Bilgin, C. Micro/nano encapsulation of some paraffin eutectic mixtures with poly (methyl methacrylate) shell: Preparation, characterization and latent heat thermal energy storage properties. *Appl. Energy* **2014**, *136*, 217–227. [CrossRef]
104. Royon, L.; Guiffant, G.; Flaud, P. Investigation of heat transfer in a polymeric phase change material for low level heat storage. *Energy Convers. Manag.* **1997**, *38*, 517–524. [CrossRef]
105. Cunha, S.; Aguiar, J.; Ferreira, V.; Tadeu, A. Mortars based in different binders with incorporation of phase-change materials: Physical and mechanical proper-ties. *Eur. J. Environ. Civ.Eng.* **2015**, *19*, 1216–1233. [CrossRef]
106. Zhang, H.; Wang, X. Synthesis and properties of microencapsulated n-octadecane with polyurea shells containing different soft segments for heat energy storage and thermal regulation. *Sol. Energy Mater. Sol. Cells* **2009**, *93*, 1366–1376. [CrossRef]
107. Liu, Z.; Yu, Z.; Yang, T.; Qin, D.; Li, S.; Zhang, G.; Haghighat, F.; Joybari, M.M. A review on macro-encapsulated phase change material for building envelope applications. *Build. Environ.* **2018**, *144*, 281–294. [CrossRef]
108. Saxena, R.; Rakshit, D.; Kaushik, S. Experimental assessment of Phase Change Material (PCM) embedded bricks for passive conditioning in buildings. *Renew. Energy* **2019**, *149*, 587–599. [CrossRef]
109. Lucas, S.; Ferreira, V.; de Aguiar, J.B. Latent heat storage in PCM containing mortars—Study of microstructural modifications. *Energy Build.* **2013**, *66*, 724–731. [CrossRef]
110. Gao, Y.; He, F.; Meng, X.; Wang, Z.; Zhang, M.; Yu, H.; Gao, W. Thermal behavior analysis of hollow bricks filled with phase-change material (PCM). *J. Build. Eng.* **2020**, *31*, 101447. [CrossRef]
111. Kuznik, F.; Virgone, J.; Noel, J. Optimization of a phase change material wallboard for building use. *Appl. Therm. Eng.* **2008**, *28*, 1291–1298. [CrossRef]
112. Keech, R. Changing phase: Are PCMs living up to their promise? *Sanctuary Modern Green Home Mag.* **2018**, *42*, 72–74.
113. Liu, L.; Chen, J.; Qu, Y.; Xu, T.; Wu, H.; Huang, G.; Zhou, X.; Yang, L. A foamed cement blocks with paraffin/expanded graphite composite phase change solar thermal absorption material. *Sol. Energy Mater. Sol. Cells* **2019**, *200*, 110038. [CrossRef]
114. Kheradmand, M.; Azenha, M.; Aguiar, J.; Castro-Gomes, J. Experimental and numerical studies of hybrid PCM embedded in plastering mortar for enhanced thermal behaviour of buildings. *Energy* **2016**, *94*, 250–261. [CrossRef]
115. Meng, E.; Yu, H.; Zhou, B. Study of the thermal behavior of the composite phase change material (PCM) room in summer and winter. *Appl. Therm. Eng.* **2017**, *126*, 212–225. [CrossRef]
116. Li, C.; Yu, H.; Song, Y. Experimental investigation of thermal performance of microencapsulated PCM-contained wallboard by two measurement modes. *Energy Build.* **2018**, *184*, 34–43. [CrossRef]
117. Hasan, M.I.; Basher, H.O.; Shdhan, A.O. Experimental investigation of phase change materials for insulation of residential buildings. *Sustain. Cities Soc.* **2018**, *36*, 42–58. [CrossRef]
118. Wang, Q.; Wu, R.; Wu, Y.; Zhao, C. Parametric analysis of using PCM walls for heating loads reduction. *Energy Build.* **2018**, *172*, 328–336. [CrossRef]
119. Mourid, A.; El Alami, M.; Kuznik, F. Experimental investigation on thermal behavior and reduction of energy consumption in a real scale building by using phase change materials on its envelope. *Sustain. Cities Soc.* **2018**, *41*, 35–43. [CrossRef]
120. Kant, K.; Shukla, A.; Sharma, A. Heat transfer studies of building brick containing phase change materials. *Sol. Energy* **2017**, *155*, 1233–1242. [CrossRef]
121. Jin, X.; Medina, M.A.; Zhang, X. Numerical analysis for the optimal location of a thin PCM layer in frame walls. *Appl. Therm. Eng.* **2016**, *103*, 1057–1063. [CrossRef]
122. Xie, J.; Wang, W.; Liu, J.; Pan, S. Thermal performance analysis of PCM wallboards for building application based on numerical simulation. *Sol. Energy* **2018**, *162*, 533–540. [CrossRef]
123. Olivieri, L.; Tenorio, J.A.; Revuelta, D.; Navarro, L.; Cabeza, L.F. Developing a PCM-enhanced mortar for thermally active precast walls. *Constr. Build. Mater.* **2018**, *181*, 638–649. [CrossRef]
124. Li, L.; Yu, H.; Liu, R. Research on composite-phase change materials (PCMs)-bricks in the west wall of room-scale cubicle: Mid-season and summer day cases. *Build. Environ.* **2017**, *123*, 494–503. [CrossRef]
125. Soares, N.; Gaspar, A.; Santos, P.; Costa, J.J. Experimental evaluation of the heat transfer through small PCM-based thermal energy storage units for building applications. *Energy Build.* **2016**, *116*, 18–34. [CrossRef]
126. Elnajjar, E. Using PCM embedded in building material for thermal management: Performance assessment study. *Energy Build.* **2017**, *151*, 28–34. [CrossRef]
127. Cabeza, L.F.; Castellón, C.; Nogués, M.; Medrano, M.; Leppers, R.; Zubillaga, O. Use of microencapsulated PCM in concrete walls for energy savings. *Energy Build.* **2007**, *39*, 113–119. [CrossRef]
128. Wang, X.; Yu, H.; Li, L.; Zhao, M. Experimental assessment on the use of phase change materials (PCMs)-bricks in the exterior wall of a full-scale room. *Energy Convers. Manag.* **2016**, *120*, 81–89. [CrossRef]

129. Soares, N.; Gaspar, A.; Santos, P.; Costa, J.J. Multi-dimensional optimization of the incorporation of PCM-drywalls in lightweight steel-framed residential buildings in different climates. *Energy Build.* **2013**, *70*, 411–421. [CrossRef]
130. Fateh, A.; Klinker, F.; Brütting, M.; Weinläder, H.; Devia, F. Numerical and experimental investigation of an insulation layer with phase change materials (PCMs). *Energy Build.* **2017**, *153*, 231–240. [CrossRef]
131. Jamil, H.; Alam, M.; Sanjayan, J.; Wilson, J.L. Investigation of PCM as retrofitting option to enhance occupant thermal comfort in a modern residential building. *Energy Build.* **2016**, *133*, 217–229. [CrossRef]
132. Barreneche, C.; Navarro, M.E.; Fernandez, A.I.; Cabeza, L.F. Improvement of the thermal inertia of building materials incorporating PCM. Evaluation in the macroscale. *Appl. Energy* **2013**, *109*, 428–432. [CrossRef]
133. Saffari, M.; de Gracia, A.; Ushak, S.; Cabeza, L.F. Passive cooling of buildings with phase change materials using whole-building energy simulation tools: A review. *Renew. Sustain. Energy Rev.* **2017**, *80*, 1239–1255. [CrossRef]
134. Biswas, K.; Abhari, R. Low-cost phase change material as an energy storage medium in building envelopes: Experimental and numerical analyses. *Energy Convers. Manag.* **2014**, *88*, 1020–1031. [CrossRef]
135. Mandilaras, I.; Stamatiadou, M.; Katsourinis, D.; Zannis, G.; Founti, M. Experimental thermal characterization of a Mediterranean residential building with PCM gypsum board walls. *Build. Environ.* **2013**, *61*, 93–103. [CrossRef]
136. Li, Y.; Nord, N.; Xiao, Q.; Tereshchenko, T. Building heating applications with phase change material: A comprehensive review. *J. Energy Storage* **2020**, *31*, 101634. [CrossRef]
137. Marin, P.; Saffari, M.; de Gracia, A.; Zhu, X.; Farid, M.; Cabeza, L.F.; Ushak, S. Energy savings due to the use of PCM for relocatable lightweight buildings passive heating and cooling in different weather conditions. *Energy Build.* **2016**, *129*, 274–283. [CrossRef]
138. Barzin, R.; Chen, J.; Young, B.; Farid, M.M. Application of weather forecast in conjunction with price-based method for PCM solar passive buildings–an ex-perimental study. *Appl. Energy* **2016**, *163*, 9–18. [CrossRef]
139. Cao, V.D.; Bui, T.Q.; Kjøniksen, A.-L. Thermal analysis of multi-layer walls containing geopolymer concrete and phase change materials for building applications. *Energy* **2019**, *186*, 115792. [CrossRef]
140. Zhou, D.; Zhao, C.; Tian, Y. Review on thermal energy storage with phase change materials (PCMs) in building applications. *Appl. Energy* **2012**, *92*, 593–605. [CrossRef]
141. Kuznik, F.; Virgone, J. Experimental investigation of wallboard containing phase change material: Data for validation of numer-ical modeling. *Energy Build.* **2009**, *41*, 561–570. [CrossRef]
142. Berardi, U.; Manca, M. The Energy Saving and Indoor Comfort Improvements with Latent Thermal Energy Storage in Building Retrofits in Canada. *Energy Procedia* **2017**, *111*, 462–471. [CrossRef]
143. Wijesuriya, S.; Tabares-Velasco, P.C. Experimental apparatus and methodology to test and quantify thermal performance of micro and macro-encapsulated phase change materials in building envelope applications. *J. Energy Storage* **2020**, *32*, 101770. [CrossRef]
144. Song, M.; Niu, F.; Mao, N.; Hu, Y.; Deng, S.S. Review on building energy performance improvement using phase change materials. *Energy Build.* **2018**, *158*, 776–793. [CrossRef]

Article

A Comparative Study on the Thermal Energy Storage Performance of Bio-Based and Paraffin-Based PCMs Using DSC Procedures

Mona Nazari Sam [1], Antonio Caggiano [1,2,*], Christoph Mankel [1] and Eddie Koenders [1]

1 Institut für Werkstoffe im Bauwesen, Technische Universität Darmstadt, 64287 Darmstadt, Germany; sam@wib.tu-darmstadt.de (M.N.S.); mankel@wib.tu-darmstadt.de (C.M.); koenders@wib.tu-darmstadt.de (E.K.)

2 CONICET, LMNI, INTECIN, Facultad de Ingeniería, Universidad de Buenos Aires, Ciudad Autónoma de Buenos Aires C1127AAR, Argentina

* Correspondence: caggiano@wib.tu-darmstadt; Tel.: +49-6151-16-22210

Received: 21 February 2020; Accepted: 1 April 2020; Published: 5 April 2020

Abstract: Thermal-Energy Storage (TES) properties of organic phase change materials have been experimentally investigated and reported in this paper. Three paraffin-based Phase Change Materials (PCMs) and one bio-based PCM are considered with melting temperatures of 24 °C, 25 °C and 26 °C. Sensible heat storage capacities, melting characteristics and latent heat enthalpies of the studied PCMs are investigated through Differential Scanning Calorimetry (DSC) measurements. Two alternative methods, namely the classical dynamic DSC and a stepwise approach, are performed and compared with the aim to eliminate and/or overcome possible measurement errors. In particular, for DSC measurements this could be related to the size of the samples and its representativity, heating rate effects and low thermal conductivity of the PCMs, which may affect the results and possibly cause a loss of objectivity of the measurements. Based on results achieved from this study, clear information can be figured out on how to conduct and characterize paraffin and bio-based PCMs, and how to apply them in TES calculations for building applications and/or simulations. It is observed that both paraffinic and bio-based PCMs possess a comparable TES capacity within the selected phase transition temperature, being representative for the human thermal comfort zone. The phase change of bio-based PCMs occurred over a much narrower temperature range when compared to the wider windows characterizing the paraffin-based materials. Bio-based PCMs turned out to be very suitable for building applications and can be an environmentally friendly substitute for petroleum-based PCMs.

Keywords: thermal-energy storage (TES); phase change materials (PCMs); latent enthalpy; melting temperature; differential scanning calorimetry; IEA method

---

1. Introduction

The global challenge to strongly cut back the use of fossil fuels with the aim to implement renewable resources and to neutralize greenhouse gas emissions make energy efficiency a key issue that is at the center of our society [1,2]. According to the EU commission, heating and cooling the residential and non-residential sector accounts for half of the EU's energy consumption, while about 84% of it is still generated from fossil resources [3]. Since the introduction of the new EU Buildings Directive 2019/2021 [4], all member states are obliged to guarantee that all new constructions are designed as "Nearly Zero Energy Buildings" (NZEBs), from the beginning of 2021. This obligation has been already applied to non-residential buildings from the beginning of 2019. Therefore, the development of new and smart energy storage solutions and technologies, with the aim to use environmental thermal energy more efficiently, and to balance out daily heating/cooling demands, are worth investigation for building applications [5,6].

One promising technique available for Thermal Energy Storage (TES) applications is by implementing Phase Change Materials (PCMs) in construction and building composites [7]. PCMs can be used to store large quantities of heat, not only through their sensible capacity, but also (sometimes predominantly) via their latent storage property [8]. As PCMs are capable of storing large amounts of latent heat at constant temperature, they are contributing to the energy efficiency and thermal comfort of residential and non-residential buildings, by balancing out daily environmental heat demands [9]. Consequently, the passive storage/release of latent heat through phase transitions from solid to liquid or vice versa, allows to save considerable amount of primary energy [10]. The principle of latent TES can be employed in a wide range of applications [11], such as, solar heating systems [12], building air conditioning [13], building envelope [14], production of energy-saving cementitious composites [15] and high porous insulation systems [16], waste heat recovery in residential and non-residential sectors [17], and many other applications [18].

Designing large-scale practical applications for passive latent thermal energy storage systems requires in-depth knowledge on the thermal characteristics of PCM before, during and after its phase change. Therefore, an accurate and correct determination of the thermal properties of PCM systems is crucial to efficiently design composite systems or devices that use latent TES [19].

Differential Scanning Calorimetry (DSC) is an effective method to characterize the thermal behavior of PCMs, and to determine their TES capacities, in terms of transition temperature, enthalpy and specific heat, and its stability throughout the various melting and crystallization cycles. The heating and cooling rates of DSC measurements are typically much faster than in real applications, while the sample mass is also very small (less than 90 mg), which might lack representativity for real size applications [20,21]. Some commercially calorimeters are already available on the market for measuring and analyzing larger sample masses with a very high precision and adopting quite low heating rates (as low as 0.01 K/min). Actually, samples with a large mass implicitly show a lower error ratio resulting from reduced sample inhomogeneity. However, larger samples may also strongly affect the course of the measured phase change temperatures (i.e., the onset temperature or the maximum one). It follows that DSC measurements can be affected by changing the heating/cooling rate and/or sample mass. They mainly influence the thermal equilibrium status in the sample, thus producing a shifting of the phase change temperature and also lead to non-realistic shapes of the heat capacity and/or enthalpy temperature responses $h(T)$ [22]. Moreover, DSC test results depend on further factors such as sample preparation, correct calibration of the experimental set-up, and many other factors that should, in fact, be standardized in order to achieve comparable and objective results, under different test environments [23]. Therefore, developing an appropriate and objective methodology, in this field, is essential to improve the accuracy of the PCM characterization procedure and to make measurement errors negligible [24].

An alternative method that overcomes the aforementioned DSC limitations (i.e., mass influence and heating rate effects) is the T-History Method (THM), which was originally developed by Yinping and Yi [25]. The method records the time–temperature evolutions of PCM samples against a well-known reference material, usually water. THM easily allows to evaluate the heat capacity, temperature of melting/crystallization, enthalpy and phase change temperature. The accuracy and soundness of this method was evaluated by many researchers, see for example [26–28]. Hong et al. [29] verified the accuracy of a modified THM for several PCMs, having different freezing patterns. A further improvement of the THM measurement technique was developed by Stankovic and Kyriacou [30]. A procedure to numerically correct the enthalpy-temperature response of PCMs obtained from THM was developed by Tan et al. [31]. Then, a critical comparison between DSC and THM was done by Rathbeger et al. [32]. The authors mainly concluded that the key difference between both methods is represented by the sample size of the investigated material, and the temperature profile to which they are subjected. THM samples are much larger than DSC samples (about 1000 times), while also constant heating and/or cooling temperatures are applied. In addition, THM minimizes heating/cooling rate effects, overcomes sample size issues and can be used for low thermal conductivity samples. The only drawback of THM, which is actually crucial for standardized measurements and test procedures, is that the accuracy of the lab results largely depends on the measuring procedures and the self-built calorimeters. For this reason, measurements from different laboratories can differ a lot and are generally not comparable [33].

The main objective of this work is to investigate the thermo-physical properties and heat storage capacity of a representative organic bio-based (non-paraffin) PCM, in terms of phase-change enthalpy, specific heat capacity and melting temperature, using different experimental setups. The influence of various parameters, involved in DSC measurements, are compared with results done on three comparable paraffin waxes. The measurements are carried out in accordance with the currently available standard of the International Energy Agency (IEA), i.e., under the Task 42 Annex 29, to characterize the thermal-energy properties of the PCMs under discussion. Heat-flux DSC dynamic measurements and DSC stepwise procedures are considered and compared to scrutinize the aforementioned aims.

To the Authors' best knowledge, only a few studies on thermal analysis (even under the still not published final version of the IEA-SHC 42 Annex 29 Standard for common PCMs) are so far available in scientific literature dealing with the use of bio-based PCM. Moreover, no Standard is currently available on the thermal analysis of organic PCMs, but there are some discussions going on in this direction. In this context, with the current paper, the authors like to contribute to this discussion by presenting the differences in DSC responses between three commonly used petroleum-based PCMs (paraffinic) and one novel bio-based PCM. Especially highlighting the differences in TES characteristics and data of various kinds of PCMs, in a general sense. Thus, this document tried to examine if the currently employed procedures can also be applied to other types of PCMs, such as the considered bio-based one investigated in this work. In addition, the presented results lead to demonstrate the feasibility of completely substituting petroleum based PCMs with a more environmentally friendly bio-based one. This aspect may open doors for a new scenario in research for developing novel PCMs, that can be employed in thermal storage materials for building construction materials, which can be more sustainable, eco-friendly and energy-saving.

The paper is structured as follows: Section 2 reports Materials and Methods of the experimental program and outlines the key thermophysical characteristics of the investigated materials and shows the overall experimental program. Then, in Section 3, the experimental results for both paraffin and bio-based PCMs are described following the dynamic DSC procedure. Section 4 outlines the results of the stepwise approach where some comparisons between the two alterative procedures are conducted to characterize the TES of the PCMs. Finally, concluding remarks and future developments of this research are addressed in Section 5.

2. Materials and Methods

In this section, the employed materials, methods and experimental program, considered for analyzing the TES properties of the selected PCMs, are presented. Particularly, two types of organic phase change materials are analyzed, namely three Paraffin Waxes and one Bio-based PCM, which have physical, kinetic, chemical and economic relevance for constructions and building application in civil engineering [34,35].

2.1. *Paraffin Wax*

Three commercial paraffin-based waxes (RT-series® by Rubitherm GmbH, Berlin, Germany) are used as reference PCMs. These commercial products, having a melting temperature T_m that ranges between 22 °C and 26 °C (suitable for enhancing the thermal comfort in building applications) are selected for this study. More specifically, RT24, RT25 and RT26 are taken into consideration which are characterized with a T_m of 24 °C, 25 °C and 26 °C, respectively.

2.2. Bio-based PCM

One bio-based PCM has also been investigated in this work as an eco-friendly alternative to the petroleum-based paraffin wax. The bio-based material, viz. PureTemp25 by PureTemp LLC (Minneapolis, MN, USA), made of natural oils with a T_m of 25 °C, is used, while its results can be directly compared to RT25. The thermal properties of the RTs and the bio-based PCM are listed in Table 1.

Table 1. Properties of RT24, 25, 26 and PureTemp25 as given by manufacturers´ datasheets [36,37].

Properties	RT24	RT25	RT26	PureTemp25
T_m (main peak) [°C]	24	25	26	25
Density liquid [kg/L]	0.77 (40 °C)	0.76 (at 40 °C)	0.75 (at 30 °C)	0.86
Density solid [kg/L]	0.88 (15 °C)	0.88 (at 15 °C)	0.90 (at 20 °C)	0.95
Specific Heat Capacity (liquid) [kJ/kg×K]	-	-	-	2.29
Specific Heat Capacity (solid) [kJ/kg×K]	-	-	-	1.99
Spec. Heat Capacity [kJ/kg×K]	2	2	2	-
Heat storage capacity [kJ/kg]	160 (16–31 °C)	170 (16–31 °C)	180 (19–34 °C)	187 *

**Measurement interval not available in the Datasheet.*

2.3. Methods

This section reports the methods used to investigate the TES properties of the PCMs presented in Sections 2.1 and 2.2. Two alterative heat-flux DSC methods (Figure 1), namely the dynamic and a stepwise method, have been considered for this purpose. The German Standards DIN 51005 [38] and DIN 51007 [39] have been considered as a reference to perform the DSC tests (generally applied for this test method to any kind of material), while, the IEA standard procedure [40] is followed to determine the heat storage capacity of PCMs.

(a)

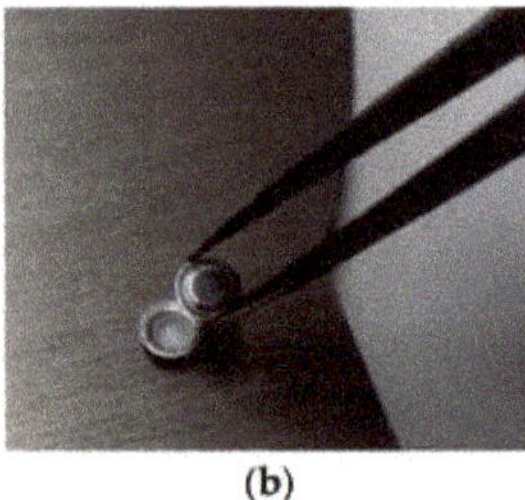
(b)

(c)

Figure 1. (**a**) Differential Scanning Calorimetry (DSC) 214 Polyma equipment from company Netzsch (Selb, Germany, with a T work range −170 °C to 600 °C — Heating/Cooling rate 0.001 K/min to 500 K/min — Indium Response Ratio > 100 mW/K — Resolution (technical) 0.1 µW — Enthalpy precision ±0.1% for indium, ± 0.05% to ±0.2% for most samples), (**b**) aluminum sample holders (maximum volume capacity of 40 µL) and (**c**) position into the DSC device.

Characterizing the PCMs was done by measuring three cycles over three different temperature ranges: (i) solid phase (−20 to 10 °C); (ii) phase change/transition (melting and/or crystallization), (10 to 30 °C); and (iii) liquid phase (30–60 °C). Figure 2 shows these three ranges that occur during the DSC heating/cooling tests, and also describes the typical enthalpy and specific heat capacity evolutions of non-isothermal PCMs.

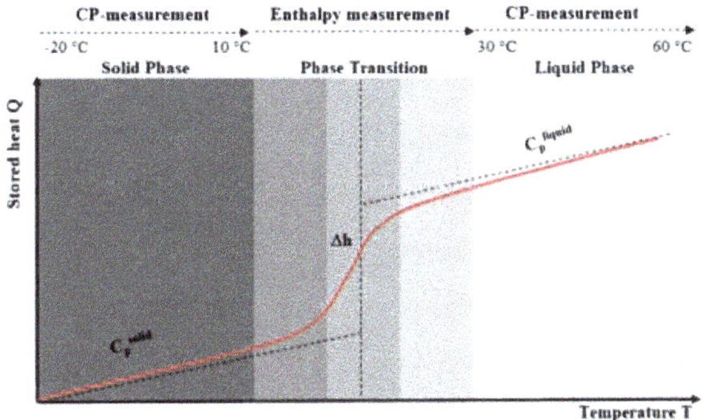

Figure 2. Separation of the measurements in 3 phases: (i) heat (sensible) capacity in solid phase, (ii) heat (latent) capacity under phase change, and (iii) heat (sensible) capacity in liquid phase according to DIN EN ISO 11357-3 [41].

It may be worth highlighting that only for the dynamic DSC tests, and following the IEA indications [40], each sample measurement consisted of 3 (DSC) cycles over the pre-defined temperature range (−20 to 60 °C). Particularly, the first cycle was performed to eliminate previous thermal histories of the specimen; the second cycle was carried out to identify the TES characteristics; and finally, the third one was done mainly to check the reproducibility of the results and the possible appearance of cyclic chemical instability of the material.

2.3.1. Dynamic DSC Method and Evaluation of TES Parameters

The most common way to operate DSC tests is by performing the experimental procedure with a constant heating/cooling rate. This is known as "dynamic DSC", since the heat transfer (energy) is evolving without a necessary thermodynamic equilibrium inside the analyzed sample. However, DSC measurements with a fast and dynamic process and occurrence of phase change phenomena can be affected by this lack of thermodynamic equilibrium, and, hence can provide non-realistic enthalpy temperature data *h (T)*. The shape of the latter can be largely affected while the melting point may be systematically shifted towards higher values for the heating cycles, or lower crystallization temperatures for the cooling process.

This will result in significant errors in the temperature-dependent measurements, such that the heat supply/release, referred to each temperature record, cannot be objectively attributed to the real TES values. For this reason, carrying out measurements with different heating rates and mass variations can provide good information about the influence of the measured variables, detects the needed thermal equilibrium status of the material and verifies the soundness of the measured data.

From a practical point of view and to avoid the above-mentioned complications, the heating (or cooling) rate must be low enough to allow the sample to be measured close to the various isothermal states and with reasonable accuracy. The procedure described in the IEA standard [40] gives some direction in this sense and it was used to control/solve heat rate issues. This was done by changing the heating rate slowly from 10.0 K/min (high rate) to 0.125 K/min (low rate), or even lower, until the temperature difference between two inflection points of consecutive enthalpy curves (i.e., the peaks of the corresponding c_p-T curves), is lower than a predefined threshold (normally fixed by the standard to 0.2 K). Thus, the maximum permissible heating rate will be the minimum heating rate of two consecutive heating curves which comply with the above criterion [40]. A heating rate of 10 K/min was adopted for the measurements in the sensible range following the DIN 51007 [39]. In these measurement conditions, large quantity samples were tested in order to achieve a proper signal from the device.

The enthalpy change (at both sensible and latent stage), $dH(T)$, of a sample, can be evaluated by integrating the heat flow registered during the DSC measurements. It can be assumed that a small variation of enthalpy dH (or the specific one, dh) is equal to a small amount of heat dQ (or the specific one, dq) added/released [42,43]. This is valid for those processes characterized at constant pressure, as it happens during DSC experiments.

Therefore, it can be written that:

$$dH = dQ \quad m^{sp} dh = m^{sp} dq \tag{1}$$

being

$$dq = c_p^{sp} \cdot dT \tag{2}$$

where m^{sp} is the sample mass, chosen as small as possible to comply with the instrument-specific requirements and to reach the state of equilibrium more quickly [39]. However, m^{sp} should be large enough to guarantee the representativity of the material sample. Furthermore, dT represents a small variation of temperature, and $c_p^{sp}(T)$ the specific heat capacity.

Based on the recorded heat flow (DSC signal), the specific heat capacity of the sample $c_p^{sp}(T)$ can be easily determined from Equation (2), as follows

$$c_p^{sp} = \frac{dq}{dT} \xrightarrow[\text{under finite } \Delta T]{} c_p^{sp} = \frac{\Delta q}{\Delta T} \tag{3}$$

while the enthalpy variations have been directly evaluated from Equation (1).

In general, changes in enthalpy (latent heat) or the specific heat capacity (sensible heat) of an examined sample are determined by recording the absorbed heat between two equilibrium states, assigned as baselines of the acquired measurement curves. It is worth highlighting that the baseline-construction due to the specific heat capacity, measurements outside its melting range, is determined by performing three measurements: (1) empty measurement, (2) calibration measurement, and (3) real sample measurement for each temperature range (more details are available in [39]). This is due to the correction of measurement results possible affected by asymmetries and to compensate device specific errors.

2.3.2. Stepwise DSC Method and Evaluation of TES Parameters

A stepwise method is used as an alternative for reducing the heating rate effect that is affecting the measurements of dynamic DSC tests. In the stepwise procedure, the net heat applied in a certain temperature interval, is the same as the amount considered in the dynamic DSC tests (see Figure 3a,b). However, the whole interval is sub-divided into small sub-steps. This will favor the accumulation time to reach the thermodynamic equilibrium, between each temperature step, and will improve the temperature resolution (indicated in Figure 3b), which was assumed to be 1 K in this study [44]. In this procedure, isothermal (equilibrium) states are always awaited between two subsequent temperature steps (i.e., 1 K). The waiting needed to reach this state of equilibrium mainly depends on the selected (temperature) step and can only be determined empirically. In this work, the waiting times were taken 10 min at the beginning of the melting and 20 min by reaching the melting peaks. More details are described in Section 4.

The total specific heat, supplied over the individual intervals, is obtained by summarizing the heat increments supplied at each sub-step $\Delta q^{\text{stepwise}}$. Thus, the experimental approach used in the stepwise method to analyze the $h(T)$ response of the materials, along with the melting peak temperature T_m, follows the same relationships as proposed for the dynamic DSC test:

$$\Delta H = m^{sp} \Delta h = m^{sp} \Delta q^{\text{stepwise}} \tag{4}$$

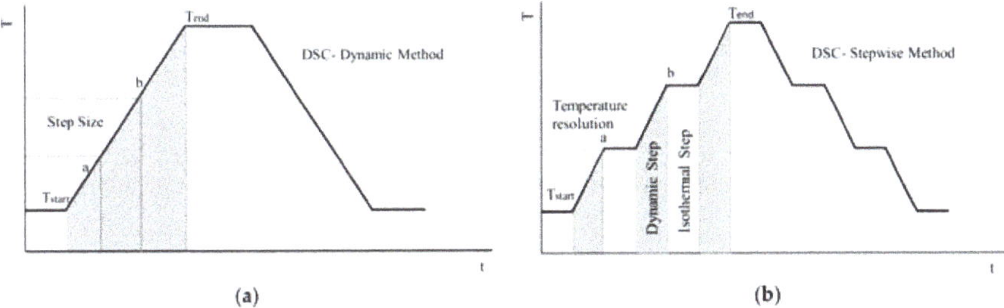

Figure 3. *T(t)* program for (**a**) dynamic DSC method and (**b**) stepwise DSC method.

2.4. Experimental Program

For the measurement of each parameter pair (c_p, h) three samples per material (paraffin-based RT24, RT25, RT26 and bio-based PureTemp25, as shown in Figure 4) were taken into account. Table 2 reports an overview of the full experimental program considered for the dynamic and stepwise method, respectively. The table gives information regarding the considered heating/cooling rates (in both dynamic and stepwise methods), which range from 10 K/min to 5 K/min, 2 K/min, 1 K/min, 0.5 K/min, 0.25 K/min, and 0.125 K/min. Furthermore, the investigated parameters, considered masses and test type of DSC method are also highlighted in the table.

Figure 4. DSC specimens for the measurement of specific heat capacity and enthalpy of Phase Change Materials (PCMs) following the dynamic and stepwise procedures.

Table 2. Samples, heating rates, investigated parameters and DSC test-type applied in this study.

Tests	**Materials (Mass Values in mg)**													
	RT24			**RT25**			**RT26**				**PureTemp25**			
	P1	P3	P3	P1	P2	P3	P4	P1	P2	P3	P1	P2	P3	P4
Dynamic Heat Rate (R), (R: 10, 5, 2, 1, 0.5, 0.25, 0.125 K/min)	✔ 18.8	-	-	✔ 17.2	-	-	-	✔ 17.7	-	-	✔ 18.9	-	-	-
Dynamic Sample Mass (R: 0.125, 0.25 K/min)	-	-	-	✔ 16.3	✔ 18.4	✔ 18	✔ 7.9	-	-	-	✔ 18.9	✔ 19	✔ 17.7	✔ 10.1
Dynamic - Enthalpy Temperature h (T) (R: 0.125 K/min)	✔ 13.3	✔ 16.5	✔ 16.9	✔ 16.3	✔ 18.4	✔ 18	-	✔ 17.7	✔ 18.5	✔ 16.1	✔ 18.9	✔ 19	✔ 17.7	-
Dynamic - Heat Capacity c_p (T) (R: 10 K/min)	✔ 14.3	✔ 16	✔ 16.1	✔ 17.2	✔ 18.4	✔ 18	-	✔ 16.4	✔ 18.9	✔ 12.4	✔ 10.5	✔ 18.9	✔ 18	-
Step-wise Heat Rate (R:2, 1, 0.5, 0.25, 0.125 K/min)	-	-	-	✔ 16.3	-	-	-	-	-	-	✔ 18.9	-	-	-
Step-wise - Enthalpy Temperature h (T) (R: VAR)	-	-	-	✔ 16.3	-	-	-	-	-	-	✔ 18.9	-	-	-

Abbreviations in table: R: heating Rate; h: enthalpy; c_p: specific heat capacity; VAR: variable; RT: RubiTherm.

3. Dynamic DSC Results

3.1. Heating Rate Effect

Dynamic DSC tests were performed for all PCMs (3 paraffin-based and 1 bio-based), considering several heating rates (ranging from 10 K/min to 0.125 K/min) and various masses. The effect of the heating rate, between 10 °C to 50 °C, was investigated for only one of the samples per each PCM type (labelled as P1 in Table 2 of either RT24, -25, -26 or PureTemp 25).

Figures 5–8 show the resulting $c_p(T)$ and $h(T)$ (latent-only) curves of the samples abovementioned. It can be observed that the melting temperature T_m, defined as the temperature at which the maximum peak of the $c_p(T)$ curve is registered, of all four samples mainly depends on the heating rate. T_m is shifting progressively to lower values with decreasing heating rates. When comparing the different DSC measurements of each sample, the influence of the heating rate on the curves becomes clearly visible. It may be worth mentioning that even the $c_p(T)$ and $h(T)$ (latent-only) curves are largely influenced by the rate of heating during the melting transition, the related measured total specific latent heat (enthalpy) is independent on the heating rate.

From the experimental results (Figures 5–8) it can be also observed that the acceptable heating rate to be selected in order to stay below the defined threshold value of 0.2 K (i.e., the difference between a high and successive smaller heating rate, as proposed by the IEA standard [40] and described in Section 2.3.1), must be 0.125 K/min (or lower) for RT25 and PureTemp 25, and 0.250 K/min (or lower) for RT24 and RT26. This would reduce as much as possible the heating rate issues for the considered PCMs.

The c_p curves of RT24, -25 and -26 with different heating rates contain mainly two (local) peaks, at different temperatures. These double-peak curves actually become more pronounced for lower heating rates. The first local one, as for each curve and considering 0.125 K/min, was registered at 6.2 °C for RT24 (Figure 5), at 17.6 °C for RT25 (Figure 6) and 17.4 °C for RT26 (Figure 7). The second local (and also maximum) peak, which is representing the melting of paraffin, while adopting a heating rate of 0.125 K/min, was achieved at 25.0 °C for RT24 (Figure 5), at 26.5 °C for RT25 (Figure 6) and 26.2 °C for RT26 (Figure 7).

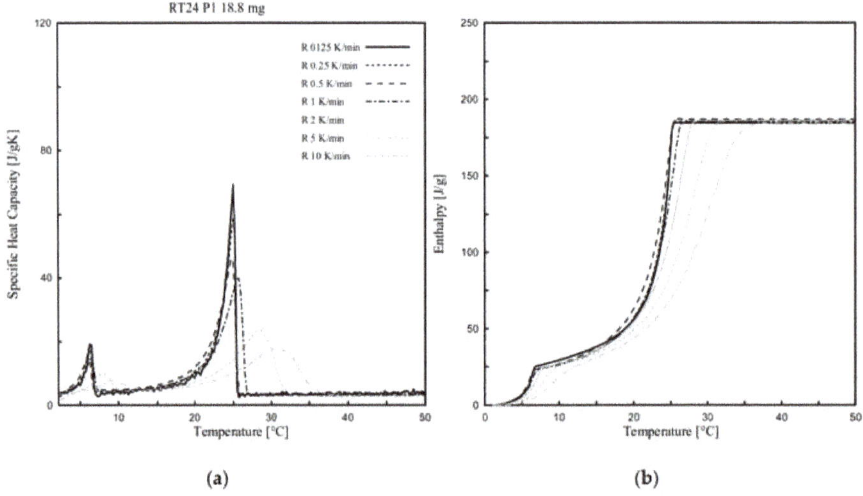

Figure 5. Dynamic DSC measurements: (**a**) $c_p(T)$ and (**b**) $h(T)$ curves (latent only) with different heating rates for RT24.

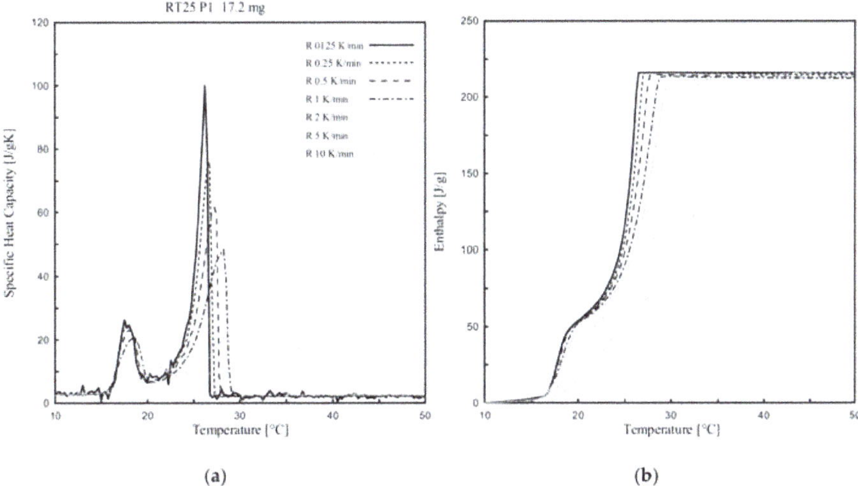

Figure 6. Dynamic DSC measurements: (**a**) $c_p(T)$ and (**b**) $h(T)$ curves (latent only) with different heating rates for RT25.

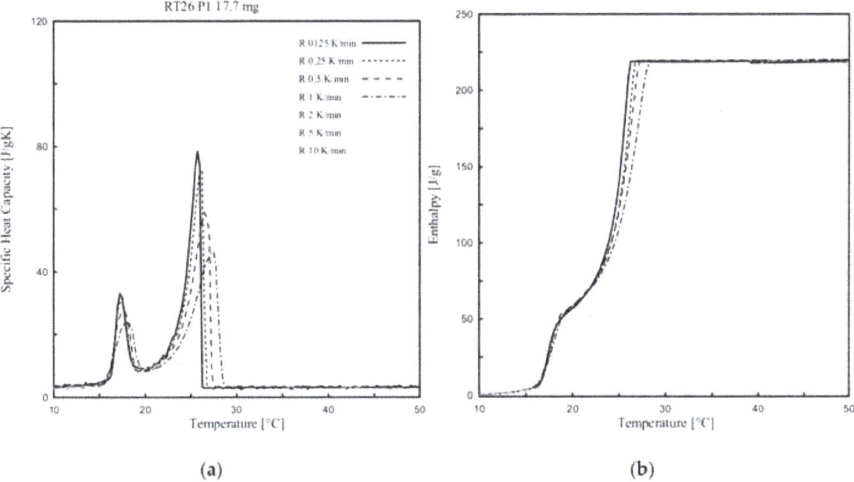

Figure 7. Dynamic DSC measurements: (**a**) $c_p(T)$ and (**b**) $h(T)$ curves (latent only) with different heating rates for RT26.

Similar behavior can be observed from the c_p- and h-T (latent-only) DSC curves of the bio-based PureTemp25. From the DSC tests, two adjacent peaks appear in the c_p-T curves. The first melting temperature of the highest peak was measured at 24.6 °C (Figure 8), while the second local (not maximum) peak was registered at a higher temperature of 26.2 °C. The actual melting peak temperature of 24.6 °C was close to the declared value in the datasheet, which is 25 °C (Table 1).

The presence of a local (not maximum) peak, that was detected for both PCM types, could be the result of either a mixture of different compounds, having different chain lengths, or a possible occurrences of solid–solid latent phase transitions prior the melting.

These results also show that the declared melting temperatures in the datasheets (Table 1) are lower than the ones measured with the dynamic DSC tests. This was expected, since measuring the exact melting peak temperature requires a heating rate that should be as low as possible, meaning theoretically nearly 0 K/min. In practice, this is not feasible, since extremely low heating rates result in very weak signals and characterized by huge noise on the measured signal and implicitly limits the choice of the lowest possible heating rate to be assigned.

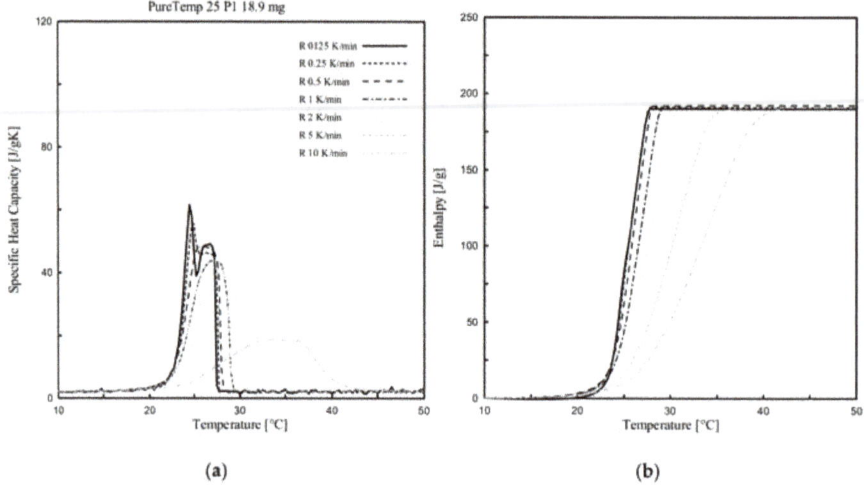

Figure 8. Dynamic DSC measurements: (**a**) $c_p(T)$ and (**b**) $h(T)$ curves (latent only) with different heating rates for PureTemp25.

3.2. Mass of Sample Effect for RT25 and PureTemp 25

The effect of the sample mass in the dynamic DSC experiments is discussed in this subsection. Paraffin-based RT25 and bio-based PureTemp 25 were considered for this aim and two masses per each PCM were investigated. Tests were performed considering different heating rates (again ranging from 10 K/min to 0.125 K/min), with temperatures raising between 10 °C and 50 °C; 7.9 mg and 16.3 mg samples were tested for RT25, while 10.1 mg and 18.9 mg for the PureTemp 25.

Figure 9 shows the comparative evolution of the melting peak temperatures, T_m, against the different heating rates by considering the aforementioned two masses (low mass 7.9 mg and high mass 16.3 mg, respectively). It can be observed that the measured melting points, through the DSC tests for the small sample mass (7.9 mg), rather quickly approximates the melting peak temperature (i.e., 26.0 °C of the lowest heating rate 0.125 K/min). Hence, T_m was 26.50 °C at 0.50 K/min and 26.25 °C at 0.25 K/min. Contrarily, the sample with the higher mass (16.3 mg) showed more influence of the heating rate: i.e., T_m = 27.25 °C at 0.50 K/min, 26.75 °C at 0.25 K/min and 26.25 °C at 0.125 K/min.

Also, for PureTemp 25 the comparison of melting peak temperatures, T_m, against the various heating rates and two different masses (10.1 mg and 18.9 mg, respectively) is presented in Figure 10. As for RT25, the sample with the lower mass (10.1 mg) showed a less influence of the heating rate on the melting peak temperature (e.g., T_m = 24.50 °C at 0.25 K/min and 24.25 °C at 0.125 K/min). For the higher mass (18.9 mg) this was, T_m = 25.00 °C and 24.50 °C, at 0.25 K/min and 0.125 K/min, respectively.

It should be noted that all curves, which were obtained from measurements with heating rates less than 0.125 K/min, had high fluctuations due to signal noises, which make the evaluation of the data and baseline construction difficult to build up. For this reason, a heating rate of 0.125 K/min was considered as the lowest possible choice at which the instrument could still detect signals that were easy to be analyzed. Moreover, 0.125 K/min, for all tests, provides a proper accuracy to fulfill the IEA standard requirements [40], see Section 2.3.1.

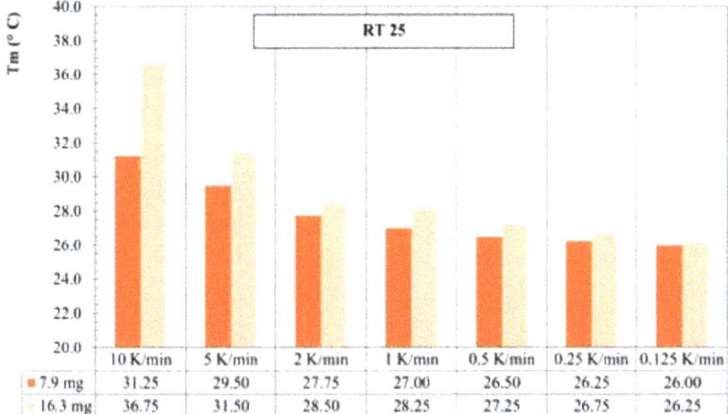

Figure 9. Melting peak temperature T_m for dynamic DSC measurements with several heating rates and 2 different masses for RT25.

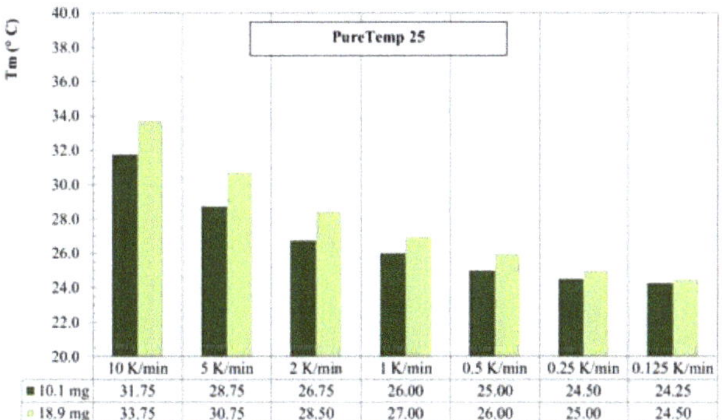

Figure 10. Melting peak temperature T_m of dynamic DSC measurements with several heating rates and 2 different masses for PureTemp25.

The results in this section showed that the selection of the "heating rate" and "sample mass" arises mainly from a compromise between accuracy of the measurements (enthalpy, specific heat capacities and melting peak values) and mitigation of the data due to heating rate effects. Besides the effect of instrument precision and adopted parameters, the measurement accuracy also depends on the representativeness of the sample, which also requires the analysis of larger masses. However, higher mass samples need a strong reduction of the heating rate for achieving a proper thermal equilibrium and valid TES analysis. From the practical point of view, very low heating rates go along with huge signal noises. Contrarily, in low mass samples, thermal equilibrium can be reached more easily with higher heating rates. Experimentally, the most representative mass should be found first, followed by choosing the proper "noiseless" heating rate.

3.3. Specific Heat Capacities, Enthalpy and Melting Points for Heating and Cooling

For each type of PCM, six samples were investigated through dynamic DSC tests under both heating and cooling. Three of them were used for $c_p(T)$ characterization, while the remaining three for $h(T)$. Before and after the non-isothermal phase change (either melting or crystallization), a heating/cooling rate of 10 K/min was adopted for all tests to minimize noise ratios in the signal [39]. In the latent stage, the considered heating/cooling rates were of 0.250 K/min for RT24 and RT26 and 0.125 K/min for RT25 and PureTemp 25. These assumptions were done following the IEA standard procedure [40] to mitigate the influence of the heating/cooling rate for each compound, as discussed in previous Section 3.1.

Figure 11 shows the results of dynamic DSC measurements done for the paraffin waxes (RT24, RT25, RT26) and the bio-based PCM (PureTemp 25), for determining the c_p [$J/g \times K$] versus T [°C] response, in both latent and sensible ranges. All DSC curves are characterized by an almost sensible behavior in the temperature range far from the melting and crystallization points (i.e., in the solid range −20 °C to 10 °C and the liquid range 30 °C to 60 °C), while a non-isothermal latent behavior appears in the phase change region (either with solid-to-liquid and liquid-to-solid responses). It can be also observed that the results show very little (almost negligible) deviation, for all tests, as indicated by the gray area. In the liquid-sensible ranges (i.e., between 30 and 60 °C), the determined specific heat capacities varied between 2.1 and 2.3 J/g × K (for both heating/cooling), for all samples RT24, RT25, RT26 and PureTemp 25. Then, in the lower temperature ranges (solid-sensible responses, between −20 and 10 °C), the c_p values varied between 1.7 and 2.0 J/g × K (for both heating/cooling), for RT24, RT25, RT26 and PureTemp 25. In the latent responses, pronounced peaks developed that represent the solid–liquid and liquid–solid phase changes of the various PCMs.

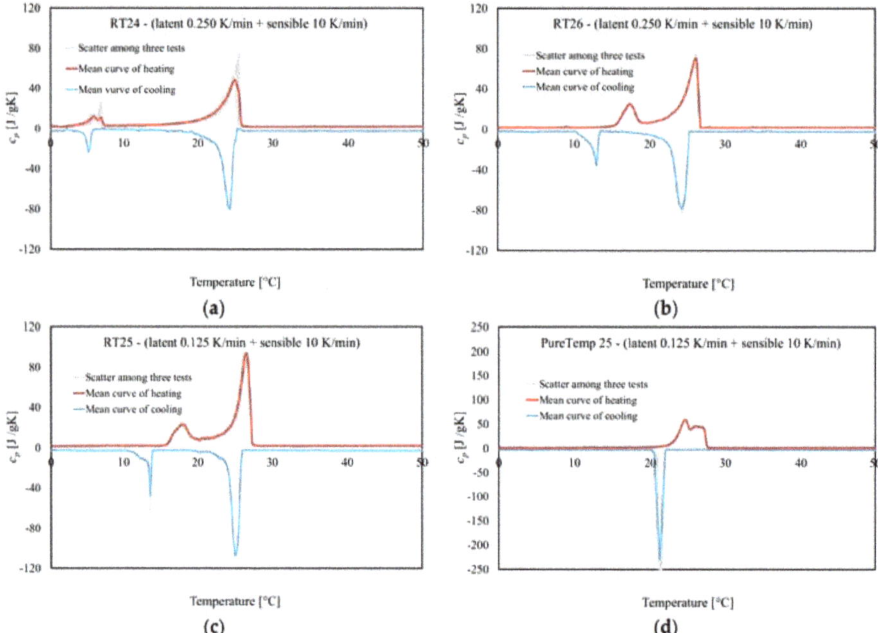

Figure 11. Specific heat capacities following the dynamic DSC tests (for heating between 0 °C and 50 °C and cooling between 50 °C and 0 °C): (**a**) RT24, (**b**) RT26, (**c**) RT25 and (**d**) PureTemp 25. Note: the total specific heat capacities have been constructed by additively linking the sensible parts with the latent ones.

In Figure 12, the resulting enthalpy curves (latent part only), measured with the dynamic DSC method, are shown for all materials. The plotted curves represent the mean values of the measurements done for three samples, over three cycles. These data mainly represent the absorbed/released thermal energies for heating and cooling, respectively, and the unitary latent heat during a phase change. The measured total and specific h of each material is shown in Figure 13. It can be observed that the absorbed heat from 0 J/g to the total specific latent heat of each PCM (i.e., 182.1 J/g for RT24, 193.3 J/g for RT25, 211.9 J/g for RT26 and 186.1 J/g for PureTemp 25) is represented by a clear non-isothermal behavior. Similar trends, in a releasing way, from the maximum values of $h(T)$ (i.e., form 178.8 J/g for RT24, 213.4 J/g for RT25, 211.4 J/g for RT26 and 186.8 J/g for PureTemp 25) to 0 J/g, can be observed for the cooling responses. From the data, it can be observed that the bio-based PureTemp 25, compared to the RT ones, is characterized by a faster acceleration of the *h(T)* response. This means that the phase change occurs in a narrower melting or crystallization range, which is often more appreciated in practical passive applications for buildings.

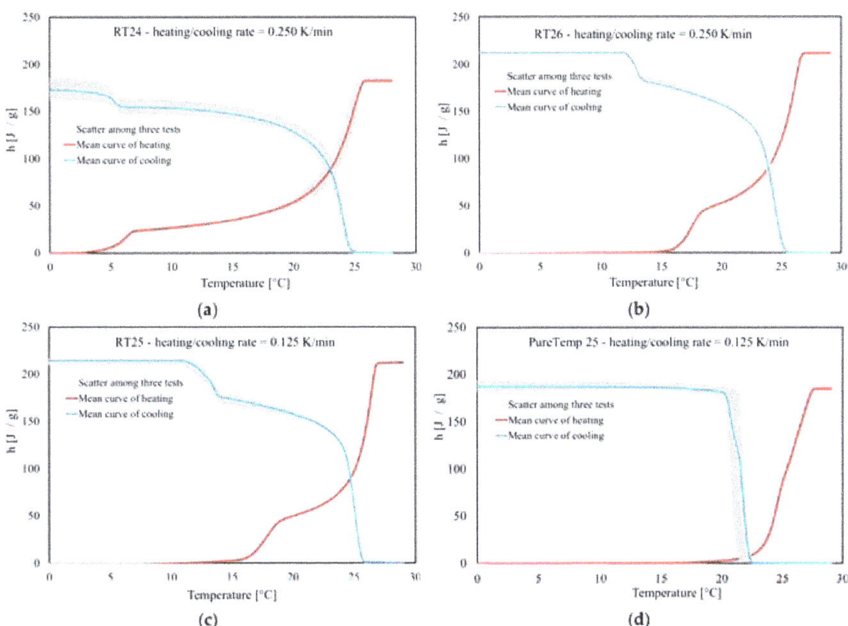

Figure 12. Enthalpy (latent-only) measurements (heating and cooling) following the dynamic DSC tests: (**a**) RT24, (**b**) RT26, (**c**) RT25 and (**d**) PureTemp 25.

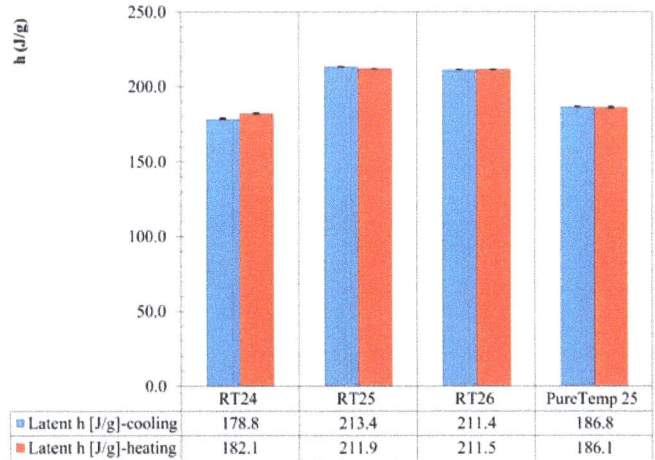

	RT24	RT25	RT26	PureTemp 25
Latent h [J/g]-cooling	178.8	213.4	211.4	186.8
Latent h [J/g]-heating	182.1	211.9	211.5	186.1

Figure 13. Latent absorbed/released h [J/g] during melting/crystallization following the dynamic DSC method.

Finally, the maximum peak temperatures (meaning those points at which the maximum c_p was registered) for either melting or crystallization of all PCMs are compared in Figure 14. The graphs further show the "melting" and "crystallization" temperatures of the three specimens, their mean values and the statistical deviation via error bars. It can be seen that the peak temperature (mean value of three specimens) at which melting happened were detected at 25.0, 26.5, 26.2 and 25.4 °C for RT24, RT25, RT26 and PureTemp 25, respectively. Similarly for the crystallization measurements (again mean value of three specimens), peak temperatures were registered at 23.8, 25.1, 24.2 and 21.1 °C for RT24, RT25, RT26 and PureTemp 25, respectively.

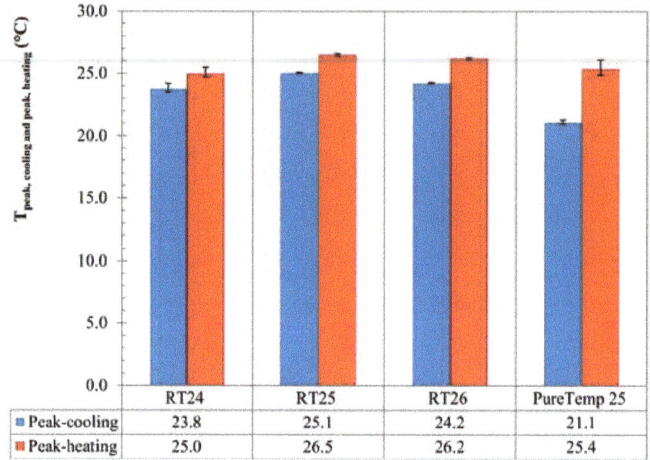

Figure 14. Comparison of the maximum peak temperatures (°C) of all samples in the melting/crystallization range.

In Figure 14 it can also be seen that there is an obvious difference between the cooling and heating peak temperature. Overheating or subcooling effects [18], occurred due to a kinetic delay in phase transformation, could be a possible explanation for the shifted heating/cooling peaks. Upon the attained results, the subcooling is lower in paraffins, also known as alkanes (about 1.2–2 K), compared to the PureTemp 25 (made of fatty acids) with about 4.3 K subcooling. These phenomena occur even if the amount of stored/released heat stayed almost the same (see Figure 13).

Finally, the $T_{onset,heating}$ and $T_{onset,cooling}$ are reported in Table 3. They are representing the extrapolated peak start temperature (defined $T_{p,ini,e}$ as sample transformation temperature by DIN 51007:2019 [39]). These temperature values represent the intersection point of the extrapolated start baseline with the tangent (or a straight) line through the linearly rising or falling part of the endothermic or exothermic peak. The onset temperatures are worth knowing since at this point the lower surface area of the sample (crucible bottom) begins to change phase, as this temperature can be determined sufficiently reproducible and is almost independent of the heating rate. All other temperatures largely depend on test conditions, masses and heating/cooling rates (as shown in Figures 5–8).

Table 3. Comparison of extrapolated peak initial temperatures (assigned to the transformation temperature) of all samples in the melting/crystallization range according to DIN 51007 [39].

PCM Type	RT24 R 0.250 K/min		RT25 0.125 K/min		RT26 0.250 K/min		PureTemp 25 0.125 K/min	
	$T_{onset, heatir}$	$T_{onset, cooling}$	$T_{onset, heatir}$	$T_{onset, heatir}$	$T_{onset, heatir}$	$T_{onset, heatir}$	$T_{onset, heatir}$	$T_{onset, heating}$
$P_{1,2,3}$	3.0	24.3	15.5	25.7	15.6	25.1	23.0	21.9

4. Stepwise DSC Results

4.1. Heating Rate Effect

Tests with different heating rates (ranging from 2 K/min to 0.125 K/min) have also been performed for the stepwise DSC method. These activities were scheduled for RT25 and PureTemp 25 under temperatures ranging between 10 and 30 °C (a range which is relevant for the phase transition).

Figure 15a and a show the stored heat, per each temperature interval (being 1 K, representing the difference between two isothermal temperature increments), while the accumulated specific enthalpy (latent part only) is shown in Figures 15b and 16b. Actually, these results represent the determined partial enthalpies and the cumulated ones between 10 °C and 30 °C from the measurements done for RT25 and PureTemp 25.

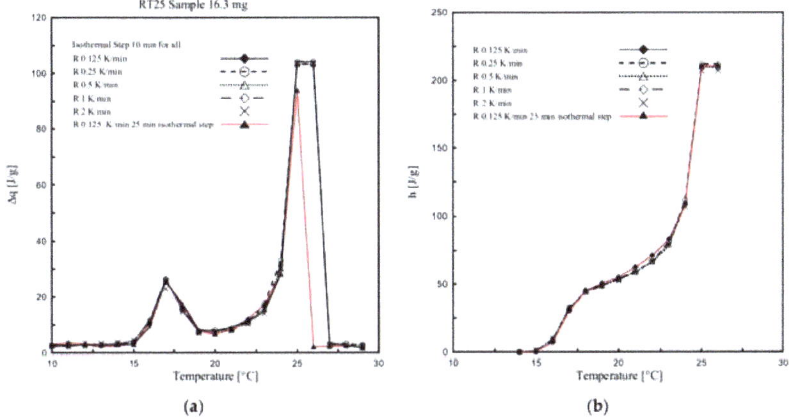

(a) (b)

Figure 15. (**a**) Δq *(T)* and (**b**) *h(T)* curves determined through stepwise DSC and considering different heating rates for RT25.

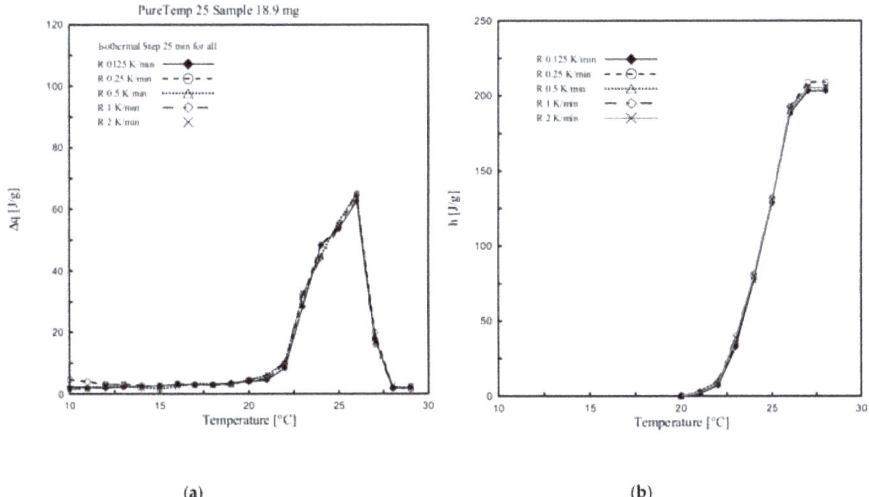

(a) (b)

Figure 16. (**a**) Δq *(T)* and (**b**) *h(T)* curves determined through stepwise DSC and considering different heating rates for RT25.

It is shown in Figures 15 and 16 that all the results, having different heating rates, are completely overlapping. Particularly, the Δq-T and h-T (dotted) curves obtained by considering different heating rates, i.e. 2, 1, 0.5, 0.25 and 0.125 K/min, show almost the same values, with a very small difference in terms of either Δq *or* h values. The enthalpy deviation between the different heating curves has a maximum of 5 J/g, and this occurred mainly close to the melting range. This difference is less than 3% of the total and specific heat absorbed in the considered temperature range. This means that with the stepwise method the influence of the heating rate, which classically affects dynamic DSC measurements, can be neglected. (see Section 3.1 where the results were obtained through dynamic DSC measurements).

It may be worth mentioning that the length of the relaxation time (namely, isothermal step) is an important parameter for the temperature resolution of the accumulated heat, especially in the temperature ranges close to the melting peak (defined as the point in which the maximum amount of stored heat can be measured). Particularly, to reach thermal equilibrium in these zones a longer isothermal step is required. This can be observed in Figure 15a where the highest peak (readable in the ranges 25 °C and 26 °C) for all tests is obtained for heating rates of 2, 1, 0.5, 0.25 and 0.125 K/min, while performed with an isothermal step of 10 min. To evaluate the sensitivity of the results regarding the duration of the applied isothermal step, an additional measurement was conducted for a heating rate of 0.125 K/min and where the isothermal step was increased from 10 min to 25 min with the aim to reach proper thermodynamic equilibrium during the phase change transition (see the red curve in Figure 15). The results clearly show the influence of this increased isothermal step.

4.2. Enthalpy and Phase Change Temperature

The enthalpy measurements, i.e., the absorbed specific heat along the considered temperature steps, and phase change temperatures obtained by an adopting isothermal step time of 25 min and various heating rates, are reported in this subsection. Both sensible and latent absorbed heat, in the range of 10 °C and 30 °C, are evaluated.

Figures 17 and 18 show a histogram of the stored heat, Δq *(T)*, between the temperature intervals, representing the temperature resolution of the acquired data, of 1 K, for the aforementioned tests. The corresponding tabular data is provided in Table 4. From the results, it can be seen that the melting of RT25 mainly starts at 15 °C and ceases at 26 °C. A baseline construction was used to evaluate and separate the sensible part form the latent one. The sensible absorbed heat was evaluated at 2.10 J/g, which was in agreement with the value declared in the datasheet, i.e., 2 J/g, (see Table 1). The melting peak temperature as indicated in Figure 17 can be appreciated in the range of 25–26 °C. The (latent) melting of the PureTemp 25 takes place in a smaller temperature interval (from 21 °C to 28 °C, see Figure 18). The absorbed sensible heat of PureTemp 25 was evaluated to 2.89 J/g while its melting point falls in the range of 25–26 °C.

The phase transition enthalpy (considering the latent part only), in the defined melting range of 15 °C to 26 °C and constructing the baseline at 15 °C, was 207.21 J/g for RT25 (see Table 4). This value is slightly different to the one measured through the dynamic DSC measurements (211.9 J/g reported in Figure 13). The total latent heat of PureTemp 25, evaluated in the temperature range between 21 °C and 28 °C, is 207.5 J/g (see Table 4), which somehow deviates from the value measured using the dynamic DSC method (186.1 J/g, showed in the Figure 13). However, it may be worth mentioning that the results obtained from dynamic DSC tests or the stepwise methods are highly depending on the baseline construction, which separates the sensible heat from the latent part. As the baseline-construction can be done more precisely for the dynamic DSC method rather than for the stepwise one, the final results of the latent heat by using both the aforementioned methods might be significantly affected.

From the results it can be stated that the stepwise DSC method is capable of cleaning and eliminating the heating rate effects, which is classically of interest for dynamic DSC measurements. It is a more appropriate method that is able to capture the total enthalpy in the defined temperature range by applying predefined temperature steps, and also for validating the melting peak (range) more precisely than with dynamic DSC tests. However, some drawbacks are still characterizing the stepwise DSC method, which is mainly that the temperature resolution adopted in these approaches is still too high. The smaller the temperature steps, the better the temperature resolution. However, the measuring time could be considerably longer as the step sizes decrease. In general, the shape characterization of $c_p(T)$ curves can be achieved faster (and more properly) by using the dynamic DSC method than the stepwise one.

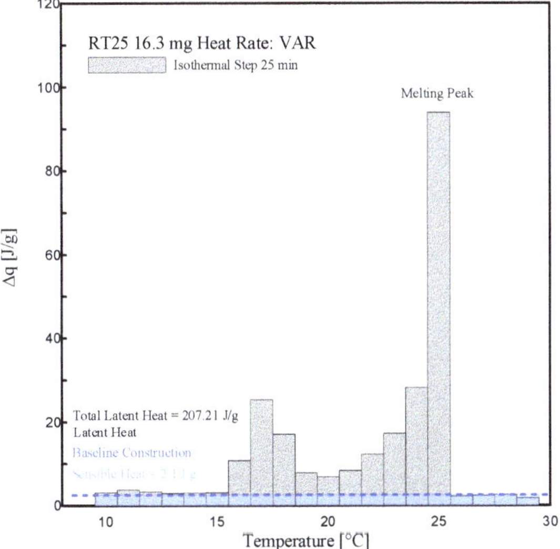

Figure 17. Histogram view of stored heat Δq *(T)* for RT25, at several temperature intervals and adopting an isothermal time step of 25 min.

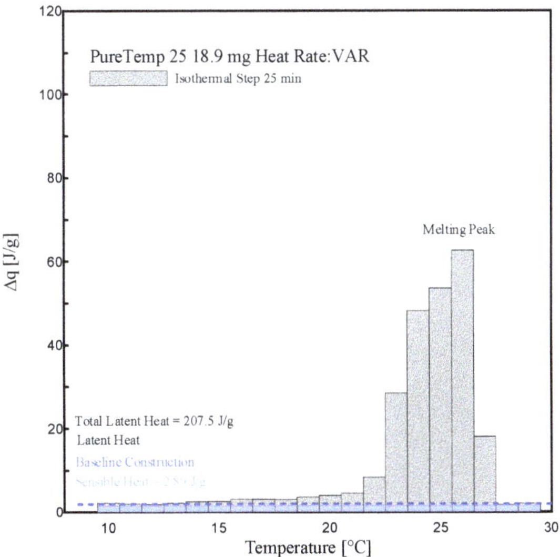

Figure 18. Histogram view of stored heat Δq *(T)* for PureTemp 25, at several temperature intervals and adopting an isothermal time step of 25 min.

Table 4. Stored heat values (sensible = S, latent = L and total = S + L) of RT 25 and PureTemp 25, as shown in Figures 17 and 18.

Temp. Range [°C]	RT25 (Mass: 16.3 mg)		PureTemp25 (Mass: 18.9 mg)	
	Δq (total: S + L) [J/g]	Δq (L) [J/g]	Δq (total: S + L) [J/g]	Δq (L) [J/g]
15 to 16	3.03	0.93	-	-
16 to 17	10.70	8.60	-	-
17 to 18	25.33	23.23	-	-
18 to 19	16.96	14.86	-	-
19 to 20	7.77	5.67	-	-
20 to 21	6.77	4.67	-	-
21 to 22	8.26	6.16	4.48	1.59
22 to 23	12.18	10.08	8.30	5.41
23 to 24	17.23	15.13	28.39	25.50
24 to 25	28.13	26.03	48.16	45.27
25 to 26	93.95	91.85	53.53	50.64
26 to 27	-	-	62.56	59.67
27 to 28	-	-	22.32	19.43
-	230.31	**207.21**	227.74	207.48

In summary, it can be observed that the temperature resolution for DSC tests depends on the amount of analyzed intervals, needed to measure the heat storage capacity, within the considered temperature range. Based on the chosen DSC method the "resolution" of the data, representing the heat supplied to the sample and the corresponding temperature range, can differ. In the stepwise method heating is applied in small intervals called "steps". Thus, the resolution is equal to the height (or length) of the temperature steps, however it is required that the start and end of each step must be in an isothermal state. For dynamic DSC tests, the interval is continuously scanned by heating of the sample. There are actually no steps waiting for thermodynamic equilibrium. This means that the heating rate needs to be low enough to assure thermodynamic equilibrium within the entire sample. Theoretically, this will occur by employing a heating rate close to 0 K/min., which is practically not feasible. This means that the heat supplied to the sample, according to each data recording (equal to the enthalpy change), cannot be fully assigned to the measured temperature. Too high heating rates may cause significant errors (less "accuracy" of the results) in the data of the heat stored as a function of temperature. However, in DSC dynamic tests the resolution of the data itself is only restricted by the data recording system.

5. Conclusions

In this work, a detailed experimental program is reported for analyzing the Thermal Energy Storage (TES) capacity of three paraffin-based and one bio-based PCM, employable for construction and buildings applications. For this aim, the paraffin-based waxes, RT24, RT25 and RT26, and an eco-friendly bio-based PCM, PureTemp 25 were examined using DSC testing equipment. All selected PCMs have melting/ crystallization temperatures within the well-known comfort zone temperature for buildings, e.g., ranging between 19 and 26 °C. Heat storage capacities, melting/ crystallization responses and enthalpies, under both sensible (solid and liquid) and latent TES responses, were investigated through DSC tests.

Two alternative methods were performed and compared, i.e., a dynamic DSC and a stepwise DSC method. Based on the results reported in this work, the following conclusions can be drawn:

- Dynamic DSC tests, performed for all 4 different PCMs and considering several heating rates (ranging from 10 K/min to 0.125 K/min), showed that the measurable melting peak temperatures T_m (indicated as that point at which the maximum peak of $c_p(T)$, or alternatively the inflection point of $h(T)$, can be registered), mainly depends on the heating rate. T_m can be shifted progressively to lower values as the heating rate decreases and vice versa;
- The $c_p(T)$ and $h(T)$ curves under dynamic DSC strongly depend on the selected heating rates. Only the related measure of the accumulated (total) specific (unitary) latent heat is independent to the heating rate. In general terms, the total achievable h is the same when different heating rates are assumed, even though the $h(T)$ curves and their slopes, $cp(T)$, are heating rate dependent;
- Dynamic DSC tests, performed by considering 2 different (low and high) sample masses, for RT25 and PureTemp 25, highlighted that the specimens with a lower mass are less affected by the heating rate. However, they suffer higher measurement noise and less representativity of the material under investigation;
- Dynamic DSC results of higher mass samples showed more stable measurements, especially in the sensible parts, and seemed to be more representative for the behavior of the investigated PCMs. However, they turned out to be more sensible to heating rate effects under latent heat storage.
- It can be concluded that the selection of the proper "heating rate" and "sample mass" (aimed at fulfilling the IEA requirements) arises from a compromise between accuracy of the measurements (i.e., enthalpy, specific heat capacities and melting point) and the mitigation of the heating/cooling rate effects.
- It can be stated that reliable and reproducible results can be achieved for characterizing the aforementioned (paraffin- and bio-based) PCMs by following the IEA standard procedure and adopting the dynamic DSC method.
- Results following the stepwise method demonstrated that both $\Delta q(T)$ and $h(T)$ results, obtained by considering several heating rates (2, 1, 0.5, 0.25 and 0.125 K/min), presented almost the same trend and similar values (with differences significantly less than 0.2 °C and/or in terms of enthalpy of max. 5 J/g between them). This means that with the stepwise method the heating rate influence, which classically affects the dynamic DSC, can be fully controlled.
- The main drawbacks of the stepwise method, however, are related to its enormously time-consuming character, imposed by the test procedure. The required length of the relaxation time (namely, isothermal step) is sometimes too high to achieve proper results, especially in the melting region.
- The temperature resolution of the stored heat is considerably higher, using the stepwise method, in comparison to the dynamic one. However, a maximum resolution of 1 K, as those employed in the stepwise method of this study, is not high enough to obtain the $c_p(T)$ curve of the material, which would require smaller discretized temperature steps, and also does not allow describing the melting temperature of the material in a concise way.
- The results obtained following both dynamic and stepwise methods are dependent on the baseline construction, which allows separating the sensible from the latent heat part. In this regard, the baseline-construction can be more precisely built in the dynamic method than the stepwise one, because more valuable measurement data can be evaluated within very small temperature steps. For this reason, the stepwise method is more appropriate to measure the total enthalpy (both sensible and latent heat) in defined temperature steps and also validating the melting peak reached from the dynamic procedure, while the dynamic measurements can ascertain the stored latent heat much more quickly and precisely. Then using the IEA Standard, dynamic measurements are less time-consuming and also more precise to characterize the melting behavior of the material.

Further experimental characterizations of the TES capacity of several other types of bio-based PCM, to be employed as environmentally friendly substitute of petroleum-based PCM in cement-based systems, are currently under development.

Author Contributions: Conceptualization, M.N.S. and A.C.; methodology: materials, methods and processing M.N.S., C.M. and A.C.; data curation, M.N.S., C.M. and A.C.; writing—original draft preparation, M.N.S. and A.C.; writing—review and editing, everybody; supervision, A.C. and E.K.; project administration, E.K.; funding acquisition, E.K. All authors have read and agreed to the published version of the manuscript.

Funding: The second author acknowledges the Alexander von Humboldt-Foundation (www.humboldt-foundation.de/) for funding his position at the WiB – TU Darmstadt under the research grant ITA-1185040-HFST-P (2CENERGY project). The support to networking activities provided by the PoroPCM Project (part of the EIG CONCERT-Japan funding, http://concert-japan.eu/) is also gratefully acknowledged. This work represents a preliminary study of the forthcoming research activities to be realized within the framework of the NRG-STORAGE project (number 870114, 2020-2024), financed by the European Union H2020 Framework under the LC-EEB-01-2019 call, H2020-NMBP-ST-IND-2018-2020/H2020-NMBP-EEB-2019, IA type.

Conflicts of Interest: The authors declare no conflict of interest.

References

1. Rashid, K.; Ellingwood, K.; Safdarnejad, S.M.; Powell, K.M. Designing Flexibility into a Hybrid Solar Thermal Power Plant by Real-Time, Adaptive Heat Integration. *Comput. Aided Chem. Eng.* **2019**, *47*, 457–462.
2. Ale, S.; Femeena, P.V.; Mehan, S.; Cibin, R. Environmental impacts of bioenergy crop production and benefits of multifunctional bioenergy systems. In *Bioenergy with Carbon Capture and Storage*; Academic Press: Cambridge, MA, USA, 2019; pp. 195–217.
3. European Commission. Available online: https://ec.europa.eu/energy/en/topics/energy-efficiency/heating-and-cooling (accessed on 4 April 2020).
4. EU-EC. Available online: https://ec.europa.eu/energy/sites/ener/files/documents/germany_de_version.pdf (accessed on 4 April 2020).
5. Da Cunha, S.R.L.; de Aguiar, J.L.B. Phase change materials and energy efficiency of buildings: A review of knowledge. *J. Energy Storage* **2020**, *27*, 101083. [CrossRef]
6. Casini, M. *Smart Buildings: Advanced Materials and Nanotechnology to Improve Energy-Efficiency and Environmental Performance*; Woodhead Publishing: Cambridge, MA, USA, 2016.
7. Cárdenas-Ramírez, C.; Gómez, M.; Jaramillo, F. Characterization of a porous mineral as a promising support for shape-stabilized phase change materials. *J. Energy Storage* **2019**, *26*, 101041. [CrossRef]
8. He, M.; Yang, L.; Lin, W.; Chen, J.; Mao, X.; Ma, Z. Preparation, thermal characterization and examination of phase change materials (PCMs) enhanced by carbon-based nanoparticles for solar thermal energy storage. *J. Energy Storage* **2019**, *25*, 100874. [CrossRef]
9. Goia, F.; Perino, M.; Serra, V. Improving thermal comfort conditions by means of PCM glazing systems. *Energy Build.* **2013**, *60*, 442–452. [CrossRef]
10. Stritih, U.; Tyagi, V.V.; Stropnik, R.; Paksoy, H.; Haghighat, F.; Joybari, M.M. Integration of passive PCM technologies for net-zero energy buildings. *Sustain. Cities Soc.* **2018**, *41*, 286–295. [CrossRef]
11. Cabeza, L.F.; Castell, A.; Barreneche, C.D.; De Gracia, A.; Fernández, A.I. Materials used as PCM in thermal energy storage in buildings: A review. *Renew. Sustain. Energy Rev.* **2011**, *15*, 1675–1695. [CrossRef]
12. Dardir, M.; Panchabikesan, K.; Haghighat, F.; El Mankibi, M.; Yuan, Y. Opportunities and challenges of PCM-to-air heat exchangers (PAHXs) for building free cooling applications—A comprehensive review. *J. Energy Storage* **2019**, *22*, 157–175. [CrossRef]
13. Sardari, P.T.; Grant, D.; Giddings, D.; Walker, G.S.; Gillott, M. Composite metal foam/PCM energy store design for dwelling space air heating. *Energy Convers. Manag.* **2019**, *201*, 112151. [CrossRef]
14. Wang, X.; Yu, H.; Li, L.; Zhao, M. Experimental assessment on the use of phase change materials (PCMs)-bricks in the exterior wall of a full-scale room. *Energy Convers. Manag.* **2016**, *120*, 81–89. [CrossRef]
15. Ramakrishnan, S.; Sanjayan, J.; Wang, X.; Alam, M.; Wilson, J. A novel paraffin/expanded perlite composite phase change material for prevention of PCM leakage in cementitious composites. *Appl. Energy* **2015**, *157*, 85–94. [CrossRef]
16. Aditya, L.; Mahlia, T.M.I.; Rismanchi, B.; Ng, H.M.; Hasan, M.H.; Metselaar, H.S.C.; Aditiya, H.B. A review on insulation materials for energy conservation in buildings. *Renew. Sustain. Energy Rev.* **2017**, *73*, 1352–1365. [CrossRef]
17. Zhu, L.; Yang, Y.; Chen, S.; Sun, Y. Thermal performances study on a façade-built-in two-phase thermosyphon loop for passive thermo-activated building system. *Energy Convers. Manag.* **2019**, *199*, 112059. [CrossRef]

18. Mehling, H.; Cabeza, L.F. *Heat and Cold Storage with PCM*; Springer: Berlin/Heidelberg, Germany, 2008.
19. Akeiber, H.J.; Hosseini, S.E.; Hussen, H.M.; Wahid, M.A.; Mohammad, A.T. Thermal performance and economic evaluation of a newly developed phase change material for effective building encapsulation. *Energy Convers. Manag.* **2017**, *150*, 48–61. [CrossRef]
20. Castellón, C.; Günther, E.; Mehling, H.; Hiebler, S.; Cabeza, L.F. Determination of the enthalpy of PCM as a function of temperature using a heat-flux DSC—A study of different measurement procedures and their accuracy. *Int. J. Energy Res.* **2008**, *32*, 1258–1265. [CrossRef]
21. Jin, X.; Xu, X.; Zhang, X.; Yin, Y. Determination of the PCM melting temperature range using DSC. *Thermochim. Acta* **2014**, *595*, 17–21. [CrossRef]
22. Barreneche, C.; Solé, A.; Miró, L.; Martorell, I.; Fernández, A.I.; Cabeza, L.F. Study on differential scanning calorimetry analysis with two operation modes and organic and inorganic phase change material (PCM). *Thermochim. Acta* **2013**, *553*, 23–26. [CrossRef]
23. Gschwander, S.; Haussmann, T.; Hagelstein, G.; Barreneche, C.; Ferrer, G.; Cabeza, L.; Hennemann, P. Standardization of PCM characterization via DSC. In Proceedings of the SHC 2015 International Conference on Solar Heating and Cooling for Buildings and Industry, Beijing, China, 19–21 May 2015; pp. 2–4.
24. Feng, G.; Huang, K.; Xie, H.; Li, H.; Liu, X.; Liu, S.; Cao, C. DSC test error of phase change material (PCM) and its influence on the simulation of the PCM floor. *Renew. Energy* **2016**, *87*, 1148–1153. [CrossRef]
25. Yinping, Z.; Yi, J. A simple method, the-history method, of determining the heat of fusion, specific heat and thermal conductivity of phase-change materials. *Meas. Sci. Technol.* **1999**, *10*, 201. [CrossRef]
26. Badenhorst, H.; Cabeza, L.F. Critical analysis of the T-history method: A fundamental approach. *Thermochim. Acta* **2017**, *650*, 95–105. [CrossRef]
27. Tan, P.; Brütting, M.; Vidi, S.; Ebert, H.P.; Johansson, P.; Kalagasidis, A.S. Characterizing phase change materials using the T-History method: On the factors influencing the accuracy and precision of the enthalpy-temperature curve. *Thermochim. Acta* **2018**, *666*, 212–228. [CrossRef]
28. Buttitta, G.; Serale, G.; Cascone, Y. Enthalpy-temperature evaluation of slurry phase change materials with T-history method. *Energy Procedia* **2015**, *78*, 1877–1882. [CrossRef]
29. Hong, H.; Kim, S.K.; Kim, Y.S. Accuracy improvement of T-history method for measuring heat of fusion of various materials. *Int. J. Refrig.* **2004**, *27*, 360–366. [CrossRef]
30. Stanković, S.B.; Kyriacou, P.A. Improved measurement technique for the characterization of organic and inorganic phase change materials using the T-history method. *Appl. Energy* **2013**, *109*, 433–440. [CrossRef]
31. Tan, P.; Brütting, M.; Vidi, S.; Ebert, H.P.; Johansson, P.; Jansson, H.; Kalagasidis, A.S. Correction of the enthalpy–temperature curve of phase change materials obtained from the T-History method based on a transient heat conduction model. *Int. J. Heat Mass Transf.* **2017**, *105*, 573–588. [CrossRef]
32. Rathgeber, C.; Miró, L.; Cabeza, L.F.; Hiebler, S. Measurement of enthalpy curves of phase change materials via DSC and T-History: When are both methods needed to estimate the behaviour of the bulk material in applications? *Thermochim. Acta* **2014**, *596*, 79–88. [CrossRef]
33. Solé, A.; Miró, L.; Barreneche, C.; Martorell, I.; Cabeza, L.F. Review of the T-history method to determine thermophysical properties of phase change materials (PCM). *Renew. Sustain. Energy Rev.* **2013**, *26*, 425–436. [CrossRef]
34. Abhat, A. Low temperature latent heat thermal energy storage: Heat storage materials. *Sol. Energy* **1983**, *30*, 313–332. [CrossRef]
35. Zhou, D.; Zhao, C.Y.; Tian, Y. Review on thermal energy storage with phase change materials (PCMs) in building applications. *Appl. Energy* **2012**, *92*, 593–605. [CrossRef]
36. RUBITHERM-RT®, Technical Datasheets of RT24-RT25-RT26. Available online: https://www.rubitherm.eu/en/index.php/productcategory/organische-pcm-rt (accessed on 5 April 2020).
37. PURETEMP 25, Technical Information, (2019 PURETEMP LLC). Available online: https://www.puretemp.com/stories/puretemp-technical-data-sheets (accessed on 5 April 2020).
38. Standard DIN 51005. Thermal analysis (TA)—general Terms, German Institute for Standardisation. Available online: https://www.techstreet.com/standards/din-51005?product_id=2074265#jumps (accessed on 5 April 2020).
39. Standard DIN 51007. Thermal analysis—Differential thermal analysis (DTA) and differential scanning calorimetry (DSC)—General Principles, German Institute for Standardisation. 2019. Available online: https://infostore.saiglobal.com/en-us/Standards/DIN-51007-2019-397024_SAIG_DIN_DIN_2718768/ (accessed on 5 April 2020).

40. Gschwander, S.; Haussmann, T.; Hagelstein, G.; Sole, A.; Hohenauer, W.; Lager, D.; Mehling, H. *Standard to determine the heat storage capacity of PCM using hf-DSC with constant heating/cooling rate (dynamic mode)— A technical report of subtask A2.1 of IEA-SHC 42 / ECES Annex 29*; IEA SHC Task 42 / ECES Annex 29, DSC 4229 PCM Standard; IEA SHC: Paris, France, 2015.
41. ISO 11357-3:2018. Plastics—Differential scanning calorimetry (DSC)—Part 3: Determination of temperature and enthalpy of melting and crystallization. March 2018. Available online: https://www.iso.org/standard/72460.html (accessed on 5 April 2020).
42. Mankel, C.; Caggiano, A.; Koenders, E. Thermal energy storage characterization of cementitious composites made with Recycled Brick Aggregates containing PCM. *Energy Build.* **2019**, *202*, 109395. [CrossRef]
43. Caggiano, A.; Mankel, C.; Koenders, E. Reviewing theoretical and numerical models for PCM-embedded cementitious composites. *Buildings* **2019**, *9*, 3. [CrossRef]
44. Hiebler, S. Kalorimetrische Methoden zur Bestimmung der Enthalpie von Latentwärmespeichermaterialien während des Phasenübergangs. Ph.D. Thesis, Technische Universität München, Munich, Germany, 2007.

Article

PCM Cement-Lime Mortars for Enhanced Energy Efficiency of Multilayered Building Enclosures under Different Climatic Conditions

Cynthia Guardia *, Gonzalo Barluenga and Irene Palomar

Department of Architecture, University of Alcala, 28801 Madrid, Spain; gonzalo.barluenga@uah.es (G.B.); irene.palomar@uah.es (I.P.)

* Correspondence: cynthia.guardia@edu.uah.es; Tel.: +34-918839239; Fax: +34-918839276

Received: 3 August 2020; Accepted: 10 September 2020; Published: 11 September 2020

Abstract: Phase change materials (PCMs) are promising materials for the energy efficiency improvement of building enclosures, due to their energy storage capacity. The thermal behaviour of a multi-layered building enclosure with five different compositions of PCM cement-lime mortars was evaluated under heating and cooling cycles. The behaviour of cement-lime mortars with 20% of microencapsulated PCM mixed with other additions, such as cellulose fibres and perlite, a lightweight aggregate (LWA), were studied under climate conditions of 15 °C–82% RH (cooling) and 30 °C–33% RH (heating) that were applied with a climatic chamber. Temperature and heat flux on both sides of the multi-layered enclosure were experimentally measured in laboratory tests. Temperature was also measured on both sides of the PCM cement-lime mortar layer. It was observed that the addition of the PCM cement-lime mortar layer delayed the heat flux through the enclosure. During a heating cycle, the incorporation of PCM delayed the arrival of the heat wave front by 30 min (8.1% compared to the reference mortar without PCM). The delay of the arrival of the heat wave front during the cooling cycle after adding PCM, compared to the reference mixture, reached 40.6% (130 min of delay). Furthermore, the incorporation of LWA in PCM cement-lime mortars also improved thermal insulation, further increasing energy efficiency of the building enclosure, and can be used not only for new buildings but also for energy rehabilitation of existing building enclosures.

Keywords: phase change materials (PCMs); energy efficiency; cement-lime mortar; experimental characterization; climatic conditions; heat flux

1. Introduction

The design of new building materials for the energy efficiency improvement of existing buildings is a hot research topic [1–3]. The high energy consumption of many dwelling units built in the XX century is a main concern due to their huge carbon footprint [4,5]. Phase change materials (PCMs) are considered as a possible solution for this problem, due to their energy storage capacity [6]. During phase change, which occurs at a selected temperature that can be designed, PCMs absorb or release heat, while the material temperature remains constant (latent heat), acting as energy deposits that can be recovered when necessary [7,8]. Among the different types of PCM [7,8], the most commonly used in construction and building materials are microencapsulated paraffin waxes, which are characterized by their thermal and chemical stability without significant changes to their properties in the temperature range used in this study. They are also commercially available at a competitive price. However, paraffin waxes present low thermal conductivity and low phase change enthalpy [7–10].

Different authors have incorporated this type of microencapsulated PCM in building materials such as cement, gypsum or cement-lime mortars [8–16]. Besides, other additions, such as fibres (cellulose) and lightweight aggregates (LWAs, e.g., perlite), can be also incorporated to improve the thermal insulation capacity of mortars [16–18]. Some authors have studied the combined use of PCMs and these additions in mortars, taking advantage of both the the energy storage capacity of PCMs and the thermal insulation capacity [16]. It was found that PCM efficiency depends not only on the amount of PCM but also on the other components of the mortars [16]. The amount of PCM incorporated modified some thermal, physical and mechanical properties of mortars. Temperature range and variation also influenced the efficiency of PCM-modified mortars [15,19–21].

The experimental characterisation of materials' properties is the first step necessary to evaluate the possibilities of their application for building purposes [1,12–14,16]. However, the actual effect of the newly designed materials for improving building enclosures can only be fully evaluated by considering the multi-layered composition of real enclosures. Accordingly, some authors have studied the behaviour of multi-layered specimens by assessing new mortars with enhanced properties under different climatic conditions [15,17–21]. Climatic chambers can be used to test specific thermal conditions, simulating real environmental conditions [15,17]. It was concluded that the PCM effect on the multilayer enclosure's behaviour depended on the climatic conditions (summer/winter) [9,14,17–25].

In this paper, a study on the behaviour of a multi-layered brick wall enclosure incorporating different cement-lime mortars with the addition of 20% of PCM microencapsulated paraffin wax, LWA (expanded perlite) and cellulose fibres was carried out under different climate conditions. The aim of the study was to evaluate the thermal behaviour of the mortar layer and the overall enclosure under different climatic conditions. For this purpose, a climatic chamber was used, heating (from 15 °C to 30 °C) and cooling (from 30 °C to 15 °C) cycles were programmed and temperature and heat flux through the specimen were evaluated.

2. Materials and Methods

Five PCM-modified cement-lime mortars were analyzed. Figure 1 shows the criteria followed to design mortar compositions, considering the combined effect of energy storage capacity supplied by PCMs with the enhanced thermal insulation provided by LWAs and fibres. The experimental program assessed physical and mechanical properties and thermal conductivity in both liquid and solid states of PCMs. The temperature of the inner enclosure layers and on both external sides of the wall was analyzed. Heat flux inside and outside the enclosure was also measured.

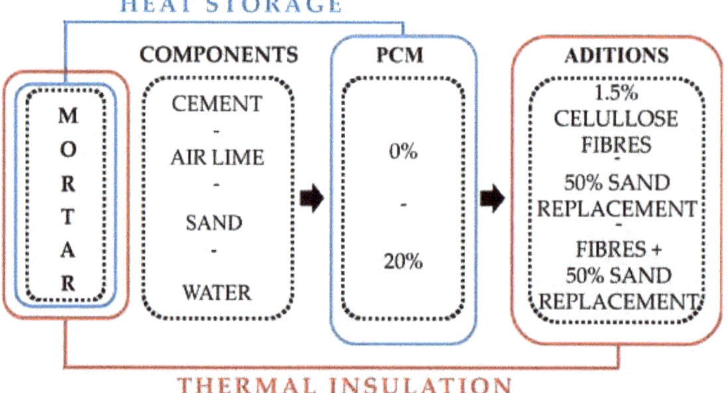

Figure 1. Composition design scheme of PCM cement-lime mortars.

2.1. Material and Mortar Compositions

The components used in the study were:

The cement used was a white type cement named BL II/B-L 32.5 N (UNE-EN 197-1) from Cementos Portland Valderrivas, Madrid, Spain.

(1) CL 90-S was the air lime added to the composition, (UNE-EN 459-1).
(2) The fine aggregate was a siliceous sand with a size between 0–4 mm.
(3) The lightweight aggregate (LWA) used was an expanded perlite (L).
(4) The fibres (F) used in this study were short cellulose fibres with a length of 1 mm—Fibracel® BC-1000 (Ø 20 µm) (Omya Clariana S.L, L'Arboc, Tarragona, Spain).
(5) The phase change material (PCM) chosen for the study was a microencapsulated paraffin wax (Micronal® DS 5040X,) with a melting point of ca. 23 ± 1 °C supplied by BASF Construction Chemicals Company, Madrid, Spain. The bulk density of this PCM is ca. 300–400 kg/m^3 and has a particle size which varies between 50–300 µm.

Table 1 presents the composition of the five mortars under study. Firstly, a reference cement-lime mortar (C) was produced. The cement:lime:aggregate ratio for this mortar was 1:0.5:4.5 by volume. Afterwards, 20% of PCM by volume of fresh mortar was added (C_{20}). In order to improve the thermal insulation capacity of the mortar, 1.5% of dry cellulose fibres considering the total fresh mortar's volume (regarding the reference mortar) was added (CF_{20}). In addition, 50% of the sand was replaced by the lightweight aggregate (CL_{20}). Finally, fibres and perlite were added (CLF_{20}) in order to design a mixture of both additions. Water to binder ratio (w/b) varied in order to get for all the mixtures a similar fresh workability. The minimum compressive strength target value was 3.5 MPa, corresponding to a CS-III grade rendering mortar (medium-high strength according to UNE-EN 998-1). The mixing process in the laboratory began blending all the dry components. Then the liquid water was added. The mixing process took a maximum of 5 min.

Table 1. Compositions of PCM cement-lime mortars, components in kg (adapted from [16]).

Components	C	C_{20}	CF_{20}	CL_{20}	CLF_{20}
BLII/B-L 32.5N	348	348	348	348	348
Air lime	55	55	55	55	55
Fine aggregates	1403	1403	1403	702	702
Fibres	-	-	0.66	-	0.66
LWA	-	-	-	94	94
PCM	-	84.6	84.6	84.6	84.6
Liquid water (*)	220	200	240	250	380
w/b (*)	0.73	0.68	0.78	0.71	0.79
D_{dry} (kg/m^3)	1400	1357	1440	868	-
D_{wet} (kg/m^3)	2264	1937	1885	1562	1561

* Fine aggregate humidity (5.3%) was also considered.

2.2. Experimental Methods

2.2.1. Hardened Properties and Thermal Parameters

The flow table test was used to measure the fresh mortar consistency, according to UNE-EN 1015-3:2000. In order to achieve a plastic consistency, the water to binder ratio was adjusted. Hardened state properties were characterised on 40 mm × 40 mm × 160 mm specimens, according to UNE-EN 1015-11. After 24 h, samples were demoulded and water cured for 28 days (21 ± 3 °C and 95 ± 5% RH). Bulk density (D) and open porosity (OP) (accessible to water) was calculated using a hydrostatic scale (UNE-EN 1015-10). According to UNE-EN 1015-19, the water vapour diffusion resistance factor (VD) was measured. VD was measured by the wet cup method. Cylindrical specimens with a 35 mm diameter and 40 ± 2 mm thickness with a saturated saline dissolution were used. After 28 days, compressive and flexural strength (CStr, FStr) were tested on standard specimens according to UNE-EN 1015-11:2000.

Thermal conductivity (λ) was measured using a thermally insulated box (hot box method). Plate samples with a size of 210 mm × 210 mm and a thickness of 24 ± 2 mm were used. Laboratory conditions during the test were 20 ± 1 °C and 50 ± 5% RH. A heat source was connected to a thermal regulator and located inside the insulated box. Sensors placed on the inner and outer surface of the sample and inside and outside the box monitored temperature (T) and relative humidity (RH) [26]. Two different temperatures were set inside the box: 25 °C and 40 °C. At 25 °C inside the box, the testing specimens remained below the PCM melting point (nominally 23 °C, although experimentally established at 22 °C [16]), while at 40 °C inside the box, all the plates were 30 °C. Thermal conductivities, λ_S (mortar's conductivity when PCM was in a solid state) and λ_L (mortar's conductivity when PCM was in a liquid state) were calculated after a thermal steady state was reached using Fourier's Law [26].

2.2.2. Climatic Chamber Testing Set-Up

A climatic chamber was used to assess the thermal behaviour of brick wall enclosures with PCM cement-lime mortars under different climatic conditions. Figure 2 shows the experimental set-up used. A multi-layered hollow brick enclosure arrangement was selected [4,5], consisting of a hollow brick wall, a gypsum-based internal rendering and an external PCM-modified coating mortar covered by a 5 cm XPS (Extruded polistyrene) insulation layer (external insulation coating system, ETICS). The sample was placed inside the climatic chamber door, and the chamber conditions were modified to produce a thermal transfer through the specimen.

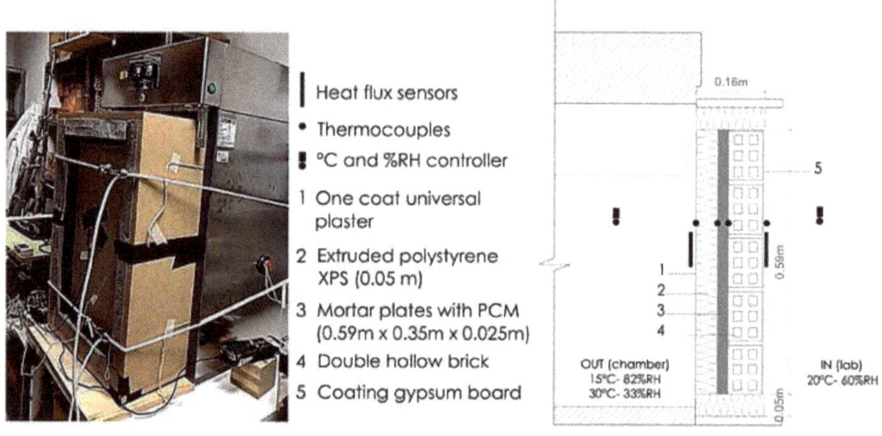

Figure 2. Climatic chamber set-up for monitoring the brick walls in different temperature conditions.

Two climatic conditions were tested in the chamber, simulating outdoor environmental conditions (OUT): a heating cycle and a cooling cycle. The heating cycle consisted of an initial stable condition of 15 °C and 82% relative humidity (RH), changing afterwards to 30 °C and 33% RH. The cooling cycle followed a reversed order, with an initial stable climatic condition of 30 °C and 33% RH, changing to 15 °C and 82% RH. Outside the chamber, the laboratory conditions (IN) remained constant at 20 ± 1 °C and 60% RH. RH inside the chamber was set to limit the water transport through the specimen. Each cycle (both heating and cooling) lasted 1400 min. During both cycles, heating and cooling, temperatures on both sides of the mortar layer, as well as at the external and internal sides of the enclosure, were monitored. Heat flux was also measured using a heat flux plate (Hukseflux HFP01 with an uncertainty degree of ±3%) for each side of the enclosure (Figure 2), registering the W/m^2 on both sides with a data logger (Hobo UX120, Bourne, MA, USA). As shown in Figure 2, heat flux sensors were adhered and hermetically sealed to the surface of the material. The temperature and RH inside and outside the climatic chamber were also monitored using thermocouples, coupled temperature and RH sensors (i-button Hygrochrom™ DS1923 (Newbury, Berkshire, UK), with an uncertainty degree of 0.02 °C and a size of Ø 17.35 mm and a thickness of 5.9 mm, and heat flux plates, located as shown in Figure 2.

3. PCM Cement-Lime Mortar Physical and Mechanical Properties

Table 2 presents the experimental results obtained for workability, physical, mechanical and thermal properties of the five mortars under study. Table 2 shows that consistency values varied between 178 mm (C) and 166 mm (C_{20}).

Table 2. Physical, mechasnical and thermal properties of the PCM cement-lime mortars (adapted from [16]).

Properties	C	C_{20}	CF_{20}	CL_{20}	CLF_{20}
Consistency (mm)	178	166	170	170	170
D (kg/m^3)	1900	1600	1660	1270	1160
OP (%)	19.56	17.72	16.77	23.33	23.09
VD	4.13	4.29	3.47	3.62	3.26
CStr (MPa)	14.33	7.17	5.83	6.00	5.33
FStr (MPa)	3.36	2.40	2.20	1.79	2.16
λ_S (W/mK)	0.23	0.20	0.30	0.29	0.23
λ_L (W/mK)	0.21	0.28	0.23	0.18	0.15

Regarding the hardened mortars' physical properties, the reference mortar (C) presented the highest bulk density (D) value, as expected, with 1900 kg/m^3, while the PCM mortar with LWA and fibres (CLF_{20}) showed the lowest, with 1160 kg/m^3. Open porosity (OP) values varied between 23.33% (CL_{20}) and 16.77% (CF_{20}). Considering the water vapour diffusion resistance factor (VD) of the mixtures under study, C_{20} showed the highest value (4.29) while CLF_{20} showed the lowest one (3.26).

Table 2 also presents the mechanical properties of the mortars. The results for compressive strength (CStr) pointed out that C presented the highest value (14.33 MPa) while CLF_{20} showed the lowest one (5.33 MPa), which agreed with the density values. Although the 20% addition of PCM in the mixtures reduced CStr, all the mortars fulfilled the minimum compressive strength target value of 3.5 MPa, corresponding to a CS-III grade rendering mortar according to UNE-EN 998-1. On the other hand, flexural strength (FStr) values ranged between 3.36 MPa (C) and 1.79 MPa (CL_{20}).

Thermal conductivity results of the PCM mortars in both the solid state (λ_S, T < 22 °C) and in the liquid state (λ_L, T > 22 °C) are also summarised in Table 2. It can be observed that λ_S varied between 0.30 W/mK (CF_{20}) and 0.20 W/mK (C_{20}). Otherwise, C_{20} presented the highest value for λ_L with 0.28 W/mK and CLF_{20} presented the lowest one with 0.18 W/mK.

As expected, the addition of 20% of PCM modified not only the thermal properties but also other hardened properties. Mechanical properties such as compressive and flexural strength decreased with the addition of PCM. On the other hand, λ_S decreased while λ_L increased when PCM was added, compared to the same mixture without PCM. These results were further analysed and discussed in a previous work [14].

4. Experimental Test Results and Thermal Analysis of Brick Wall Enclosures with PCM Mortars

The brick wall enclosures with PCM mortars were evaluated using the experimental set-up shown in Figure 2 and following the test procedure for a heating and a cooling cycle. For each sample, the thermal performance of the PCM mortar layer and the overall enclosure thermal performance were monitored during the test.

4.1. Thermal Behaviour of the PCM Cement-Lime Mortar Layer Inside the Enclosure Solution under Heating and Cooling Cycles

4.1.1. Heating Cycle

Table 3 shows the experimental results obtained during the heating cycle on both sides of the mortar layer, measured inside the enclosure. The initial and final temperature on both sides of the layer, the outer side in contact with the XPS insulation layer and the inner side in contact with the brick wall, the average values of the mortar layer and the difference in temperature between both sides are summarised.

Table 3. Temperatures on both sides of the PCM cement-lime mortar layer measured inside the enclosure during the heating cycle (from 15 °C to 30 °C).

Temperatures	C	C_{20}	CF_{20}	CL_{20}	CLF_{20}
Initial T_i (°C)					
Outer side (XPS)	19.60	19.00	20.80	19.00	20.30
Inner side (B)	21.40	20.60	21.30	20.70	20.60
Final T_f (°C)					
Outer side (XPS)	24.23	23.50	25.50	23.40	25.20
Inner side (B)	22.22	21.60	25.00	21.30	24.80
Layer average T_i	20.50	19.80	21.05	19.85	20.45
Layer average T_f	23.23	22.55	25.25	22.35	25.00
T difference T_i	1.80	1.60	1.30	1.70	0.30
T difference T_f	2.01	1.90	0.50	2.10	0.40

Initial temperatures on the outer side (XPS) of the mortar layer varied between 20.80 °C (CF_{20}) and 19.00 °C (C_{20} and CL_{20}), while on the inner side (B), temperatures ranged between 21.40 °C for C and 20.60 °C for C_{20} and CL_{20}. Regarding final temperatures (climatic chamber conditions of 30 °C and 33% RH), CF_{20} was the mixture with the highest temperature on the XPS side (25.50 °C) and CL_{20} presented the lowest (23.40 °C). On the brick wall side, the temperature varied between 25.00 °C (CF_{20}) and 21.30 °C (CL_{20}).

The average temperature of the layer can give an estimation of the solid or liquid state of PCMs. As verified in a previous study, the phase change temperature of the PCM used was 22 °C [16]. Table 3 shows that the initial average temperatures (15 °C inside the chamber) remained under 22 °C in all cases. Therefore, it can be assumed that the PCM was in a solid state. On the other hand, as the average temperature at the end of the heating cycle was above 22 °C (Table 3), it can be said that during each heating cycle, the PCM changed from solid to liquid state in all cases.

With regard to the temperature differences between both sides of the mortar plates (XPS and brick wall), it can be observed that initially the values ranged between 1.80 °C (C) and 0.30 °C (CLF_{20}), while at the end they varied between 2.10 °C (CL_{20}) and 0.40 °C (CLF_{20}).

Figure 3 presents the temperature increases on both sides of the mortar layer. Figure 3a,b compare the samples with and without PCM (C and C_{20}) and Figure 3c,d plot the temperature curves of samples with PCM (C_{20}) and the other components (CF_{20}, CL_{20} and CLF_{20}).

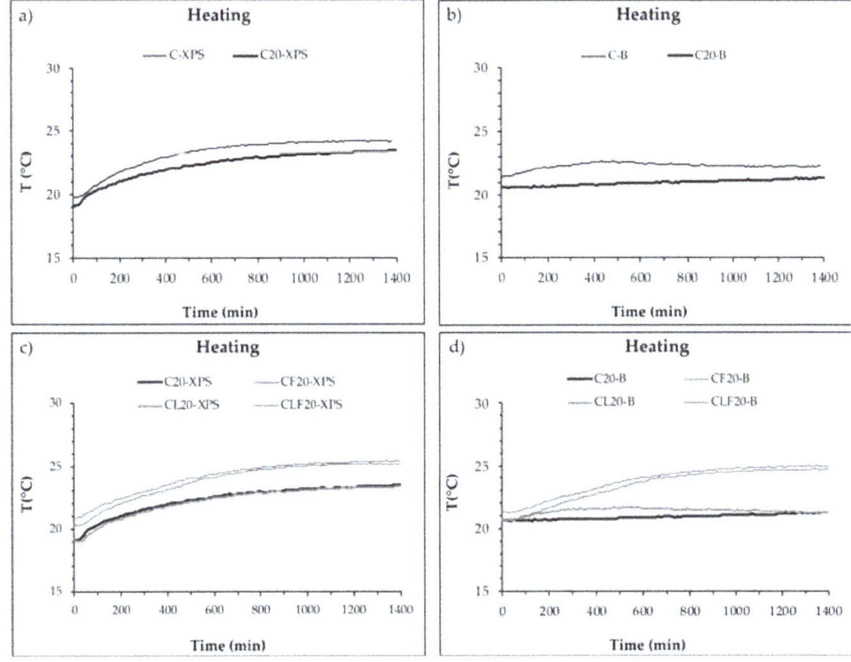

Figure 3. Temperature on both sides of the PCM cement-lime mortar layer during a heating cycle. (**a**) Inner side (XPS) of C and C_{20} plates. (**b**) Outer side (B) of C and C_{20} plates. (**c**) Inner side (XPS) of C_{20}, CF_{20}, CL_{20} and CLF_{20} plates. (**d**) Outer side (B) of C_{20}, CF_{20}, CL_{20} and CLF_{20} plates.

As expected, Figure 3a shows how the temperature on the side of the enclosure in contact with the heating source (XPS) increased continuously until the stabilisation of the system (final temperature). The incorporation of 20% of PCM reduced the slope of the curve, reducing the final temperature on the surface of the mortar layer. The same effect can be observed on the inner side of the plate in contact with the brick wall (B) of the plates. Accordingly, it can be said that PCM reduces the temperature on both sides of the mortar layer. The average temperature of the layer was therefore reduced, although the difference between them was similar (Table 3).

The enclosures with PCM mortars also showed a temperature increase due to the effect of the temperature increase inside the climatic chamber. When compared (Figure 3c,d), two groups of mixtures were identified. C_{20} and CL_{20} showed final temperatures under 23 °C and a slower increase compared to the mixtures with cellulose fibres (CF_{20} and CLF_{20}), which showed temperatures over 25 °C. Temperature differences between both sides of the C_{20} and CL_{20} plates were 2 °C, while temperature differences between CF_{20} and CLF_{20} were 0.5 and 0.4 °C, respectively. These differences were more significant on the inner side of the enclosure at the end of the test. Consequently, the addition of cellulose fibres increased the overall temperature of the mortar layer.

4.1.2. Cooling Cycle

Temperature was monitored on both sides of the mortar layer during the cooling cycle, which was tested after the heating cycle. Table 4 shows the experimental results obtained on both sides of the mortar layer. The initial and final temperature on both sides of the layer, the average values of the layer and the difference in temperature between both sides are summarised.

Table 4. Temperatures on both sides of the PCM cement-lime mortar layer measured inside the enclosure during the cooling cycle (from 30 °C to 15 °C).

Temperatures	C	C_{20}	CF_{20}	CL_{20}	CLF_{20}
Initial T_i (°C)					
Inner side (B)	22.30	21.60	25.20	21.50	23.80
Outer side (XPS)	24.10	23.90	25.60	23.90	25.40
Final T_f (°C)					
Inner side (B)	21.50	20.90	21.70	20.90	21.40
Outer side (XPS)	21.50	19.80	21.40	19.70	21.00
Layer average T_i	23.20	22.75	25.40	22.70	24.60
Layer average T_f	21.50	20.35	21.55	20.30	21.20
T difference T_i	1.80	2.30	0.40	2.40	1.60
T difference T_f	0	1.10	0.30	1.20	0.40

Initial temperatures on both sides, average temperature of the mortar layer and the difference in temperature were slightly different compared to the end of the previous heating cycles due to the thermal inertia of the enclosure specimen. The initial average temperature was over 22 °C for all mixtures, so PCM was in a liquid state.

Regarding the final temperature, the C mixture presented the highest value (21.50 °C) and CL_{20} the lowest one (19.70 °C) on the XPS side, while on the B side, the temperature ranged between 20.90 °C (C_{20} and CL_{20}) and 21.70 °C (CF_{20}). As the final average temperature was under 22 °C in all cases, PCM was in solid state after the cooling cycle. As occurred during the heating cycle, it can be stated that during the cooling cycle for each mixture, the PCM changed its phase from solid to liquid.

Table 4 shows the values of temperature differences between both sides of the plates at the beginning and end of the cooling cycle. CL_{20} was the mixture that presented the highest value with 2.40 °C, while CF_{20} presented the lowest one with 0.40 °C. The final values ranged between 1.20 °C (CL_{20}) and 0 °C (C).

The temperature evolution on both sides of the mortar layer, inside the enclosure, during a cooling cycle test are plotted in Figure 4: Figure 4a,b show the temperature development during cooling cycles of mortars without PCM (C) and with 20% of PCM (C_{20}) and Figure 4c,d plot the temperature of mixtures with PCM (C_{20}) and with the other components (CF_{20}, CL_{20} and CLF_{20}).

The results obtained showed that there was no significant difference when PCM was incorporated into the mortar (Figure 4a,b). However, when cellulose fibres or LWA were added, some differences arose. Mortar with LWA only did not show significant differences in C_{20}, although the use of cellulose fibres (both with or without LWA) increased the slope of the temperature curve, especially on the inner side of the enclosure (Figure 4d).

Summarising the results measured for a cooling cycle, the addition of PCM increased the temperature difference between the outer and the inner side of the mortar layer, LWA did not produce significant changes in temperature and cellulose fibres reduced the difference. C_{20} and CL_{20} showed a final temperature difference of over 1 °C, while for CF_{20} and CLF_{20} the difference was under 0.5 °C.

For both heating and cooling cycles, mortars with LWA and 20% of PCM recorded the highest temperature differences between the outer and inner sides of the mortar layer inside the enclosure. Considering the differences between the initial and final temperatures on both sides of the mortar layer, PCM showed greater influence during the cooling cycle than during the heating cycle. On the other hand, the addition of fibres reduced the effect of PCM and PCM plus LWA on the temperature changes produced by the heating and cooling cycles.

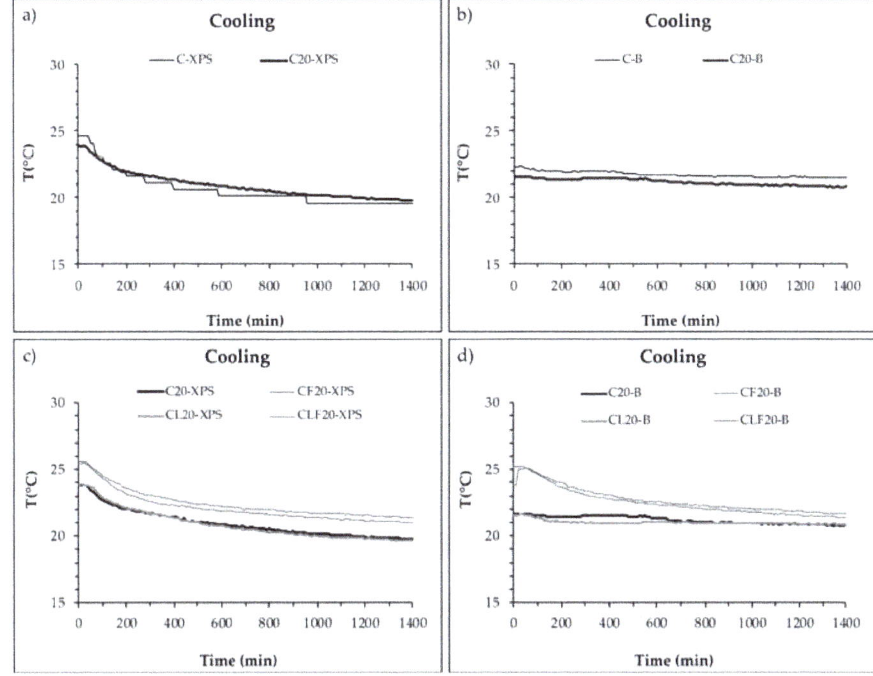

Figure 4. Temperature on both sides of the PCM cement-lime mortar layer during a cooling cycle. (**a**) Inner side (XPS) of C and C_{20} plates. (**b**) Outer side (B) of C and C_{20} plates. (**c**) Inner side (XPS) of C_{20}, CF_{20}, CL_{20} and CLF_{20} samples. (**d**) Outer side (B) of C_{20}, CF_{20}, CL_{20} and CLF_{20} samples.

4.2. *Effect of the PCM Cement-Lime Mortar Layer on the Overall Thermal Performance of the Brick Wall Enclosure under Heating and Cooling Cycles*

Temperature and heat flux (HF) were measured for enclosure specimens with an intermediate mortar layer with PCM, LWA and cellulose fibres. Heating and cooling cycles were applied with a climatic chamber and the results were compared.

4.2.1. Heating Cycle

The experimental results of temperature and heat flux on both sides of the enclosure during a heating cycle are plotted in Figure 5 and the main data are summarised in Table 5. Figure 5a,c relate the temperature evolution of mortars with and without PCM while Figure 5b,d compare the heat flux (HF) evolution during a heating cycle (from 15 °C to 30 °C). Figure 5a compares the temperature curve of enclosures with mortar layers without and with PCM, C and C_{20}, respectively. During the test, the temperature on the external side of the enclosure increased until 30 °C was reached at 200 min, remaining constant until the end of the test. Temperatures on the inner side of the wall (in laboratory conditions) were almost constant until 200 min, increasing slightly afterwards due to the arrival of the thermal wave front. It can be observed that the addition of PCM to the cement-lime mortar layer reduced the inner temperature by 1 °C compared to the same mortar without the addition of PCM.

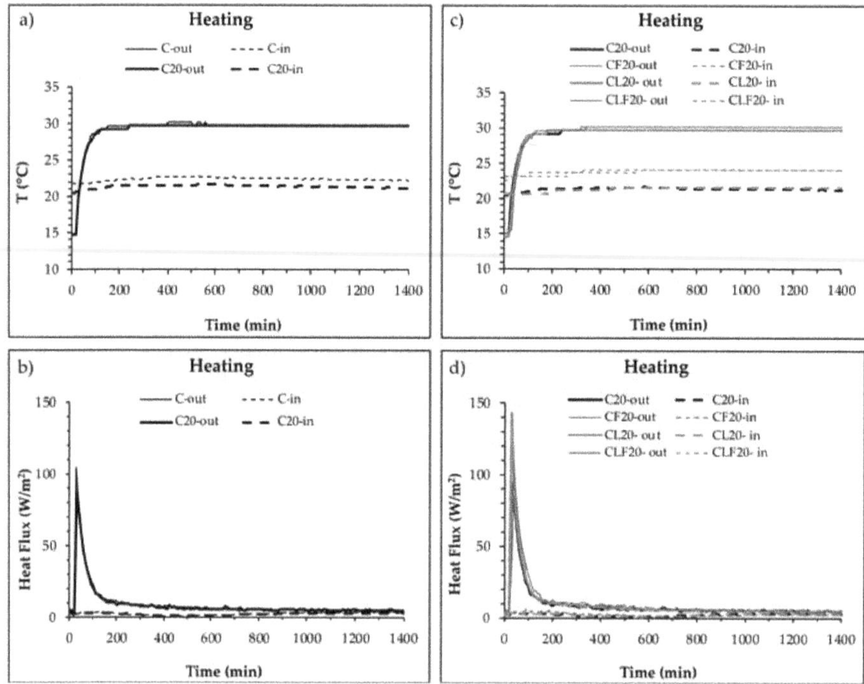

Figure 5. Temperature and heat flux monitored on the inner and outer sides of enclosures with a PCM cement-lime mortar layer during a heating cycle (from 15 °C to 30 °C). (**a**) Temperature of enclosures with and without PCM mortar layer (C and C_{20}). (**b**) Heat flux of enclosures with and without PCM mortar layer (C and C_{20}). (**c**) Temperature of enclosures with different types of PCM mortar (C_{20}, CF_{20}, CL_{20} and CLF_{20}). (**d**) Heat flux of enclosures with different types of PCM mortar (C_{20}, CF_{20}, CL_{20} and CLF_{20}).

Table 5. Summary of temperature and heat flux main results on both sides of the brick wall enclosure with a PCM cement-lime mortar layer during a heating cycle.

Thermal Parameters	C	C_{20}	CF_{20}	CL_{20}	CLF_{20}
Initial inner T_i (°C)	21.13	20.50	23.11	20.63	22.61
Final inner T_f (°C)	22.63	21.20	24.11	21.63	24.11
T_{inner} difference (°C)	1.5	0.7	1	1	1.5
Max. outer HF (W/m^2)	105	90	95	140	110
Inner infl. point (min)	370	400	580	440	540
Final outer HF (W/m^2)	3.24	4.60	3.92	3.15	4.47
Final inner HF (W/m^2)	1.82	3.15	3.15	2.55	2.55

Figure 5c compares the temperature of enclosures with PCM mortars with different compositions (C_{20}, CF_{20}, CL_{20} and CLF_{20}). As expected, no difference in the outer temperature was observed. However, two groups of inner temperature curves were identified, related to whether they had cellulose fibres (CF_{20} and CLF_{20}) or not (C_{20} and CL_{20}). Enclosures with a PCM mortar layer with fibres reached 3 °C above the inner temperature of enclosures without fibres, which barely reached 21 °C.

The heat flux measured on both sides of the enclosures during a heating cycle is plotted in Figure 5b (mortar with and without PCM) and in Figure 5d (PCM mortars with different compositions). A general trend can be observed in the outer HF, with a sharp initial increase until a peak value, followed by a fast decrease and then a slow decrease until a final stabilisation at a steady-state heat flux.

Figure 5b,d also record the heat fluxes on the inner side of the enclosure. The curves followed a general trend, where the heat flux showed a slight decrease until the arrival of the thermal wave front and a moderate increase afterwards until reaching the steady-state stabilisation value.

Table 5 summarises temperature values on the inner side of the enclosure. The initial temperature varied between 23.11 and 20.50 °C, while the final temperature ranged between 21.20 and 24.11 °C (CF_{20} and C_{20}, respectively). Enclosures with PCM mortar with cellulose fibres (CF_{20} and CLF_{20}) recorded the highest values of final temperature, which can be considered the worst scenario, while C_{20} and CL_{20} showed the most advantageous behaviour.

The peak HF values measured on the outer side of the wall (in direct contact with the climatic chamber conditions) are also presented in Table 5. The maximum peak HF value (140 W/m^2) was measured for PCM mortar with lightweight aggregate (CL_{20}) due to the lower conductivity provided by the LWA (Table 2). Accordingly, the combination of PCM and LWA showed the most advantageous behaviour, delaying the advance of the heat wave front through the enclosure.

Table 5 also presents the time of arrival of the heat wave front to the inner side of the enclosure (inner HF inflexion point). It can be observed that the enclosures with lower thermal conductivity (λ_L in Table 2) delayed the arrival of the heat wave to the inner side, increasing the effective thermal inertia of the enclosure.

The final HF measured after thermal and HF stabilisation showed small differences among the enclosures tested, ranging between 3.15 and 4.60 W/m^2 on the outer side and between 1.82 and 3.15 W/m^2 on the inner side. The small difference between the outer and the inner sides can be attributed to mass transfer related to water movement through the enclosure.

Based upon the temperature and heat flux results during the heating cycle, some analyses can be done. It can be seen that C presented an initial temperature 0.66 °C higher than C_{20}, although both constructive systems were stabilised under the same climatic conditions both inside and outside. Likewise, C and C_{20} presented great differences in the temperature increase during the test, as C increased its temperature 1 °C more than C_{20}, which could be related to the addition of 20% of PCM to the mortar. Regarding the heat flux measured on the inner side, the incorporation of PCM delayed by 30 min the arrival of the heat wave front, delaying the inflexion point from 370 to 400 min, which could be related to the heat storage capacity of PCM.

Considering mortars with 20% of PCM, two groups were identified related to the presence of cellulose fibres. Mixtures with fibres (CF_{20} and CLF_{20}) showed an initial temperature 2–3 °C higher than mortars without fibres (C_{20} and CL_{20}). This difference remained almost constant during the test. Regarding the HF peak, the compositions with lightweight aggregate showed the largest values (140 W/m^2 and 110 W/m^2) and, therefore, the larger delay in the heat wave front. It can be seen that the reference mixture (C) also showed similar values due to its higher density (Table 2) and, consequently, its higher thermal inertia.

Finally, the incorporation of lightweight aggregates and cellulose fibres also produced an extra delay in the arrival of the heat wave front to the inner side of the enclosure (Table 5 and Figure 5d). Accordingly, the delay above 400 min can be attributed to the other components beyond the effect produced by PCM alone. The delay was longer for mixtures with fibres than mixtures with LWA.

4.2.2. Cooling Cycle

The experimental results of temperature and heat flux on both sides of the enclosure during a cooling cycle (from 30 °C to 15 °C) are plotted in Figure 6 and the main data are summarised in Table 6. Figure 6a,c relate the temperature evolution of mortars with and without PCM, while Figure 6b,d compare the heat flux (HF) evolution during a heating cycle.

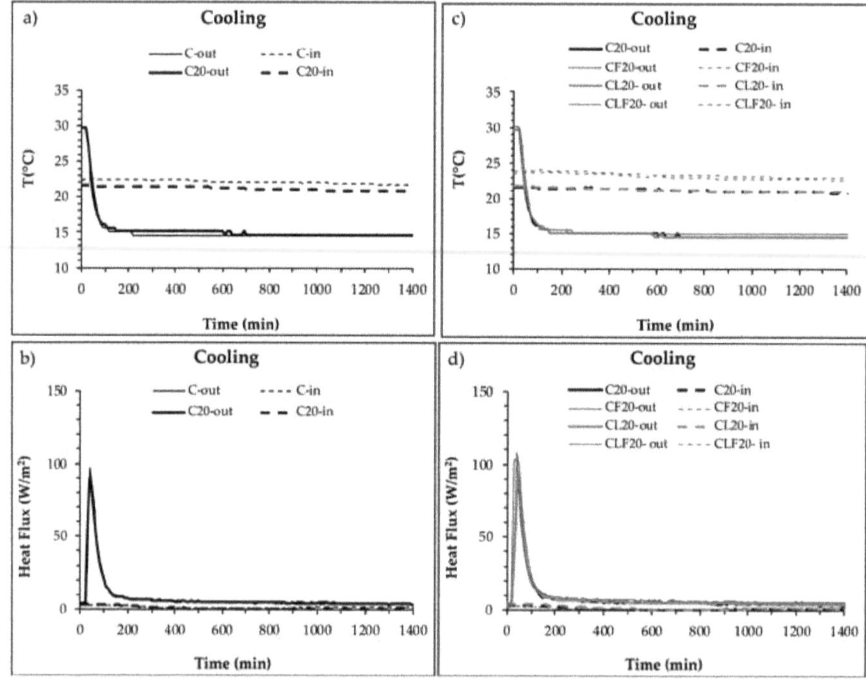

Figure 6. Temperature and heat flux monitored on the inner and outer sides of enclosures with a PCM cement-lime mortar layer during a cooling cycle (from 30 °C to 15 °C). (**a**) Temperature of enclosures with and without PCM mortar layers (C and C_{20}). (**b**) Heat flux of enclosures with and without PCM mortar layers (C and C_{20}). (**c**) Temperature of enclosures with different types of PCM mortar (C_{20}, CF_{20}, CL_{20} and CLF_{20}). (**d**) Heat flux of enclosures with different types of PCM mortar (C_{20}, CF_{20}, CL_{20} and CLF_{20}).

Table 6. Summary of temperature and heat flux main results on both sides of the brick wall enclosure with PCM cement-lime mortar layer during a cooling cycle.

Thermal Parameters	C	C_{20}	CF_{20}	CL_{20}	CLF_{20}
Initial inner T_i (°C)	22.63	22.13	24.11	21.63	24.11
Final inner T_f (°C)	21.63	21.13	23.61	20.63	23.11
T_{inner} difference (°C)	1	1	0.5	1	1
Max. outer HF (W/m^2)	100	90	110	95	105
Inner infl. point (min)	320	450	270	540	270
Final outer HF (W/m^2)	3.24	3.24	3.92	4.47	5.13
Final inner HF (W/m^2)	1.82	1.32	1.82	1.87	1.87

Figure 6a compares the temperature curve of enclosures with mortar layers without and with PCM, C and C_{20}, respectively. During the test, the temperature on the external side of the enclosure decreased until 15 °C was reached at 200 min, remaining constant until the end of the test. Temperatures on the inner side of the wall (in laboratory conditions) were almost constant until 200 min, slowly decreasing slightly afterwards due to the arrival of the thermal wave front. The addition of PCM to the cement-lime mortar layer reduced the inner temperature by 1 °C compared to the same mortar without the addition of PCM.

Figure 6c compares the temperatures of enclosures with PCM mortars with different compositions (C_{20}, CF_{20}, CL_{20} and CLF_{20}). The temperatures measured on the outer side followed the same curve in all cases, as expected. On the inner side, as happened for the heating cycle, two groups of temperature curves can be seen, corresponding to the incorporation of cellulose fibres (CF_{20} and CLF_{20}), which remained roughly 2 °C above the other mixtures.

The heat flux measured on both sides of the enclosures during the heating cycle is plotted in Figure 6b (mortar with and without PCM) and in Figure 6d (PCM mortars with different compositions). The HF curves showed a general pattern similar to that observed for the heating cycle, although the heat moved to the opposite direction in this case. The curves of the outer HF began with a sharp initial increase until a peak value, followed by a fast decrease and a final slow decrease afterwards until a stabilisation at a steady-state heat flux.

Figure 5b,d also record the heat fluxes on the inner side of the enclosure. The curves followed a general trend, where the heat flux showed a slight decrease until the arrival of the thermal wave front and a moderate increase afterwards until reaching the stabilisation steady-state value. In the case of the cooling cycle, the addition of PCM did not cause significant changes in the thermal capacity of the mortar when it changed from liquid to solid phase. On the other hand, it can be observed that mixtures with fibres presented a higher initial temperature in comparison with the mixtures without them. This trend continued over time (Figure 6d).

Table 6 summarises temperature values on the inner side of the enclosure. The initial temperature varied between 24.11 °C (CF_{20} and CLF_{20}) and 21.63 °C (CL_{20}), while the final temperature ranged between 20.63 and 23.63 °C (CL_{20} and CF_{20}, respectively). Enclosures with PCM mortar with cellulose fibres (CF_{20} and CLF_{20}) recorded the highest values of final temperature.

The peak HF values measured on the outer side of the wall (in direct contact with the climatic chamber conditions) are also presented in Table 6. The maximum peak HF value (110 W/m^2) was measured for PCM mortar with cellulose fibres (CF_{20}) because it also showed the highest initial temperature. However, when the time to reach the inflection point on the inner side of the enclosure, corresponding to the time of arrival of the heat wave front, was considered, CF_{20} did not show the longest delay. As happened for the heating cycle, C_{20} and CL_{20} showed the longest delays. It can be observed that the enclosures with lower thermal conductivity (λ_L in Table 2) delayed the arrival of the heat wave to the inner side, increasing the effective thermal inertia of the enclosure. In this case, PCM showed a greater effect (130 min) than LWA (90 min).

The final HF measured after thermal and HF stabilisation showed small differences among the enclosures tested, ranging between 3.24 and 5.13 W/m^2 on the outer side and between 1.32 and 1.87 W/m^2 on the inner side. The small difference between the outer and the inner sides can be attributed to mass transfer related to water movement through the enclosure.

Based upon the temperature and heat flux results during the cooling cycle, it can be seen that the inner temperature depended more on the initial inner temperature rather than on the mortar composition. Regarding the heat flux measured on the inner side, the incorporation of PCM delayed by 130 min the arrival of the heat wave front, due to the heat storage capacity of the PCM, while LWA supplied 90 extra minutes of delay. On the contrary, cellulose fibres did not increase the positive effect of the PCM.

5. Conclusions

An experimental study to evaluate the behaviour under different climatic conditions of a new enclosure solution containing microencapsulated phase change material cement-lime mortars with LWA and fibres inside was carried out. The experimental program assessed the thermal behaviour of the mortar layer and the overall enclosure by measuring the temperature and heat flux during heating and cooling cycles. The main conclusions of this study were:

(1) The addition of PCM to a conventional cement-lime mortar modified the physical, mechanical and thermal properties of the mortar, reducing density and strength, while increasing heat storage capacity.

(2) The addition of lightweight aggregates and fibres also modified mortar heat transfer capacity, increasing some properties already improved by PCM.
(3) The thermal behaviour of the PCM cement-lime mortars depended not only on the composition but also on the climatic conditions to which they were subjected.
(4) The addition of cellulose fibres facilitated the heat/cold transfer through the mortar layer, increasing or decreasing the average temperature of the mortars, which can be less favourable, especially in heating conditions.
(5) The addition of PCM delayed by 30 min the arrival of the heat wave front (8.1%) during the heating cycle. During the cooling cycle, the addition of PCM delayed by 130 min (40.6%) the arrival of the heat wave front compared to the reference mixture without PCM.
(6) LWA reduced thermal conductivity, increasing thermal insulation capacity and, therefore, producing an advantageous coupled effect with PCM energy storage capacity. Consequently, the combined use of PCM and LWA produced a remarkable delay of the heat wave front on the inner side of the enclosure in both heating (19%) and cooling conditions (68.7%), compared to the reference mixture.
(7) The combination of cellulose fibres and PCM showed a reduced synergic effect, but only in heating conditions.

Author Contributions: Conceptualisation, C.G., G.B., I.P.; methodology, C.G., G.B., I.P.; formal analysis, C.G., G.B.; data curation, C.G.; investigation, C.G., G.B., I.P.; writing—original draft preparation, C.G., G.B.; writing—review and editing, C.G., G.B., I.P.; supervision, G.B.; project administration, G.B., I.P.; funding acquisition, I.P. All authors have read and agreed to the published version of the manuscript.

Funding: Financial support for this research was provided by the Reasearch Program for the Promotion of Young Researchers, co-funded by Comunidad de Madrid and the University of Alcala (Spain), as part of the project IndoorComfort (CM/JIN/2019-46).

Acknowledgments: Some of the components were supplied by BASF Construction Chemicals España S.L., Omya Clariana S.L. and Cementos Portland Valderrivas.

Conflicts of Interest: The authors declare no conflict of interest.

Nomenclature

PCM	Phase change material	RH	Relative humidity
F	Cellulose fibre	T_i	Initial temperature
LWA	Lightweight aggregate	T_f	Final temperature
D	Bulk density	XPS	Insulation layer
OP	Open porosity	B	Brick layer
VD	Water vapour resistance factor	HF	Heat flux

References

1. Palomar, I.; Barluenga, G.; Puentes, J. Lime–cement mortars for coating with improved thermal and acoustic performance. *Constr. Build. Mater.* **2015**, *75*, 306–314. [CrossRef]
2. Sala, E.; Zanotti, C.; Passoni, C.; Marini, A. Lightweight natural lime composites for rehabilitation of Historical Heritage. *Constr. Build. Mater.* **2016**, *125*, 81–93. [CrossRef]
3. Bentchikou, M.; Guidoum, A.; Scrivener, K.; Silhadi, K.; Hanini, S. Effect of recycled cellulose fibres on the properties of lightweight cement composite matrix. *Constr. Build. Mater.* **2012**, *34*, 451–456. [CrossRef]
4. Terés-Zubiaga, J.; Martín, K.; Erkoreka, A.; Sala, J.; Escudero, K.M. Field assessment of thermal behaviour of social housing apartments in Bilbao, Northern Spain. *Energy Build.* **2013**, *67*, 118–135. [CrossRef]

5. Terés-Zubiaga, J.; Campos-Celador, A.; González-Pino, I.; Escudero-Revilla, C. Energy and economic assessment of the envelope retrofitting in residential buildings in Northern Spain. *Energy Build.* **2015**, *86*, 194–202. [CrossRef]
6. Ryms, M.; Januszewicz, K.; Kazimierski, P.; Zaleska-Medynska, A.; Klugmann-Radziemska, E.; Lewandowski, W.M. Post-Pyrolytic Carbon as a Phase Change Materials (PCMs) Carrier for Application in Building Materials. *Materials* **2020**, *13*, 1268. [CrossRef] [PubMed]
7. Sharma, A.; Tyagi, V.; Chen, C.; Buddhi, D. Review on thermal energy storage with phase change materials and applications. *Renew. Sustain. Energy Rev.* **2009**, *13*, 318–345. [CrossRef]
8. Cabeza, L.F.; Castell, A.; Cabeza, L.F.; De Gracia, A.; Fernandez, A.I. Materials used as PCM in thermal energy storage in buildings: A review. *Renew. Sustain. Energy Rev.* **2011**, *15*, 1675–1695. [CrossRef]
9. Ryms, M.; Klugmann-Radziemska, E. Possibilities and benefits of a new method of modifying conventional building materials with phase-change materials (PCMs). *Constr. Build. Mater.* **2019**, *211*, 1013–1024. [CrossRef]
10. Rao, V.V.; Parameshwaran, R.; Ram, V.V. PCM-mortar based construction materials for energy efficient buildings: A review on research trends. *Energy Build.* **2018**, *158*, 95–122. [CrossRef]
11. Jayalath, A.; Nicolas, R.S.; Sofi, M.; Shanks, R.; Ngo, T.; Aye, L.; Mendis, P. Properties of cementitious mortar and concrete containing micro-encapsulated phase change materials. *Constr. Build. Mater.* **2016**, *120*, 408–417. [CrossRef]
12. Pavlík, Z.; Fořt, J.; Pavlíková, M.; Pokorný, J.; Trník, A.; Černý, R. Modified lime-cement plasters with enhanced thermal and hygric storage capacity for moderation of interior climate. *Energy Build.* **2016**, *126*, 113–127. [CrossRef]
13. Lucas, S.S.; Ferreira, V.; Aguiar, B. Latent heat storage in PCM containing mortars—Study of microstructural modifications. *Energy Build.* **2013**, *66*, 724–731. [CrossRef]
14. Cunha, S.; Lima, M.; Aguiar, B. Influence of adding phase change materials on the physical and mechanical properties of cement mortars. *Constr. Build. Mater.* **2016**, *127*, 1–10. [CrossRef]
15. Mankel, C.; Caggiano, A.; Ukrainczyk, N.; Koenders, E. Thermal energy storage characterization of cement-based systems containing microencapsulated-PCMs. *Constr. Build. Mater.* **2019**, *199*, 307–320. [CrossRef]
16. Guardia, C.; Barluenga, G.; Palomar, I.; Diarce, G. Thermal enhanced cement-lime mortars with phase change materials (PCM), lightweight aggregate and cellulose fibers. *Constr. Build. Mater.* **2019**, *221*, 586–594. [CrossRef]
17. Palomar, I.; Barluenga, G.; Ball, R.J.; Lawrence, M. Laboratory characterization of brick walls rendered with a pervious lime-cement mortar. *J. Build. Eng.* **2019**, *23*, 241–249. [CrossRef]
18. Wi, S.; Yang, S.; Park, J.H.; Chang, S.J.; Kim, S. Climatic cycling assessment of red clay/perlite and vermiculite composite PCM for improving thermal inertia in buildings. *Build. Environ.* **2020**, *167*, 106464. [CrossRef]
19. Alonso, C.; Oteiza, I.; García-Navarro, J.; Martín-Consuegra, F. Energy consumption to cool and heat experimental modules for the energy refurbishment of façades. Three case studies in Madrid. *Energy Build.* **2016**, *126*, 252–262. [CrossRef]
20. Arıcı, M.; Bilgin, F.; Nižetić, S.; Karabay, H. PCM integrated to external building walls: An optimization study on maximum activation of latent heat. *Appl. Therm. Eng.* **2020**, *165*, 114560. [CrossRef]
21. Rathore, P.K.S.; Shukla, S.K. An experimental evaluation of thermal behavior of the building envelope using macroencapsulated PCM for energy savings. *Renew. Energy* **2020**, *149*, 1300–1313. [CrossRef]
22. Fachinotti, V.; Bre, F.; Mankel, C.; Koenders, E.A.B.; Caggiano, A. Optimization of Multilayered Walls for Building Envelopes Including PCM-Based Composites. *Materials* **2020**, *13*, 2787. [CrossRef] [PubMed]
23. Kishore, R.A.; Bianchi, M.V.; Booten, C.; Vidal, J.; Jackson, R. Optimizing PCM-Integrated Walls for Potential Energy Savings in U.S. Buildings. *Energy Build.* **2020**, *226*, 110355. [CrossRef]
24. Qiao, Y.; Yang, L.; Bao, J.; Yang, L.; Liu, J. Reduced-scale experiments on the thermal performance of phase change material wallboard in different climate conditions. *Build. Environ.* **2019**, *160*, 106191. [CrossRef]

25. Khan, R.J.; Bhuiyan, Z.H.; Ahmed, D.H. Investigation of heat transfer of a building wall in the presence of phase change material (PCM). *Energy Built Environ.* **2020**, *1*, 199–206. [CrossRef]
26. Herrero, S.; Mayor, P.; Olivares, F.H. Influence of proportion and particle size gradation of rubber from end-of-life tires on mechanical, thermal and acoustic properties of plaster–rubber mortars. *Mater. Des.* **2013**, *47*, 633–642. [CrossRef]

Article

Coherent Investigation on a Smart Kinetic Wooden Façade Based on Material Passport Concepts and Environmental Profile Inquiry

Amjad Almusaed [1,*], Ibrahim Yitmen [1], Asaad Almsaad [2], İlknur Akiner [3] and Muhammed Ernur Akiner [4]

1 Department of Construction Engineering and Lighting Science, Jönköping University, 551 11 Jönköping, Sweden; ibrahim.yitmen@ju.se
2 Department of Engineering and Chemical Sciences, Karlstad University, 651 88 Karlstad, Sweden; asaad.almssad@kau.se
3 Department of Architecture, Akdeniz University, Antalya 07058, Turkey; ilknurakiner@akdeniz.edu.tr
4 Vocational School of Technical Sciences, Akdeniz University, Antalya 07058, Turkey; ernurakiner@akdeniz.edu.tr
* Correspondence: amjad.al-musaed@ju.se; Tel.: +46-700451114

Abstract: Wood is one of the most fully renewable building materials, so wood instead of non-renewable materials produced from organic energy sources significantly reduces the environmental impact. Construction products can be replenished at the end of their working life and their elements and components deconstructed in a closed-loop manner to act as a material for potential construction. Materials passports (MPs) are instruments for incorporating circular economy principles (CEP) into structures. Material passports (MPs) consider all the building's life cycle (BLC) steps to ensure that it can be reused and transformed several times. The number of reuse times and the operating life of the commodity greatly influence the environmental effects incorporated. For a new generation of buildings, the developing of an elegant kinetic wooden façade has become a necessity. It represents a multidisciplinary region with different climatic, fiscal, constructional materials, equipment, and programs, and ecology-influencing design processes and decisions. Based on an overview of the material's environmental profile (MEP) and material passport (MP) definition in the design phase, this article attempts to establish and formulate an analytical analysis of the wood selection process used to produce a kinetic façade. The paper will analyze the importance of environmentally sustainable construction and a harmonious architectural environment to reduce harmful human intervention on the environment. It will examine the use of wooden panels on buildings' façades as one solution to building impact on the environment. It will show the features of the formation of the wooden exterior of the building. It will also examine modern architecture that enters into a dialogue with the environment, giving unique flexibility to adapt a building. The study finds that new buildings can be easily created today. The concept of building materials passport and the environmental selection of the kinetic wooden façade can be incorporated into the building design process. This will improve the economic and environmental impact of the building on human life.

Keywords: double skin; environmental profile; material passport; kinetic façade; wooden façade

Citation: Almusaed, A.; Yitmen, I.; Almsaad, A.; Akiner, İ.; Akiner, M.E. Coherent Investigation on a Smart Kinetic Wooden Façade Based on Material Passport Concepts and Environmental Profile Inquiry. *Materials* **2021**, *14*, 3771. https://doi.org/10.3390/ma14143771

Academic Editors: Sukhoon Pyo and Antonio Caggiano

Received: 4 May 2021
Accepted: 2 July 2021
Published: 6 July 2021

1. Introduction

The construction industry accounts for more than 30 percent of the extraction of natural resources and 25 percent of the world's solid waste since the construction industry mostly follows a sequential business paradigm of "taking, manufacturing, disposing of," using and disposing of goods at the end of their lives, as they are put together for the one-off and not reused [1]. The industry has been changing its paradigms over recent decades by implementing a circular economy model to maintain closed-loop materials with the highest capacity for minimizing waste production and resource mining for the construction industry [2]. In particular, it will continue to be a paradigm change in the industry as a whole. According to the Ellen MacArthur Foundation (EMF), which works in accelerating the transition to a circular economy, the end-of-life goods and building materials, and components, can be reused, used as resource banks for new buildings, and retained as the circular economy's general perspective in a closed-loop segment or material [3]. However, for greater market acceptance, this definition requires knowledge development and tools [4], particularly in the building sector, in which innovational engineering design takes more time [5] since buildings are always one-of-a-kind designs together with the wide supply chain, which adds to the difficulty [6]. The design of a dynamic (kinetic) façade is one of the most interesting solutions in the structure, adapting to changes in the outside environment.

The kinetic façade is a breakthrough in buildings' architecture [7]. For the Scandinavian region, kinetic façades are a new trend, in which they can currently be counted on the fingers of a hand [8]. If we talk about greening, today it is relevant all over the world; there are more and more green areas in large cities. Dynamic façades adapt the space according to the needs of the people.

Buildings with variable façades can certainly be called a new era in architecture. The kinetic façade is a breakthrough trend in building architecture. The façade is in motion; here is a more accurate description of it: A kinetic façade is a building cladding in constant motion from an engineering perspective under the influence of nature or mechanics forces [9,10]; this is a constantly changing façade pattern [11]. Therefore, selecting appropriate materials is a critical and meticulous step to achieve the building façade design's desired effect [12].

Different materials can convey different feelings on the exterior of the building. A tight grasp of the materials' properties can create an extraordinary and intriguing façade effect when using materials. The influence of materials and craftsmanship on the design of building façades is very important. In recent year primitive natural building materials to the widely used building materials such as woods, stones, and reeds have formed the architectural style characteristics of various periods.

Although kinetic façades are quite new, the types they take are already very diverse. Every style, including triangular elements to giant sunshades, wooden frames, and projected animations, serves a specific function and takes on a distinct form [13].

Consisting of qualitative knowledge, also the quantitative database of a structure's material property, the material passport (MP) presents components found in designs and demonstrates their recyclability and environmental effects [14]. Today, the structure of kinetic wooden materials for the façade in a new generation of building conception is supported by strong arguments, innovations, and improvements introduced in the recent period, helping promote ecological material for new construction worldwide [15]. The research in the kinetic wooden façade as part of sustainable environment buildings arouses great interest in the construction sector. Thus, we will see opportunities to develop wood construction research considerably at our universities in the next few years. It allows us to establish construction technology as a research subject at our institutions and strengthen the relation with city municipalities in usable form. BEAT 2000 is a systematic methodology of environmental assessment and measurement of environmental impacts of building materials in their life cycles [14] based on the SBI (DK) simulation programmer used to calculate the environmental profile.

The objective of this research is to address questions such as:

- What is the role of the kinetic façade in modern architecture today?
- How does the kinetic wooden façade contribute to modern building conception?
- Why is the material passport system required in creating the ecological kinetic façade?
- What is the evaluation of the environmental profile impact on the wooden façade?

The analysis will review the most appropriate building material passports used in a smart wooden kinetic façade to confirm the environment's natural material. First, this research attempts to classify the criteria that have an especially strong effect on material selection through material passports. The second goal is to assess various architecture systems and their basic requirement in light of the chosen solution requirements. Finally, this investigation will focus on using wooden and ecological materials to build façade models and technology.

The research will provide an efficient material passport (MP) and a smart technology for innovative kinetic wooden façade architecture, where modern technical components can be integrated into façade elements. Almost all façade materials and systems are suitable for use combined with different kinds of wood to express the designer's creation intent and match the environment [16]. Temperature and humidity can be effectively adjusted under proper insulation by cooperating with board and membrane materials with different building physical properties [16]. The building's energy consumption can be reduced to the lowest level under the premise of ensuring the indoor environment's comfort. Ruggiero et al. [17] supposed that over the last 40 years, reforming in the building sector has had an extraordinary evolution, which has led, from the intuition of the possibility to import models and methodologies from the industrial sector an environmental and safety building component. The most usable material in history was wood. Waste-free properties, thermal efficiency, durability, original texture, an advantage in handling, and several additional wood attributes contribute to a comfortable and relaxing living environment [18]. Munir et al. [19] conclude that the wood's raw, brittle, and moisture-absorbing properties are often misunderstood because of its organic, porous, and moisture-absorbing surface, where the organic nature of wood makes it environment-friendly. Regarding the environmental and energy considerations, the utilization of wood in design is regarded as the best material for addressing these problems. It is no coincidence that existing structures were made of solid wood to enhance their structural and architectural qualities. Wood has lower thermal conductivity than many construction materials and is ideal for an energy-efficient design [20]. Wood buildings behave similarly to passive solar houses by absorbing and storing heat in the wood pulp [20]. Sekularac et al. [21] think that the wood is an element of façade cladding in modern architecture, and the research is intended to expand knowledge of the possibilities and limitations and create the foundation for their correct wider use. Sekularac et al. [21] consider that the wood and wood-based components of the building are used as double skin layers in the façade, where the temperature, solar radiation, and wind have a certain impact on the architectural presence of a structure. The wooden façade has been common material used in conjunction with ecological building materials to express the wooden structure's natural texture directly. Solovev [22] believes that the construction material's choosing process is the most important decision and has long-term consequences for the structure's owner.

Zhukov et al. [23] are confident that the concept of an objective selection of construction materials should include requirements for materials, building systems in which these materials are used, work technology, architectural and planning solutions, and engineering support systems. Khoshnava et al. [24] consider that the important factor in selecting building material is to be an environmentally safe material for a toxic-free environment. That will have a positive impact on humans and the environment. Lawson [25] confirms that the safety assessment considers the impact of material on the environment in all its life cycles. Haupt and Hellweg [26] think the most important indicators of the material's environmental friendliness are the possibility of recycling, energy consumption, environmental friendliness of production, and operational characteristics. Wood occupies a special and important place among building materials, having an undoubted priority in "sustainable architecture" [27].

Assefa et al. [28] think that after experiencing wind and rain and other climatic conditions, the façade presents a special texture without losing its function. The effect of a building on the atmosphere is determined by the materials used and the energy sources used. Wooden kinetic façades may be used with or without surface protection. Shahda [29] suggests that the change in building technology was from traditional building technology to smart, sustainable architecture, expanded use of environmentally safe materials in design, and environmentally friendly wood conservation solutions in his paper. Fakourian and Asefi [30] consider that buildings with a kinetic wooden façade are climate-smart, not least because they bind carbon dioxide and prevent it from being released.

At the "World Congress of Architects" in 1993, Thayer [31] proposed that the architectural climate in general and structures are among the most significant components in the detrimental human impact on the natural environment in the "Declaration Interdependence for a Sustainable Future".

2. Façade Analysis within the Thematic Area

2.1. Wooden Façade and Building Material Passport (BMP)

Choosing passport materials for creating a competent architectural element with suitability for a building component or category becomes required. It depends on creating an objective. There is still no clear technical regulation within the EU that would comprehensively regulate the building material passport (BMP), which describes the suitability, quality, and safety requirements in different buildings and programs. All building materials require a mandatory assessment of compliance with competency, suitability, and safety requirements. A structured description of building materials allows more successful working in the ABC industry, developing the concept of efficiency and bio-economy in the construction sector [32]. The wide range of building materials put into circulation at the moment is so extensive that it is often impossible to do without difficulties in determining the composition of the mandatory accompanying classifications and information. Quality and material classification are primary details characterizing and describing building materials' products [33]. To some extent, a technical passport of products, which is an integral part of the accompanying information, can act as a passport for the quality of building materials. Still, this phrase is generally understood as any material description that testifies building materials' quality, suitability, and safety.

Companies' economic models are transformed as circular economy practices that are implemented. If data are systematized and optimized, it becomes easier to adapt, add value, and implement energy-efficient and recycling initiatives in the building industry. BMPs are instruments for incorporating the circular economy into residential design. Creating more effective and resilient ecosystems, they would be crucial in preserving and delivering knowledge to consumers in company supply chains. Resources' worth and useful life are maintained, repaired, or even improved by locating them in a database, transferring them, and reusing them. Munaro et al. [34] propose a BMP model for Brazil's wood-frame structure, including general awareness, security, preservation, use, service, installation directions, reuse, and product support history. Centered on Munaro's BMP concept, this analysis proposes a MP for wood façades.

2.2. Wooden Façade and Environmental Profile Analysis (EPA)

Environmental considerations are on the way to becoming an integral part of the design process when creating the world's architecture and construction. In the 1990s of the last century, the environment finally came on architecture's main agenda [35]. However, the problem is not infrequently handled at a somewhat naive level, focusing on the signal value rather than concrete environmental results. The environment is something with nature; so-called natural materials are preferred when environmentally friendly. In reality, all materials, even plastic, come from nature. Simultaneously, all common building materials have undergone a processing process, i.e., they are not natural in the sense of the original [14]. However, it is often meant as wood or green areas. Philosophically, it is about different ways of understanding the world.

On the other hand, the romantic wants to emphasize the sensual and the thoughts and feelings it triggers in the viewer. The engineer and the architect fill different roles in the construction. The engineer is educated in a scientific tradition and must strive for "objective truth". It is something of a mouthful about contemporary construction's multifaceted reality, even when limited to the physical field. Therefore, the engineer must specialize.

With the specialization, the overview and understanding of the built whole are weakened, just as the risk of using agreed-upon terminology increases. It requires comparing the material-related environmental impacts with the selected building material passport. It is necessary to enhance the choice process of material and its overall environmental impact to get objective results to use numbers, diagrams, words, and pictures to describe the topic.

3. Kinetic Façade Role in Modern Building Design

An ecological kinetic façade with a new modular building structure means a façade system with a modular preassembled construction adjusted to various required conditions in different sites and positions [36].

Kinetic architecture is the art and science of constructing buildings so that structural elements can move relative to each other without disrupting the building's overall integrity. Kinetic factors affect how the building panels move, fold, rotate, and transform, solving various climatic and aesthetic problems [37]. The visual transformation in this architecture direction is not hidden between the internal engineering communications [38]. The process of changing the façade of kinetic buildings is visible to everyone—if you need to hide the room from the sun, then the whole house will "take" this in. In the early 20th century, architects began to explore the possibility of introducing kinetics elements into buildings. The understanding was formed that movement in architecture can be produced with engines mechanically or using people, air, water, and other kinetic forces. For example, the wooden kinetic façade can include massive wooden elements supported by separate frames from the outer wall. According to the façade orientations, the façade's kinetic aspects are programmed to reduce sunlight's influence.

Every year, dozens of new original designs of dynamic façades and building envelopes appear globally, allowing in time to change buildings' appearance and perform several additional functions to regulate lighting, heat protection of a building, and air exchange of premises [39]. Architectural structures are considered static objects, but most have special equipment that lets the building adjust to varying settings. Controls and digital technology are revolutionizing our lives, automating nearly every aspect of our lives. These innovations are increasingly being used in building architecture and construction [40]. These involve movable partitions, walls, active ventilation openings, curtains, screens, and blinds, and the mechanized sections of the system that enable it to respond to changing external environmental conditions and human behavior. Kinetic façades as controlled dynamic structures are also found in modern building systems in most countries [41]. The change in the position of these structures is due to certain factors: if it is necessary to increase energy efficiency, when the temperature inside the building fluctuates (i.e., based on the microclimate of the room), when climatic conditions change, for artistic reasons, which attracts more people to structures and spaces. The era of responsive building components and dynamic architecture that respond to consumer demands rapidly evolved from the early 20th century to the last transformable façade erected based on algorithmic control that relies on climate data and sunlight. These responsive components are high-tech systems that use networked sensors and actuators to monitor environmental parameters and automate functional building elements' control.

3.1. How Does the Smart Kinetic Wooden Façade (SKWF) Contribute to Modern Building Conception?

A smart, efficient kinetic wooden façade is a high-tech project, where a high-tech product, in general, is a data processing object, with several interactive functions. The work will contribute to reaching the global goals for sustainable development. Relevant objectives are minimal resource usage throughout the whole product life cycle, increased prosperity by making technology and products available worldwide, and life-long learning in the industry through flexible education alternatives [21]. The importance of a smart kinetic wooden façade (SKWF) as a system that comes directly in energy performances and healthy buildings is high, where it becomes required to develop this element according to EU standards. The study will investigate the essential effect of using this system in the city façade and the environmental benefit. The study will also open a way to activate the passport concept in the design process as a sustainability tool in building design conception. Wooden kinetic façades can be largely prefabricated.

Modern architecture and technologies guarantee a high degree of precision, a basic requirement for quality construction. The assembly is thus considerably reduced, and the construction period is much shorter. Wood is a natural material. Untreated, it withstands changes in color and surface structure due to climatic influences. The natural color of fresh wood is not durable to any wood species used outdoors [42]. These changes, however, do not affect the strength of the wood in any way but are rather signs of aging of the living material. As a material for constructing façades, wood will be suitable for all types of facilities, including residential buildings. By choosing the type of wood, which corresponds to the environmental profile, and the installation and the surface treatment, the wooden façade can be completely customized, even in colors. Thus, no two wooden façades are alike. The materials selected for the wooden façade can be a special environmental characteristic. The benefits of using environmentally friendly construction materials include practical recycling choices and wood as the building material with the lowest carbon footprint [43]. The new type of ecological materials and technologies appear that make it possible to embody the most daring ideas because non-standard solutions increase the urban environment's aesthetic appeal. All types of wood used for façades have a long life, and in the end, they are easy to recycle, as described in the environmental profile factors.

3.2. *Bionic and Bioclimatic Concepts in the Adaptive Ecological Kinetic Façade*

Dynamic façades adapt the building to the time of day, weather, and light level. The building lives and exists as part of nature, wakes up at dawn, protects residents on a hot afternoon from bright sunlight, saves energy, and even replenishes its reserves. New systems are currently being actively developed to cover a building from excessive sun and regulate its temperature. Bionics is a growing industry in architecture and construction, and many bio-inspired adaptable façades have gone from concept to reality [44]. It is necessary to establish a more comprehensive, systematic, and rational "transfer" process from nature to the enclosing structures to achieve a thorough application of bionics in architecture, potentially influencing the efficiency of life. Adapting the building envelope to the external climate and user requirement and providing the desired indoor temperature can be learned from nature. Bio-adaptive enclosing structures have great potential in reducing energy consumption and providing a comfortable operating environment [45]. Biological adaptation is the ability of a system to adapt, that is, to meet specified requirements, including when environmental conditions change. Building shells are enclosing structures that can independently react to changes in their atmosphere, such as solar radiation, wind speed and direction, air temperature, and precipitation [46]. As a result, when compared to conventional static buildings, energy demand can be reduced because useful energy sources are only used when they are needed.

Bio-adaptable façades act as a kind of climate mediator between comfort requirements and environmental conditions. Façades with the built-in function of bio-adaptiveness can be designed directly for a specific user. The investigation on the systems of bio-adaptive kinetic façades can be based on various world experiences, where it can highlight the latest trends in using this type of façade in modern buildings and projects. The hypothesized adaptive kinetic façade concept is that this system's application and viability are possible in the climate of north Europa. The outer shell will open and close depending on weather conditions, regulate the temperature and humidity, and create the necessary ventilation. Thus, an ideal microclimate will be created and maintained at any time of the year, regardless of climatic conditions.

3.3. *Macro-Climatic Action and Wooden Façade Reaction*

An increase in wood moisture content above 20–23% inevitably increases fungal attack risk [43]. With drops in humidity and temperature (when the weather changes), wood deforms. Its shrinkage and swelling, alternating, lead to warping and cracking through which water enters the wood structure [47]. Ultraviolet radiation is a destroyer of wood lignin, which binds cellulose and is the main building substance. The primary signs are wood darkening. With longer exposure to the sun, the wood acquires a gray color; small cracks appear in it, and water accumulates (precipitation), which gives an impetus for the reproduction of fungi or mold. Materials for protecting the façade surfaces of wooden houses from atmospheric influences must be elastic and resistant to external forces. On another hand, the building orientation plays an important role in the relative proportion of the energy gained from the outer climate as shown in Figure 1.

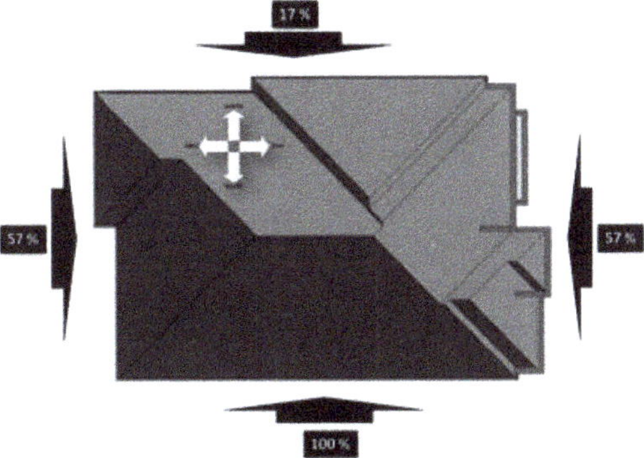

Figure 1. The relative proportion of the energy gained from the outer climate on the building walls from different sides of the world [48].

4. Wooden Materials Use in Wooden Façade

The European timber industry is strongly committed to sustainable development, especially as their raw material comes from sustainably managed forests. As the European Commission stated several years ago, "wood and timber products play an important role in mitigating climate change by absorbing and retaining carbon from the atmosphere" [49]. For a better comprehension of eco-friendly wood materials, and the recycling process, it becomes required to understand the physical and chemical properties of wood and recycled materials, as well as the interactions between wood, recycled material, and adhesive and technological conditions [50].

4.1. Environmental Profile (EP) for a General Carpentry Material

Carpentry or carpentry-work-construction work on the manufacture of wooden structures and parts is characterized by less careful wood processing. Carpentry work includes work on the construction of wooden walls, façades. By carpentry, wood is meant here as building wood exterior cladding, such as panel units for a kinetic façade) as well as veneer and chip products. The energy consumption for manufacturing is modest. As wood absorbs CO_2 during growth and is therefore considered a CO_2-neutral material, there are obvious environmental benefits from using it [51]. It has clear growth rings due to the large color difference between light springwood and dark autumn wood, selected as the color for the kinetic façade. With constructive wood protection, the use of heartwood and regular surface treatments, doors, and pine windows can last 90–120 years, and in a dry environment, the wood lasts 120–1000 years [52]. Depending on the environmental profile (EP) of 1 m^2 of this material, with a thickness of 21 mm, where the estimated cycle life is 50 years, wood has a modest thermal conductivity [53].

4.2. Larix Wood Cladding

Larix belongs to a fast-growing and durable species: some of them live 700–900 years. A coniferous tree, up to 50–80 m high, shedding foliage for the winter, light-loving, and frost-resistant, grows throughout the Northern Hemisphere of our planet. Deciduous forests can withstand temperatures down to −60 °C [54]. Larix is among the hardest and heaviest coniferous species, with a very large heartwood proportion. The crucial environmental force of larch wood is that the heartwood does not need impregnation for exterior use. Larix wood is durable and very difficult to ignite. Its biggest physical and aesthetic weakness is the appearance of large and rather dark lumps. The heartwood has poor permeability and therefore moisturizes only to a limited extent by brief water exposure. Small dimensions, such as the environmental profile of 1 m^2 Larix wood, with a thickness of 21 mm, where the estimated cycle life is 65 years, are recommended due to the lark's tendency to twist and bend [53].

4.3. Cedarwood Cladding

Cedarwood is valuable because it does not rot in water, is not subject to fouling by algae and mollusks, and is not damaged by termites. Therefore, the red cedar is one of the most favorite options for the kinetic wooden façade (KWF). The wood of European origin is called thuja. European thuja is fast-growing and weaker than North American wood [55]. Western Red cedar grows mixed with other coniferous species in the Western US and Canada [56]. Cedar is a very light wood with modest compressive and bending strength, and therefore not suitable as a construction wood. However, it is among the most durable woods for outdoor use, partly due to a high content of essential oil with a moisture-repellent effect and partly due to the fungicidal Thujaplicin. In Northern Europe, cedar has been used more frequently during the 1990s, especially for exterior cladding of façades. Therefore, it is a suitable material for a wooden façade. The environmental profile (EP) of 1 m^2 of cedarwood, with a thickness of 21 mm, has an estimated cycle life of 75 years [53].

4.4. Fiberboards

Paraffin is usually added to the wood pulp to improve the water-repellent properties of fiberboard boards. The boards' strength can be increased by binding agents such as starch, rosin, and synthetic resins. Fiberboard boards are faced with natural wood veneer, paper, fabric, plastic, fiberglass, metal, and cork. The wood fiberboards are cut and applied to stains on one side to distinguish them from ordinary softwood boards [57]. The plates are wind- and moisture-tight but open to vapor diffusion. The wax-impregnated boards are cleaner to work with than similar asphalt-impregnated ones. The panels can emit very small amounts of formaldehyde over time, on a par with ordinary planed wood. Fiberboard with the above properties is used as a wind barrier behind a ventilated exterior cladding. Today the eco-friendly materials can be used frequently in a new generation of building materials, where the environmental aspects of the various board materials are just as different as their properties and applications. Eco-friendly fiberboard panels with acceptable physical and mechanical properties are in accordance with European standards [58].

Fiberboard is disposed of by incineration. Standard fiberboard boards are divided into two main classes:

- porous
- solid

In terms of its basic properties, the fiberboard material is comparable to wood since it retains all the useful qualities of wood, for example, strength, toughness; moreover, fiberboard is a warm material. Furthermore, the environmental profile (EP) of 1 m^2 of fiberboard, with a thickness of 19 mm, has an estimated cycle life of 100 years [53].

4.5. Chipboard

Chipboard is a composite sheet material made by hot pressing of wood particles, mainly shavings, mixed with a binder of non-mineral origin with the introduction, if necessary, of special additives [59]. Particle board consists of pine or spruce shavings, possibly birch shavings such as cover layers, ureal glue, or phenol glue, as well as small amounts of wax. Slabs with a bulk density of at least 600 kg/m^3 are used for building purposes [60]. Particle board is very sensitive to moisture, partly because wood chips absorb moisture to a much greater degree than, for example, defibrated wood material, and partly because the adhesives used are not or only partially moisture resistant. Thus, among the materials used for construction and furniture production, the chipboard takes an important place. The estimated cycle life is 50 years for the environmental profile (EP) of 1 m^2 of chipboard, with a thickness of 21 mm [53].

4.6. Plywood

Plywood consists of glued together with thin wooden boards. The middle veneer layers are often spruce, possibly pine, and of Scandinavian origin. Good sorts are used, and the trunk's best parts are used for veneers [61]. The outer cover layers can be beech, birch, or other wood types, depending on where and whether the plywood is used visibly. The purpose of cross-laying veneer layers is to produce a board that does not sag or shrink significantly and has great strength concerning its weight. Depending on the type of glue, plywood can be water- and boiling-resistant. Plywood treated with flame retardants and impregnated against fungus can also be produced. Plywood, also known as a wood-laminated sheet, is a multi-layer construction material created by gluing specially designed veneer [62]. Plywood is an inexpensive finishing material suitable for exterior wall decoration. It can be used for roughing under tiles or for finishing. Depending on the sheets' brand, such a surface may require protection with varnish or paint, or not require additional processing at all. Plywood is used in construction because of its strength and durability, and versatility. Plywood can be made also from hardwoods, softwoods, bamboo, or a combination of different woods. The sustainability of plywood is determined not only by how the wood is being sourced but also by the manufacturing process [63]. The environmental profile (EP) of 1 m^2 of plywood, with a thickness of 15 mm, has an estimated cycle life of 50 years [54].

5. Results

5.1. Material Passport (MP) for Wood Façade

The Material Passport application for the wood façade is investigated using the Mnaro et al. [35] model. General data, security, sustainability, usage and service, disassembly guide, reuse, history, and other details are required for passports. Product definition, manufacturer, device structure, usage recommendations and restrictions, technological assessment, structural efficiency, impact tolerance, durability against the xylophage species, water tightness, thermal and acoustic efficiency, and system reliability are the general data collected. The data contained in terms of protection measures include fire resistance and fire response assessment. The details gathered for sustainability include the implementation/execution protocol, transportation, assembly mechanism, and component assessment methods. The criteria for evaluating the wood frame system's material and component characteristics are indicated in the National System of Technical Assessment documents, SİNAT-005/2017 [64].

5.2. The Indicators of the Environmental Profile (EP)

Sustainable development means increased welfare that considers the earth's ecosystems and the inventory of renewable and non-renewable natural resources. Therefore, environmental evaluations of the built environment are becoming increasingly common [65]. An environmental profile is shown for the buildings' materials and constructions per m^2 floor area for buildings. All results shown in the environmental profile are expressed as annual values. In the calculation process, each material's environmental impact is divided by its estimated lifetime and then summed up as part of the home's total environmental impact. An environmental profile is also displayed, which compares the annual environmental impact per m^2 of floor area from the building's heating and building materials. This profile includes only energy consumption and the greenhouse effect. Here you can see how large a share of the total environmental impact is due to the materials. Finally, a layer cake diagram is shown, which compares the greenhouse effect's environmental impact on the building's wooden façade. BEAT 2000 is a suitable tool that can be used immediately for energy and environmental assessment of any environmental building analysis. It has expanded the database, especially for alternative energy-saving solutions, and the materials included could reduce the time consumption by defining the untraditional and unusual building parts that appear during renovation.

The environmental profile (EP) consists of seven environmental indicators (see Figure 2) covering all significant physical environmental impacts and effects. Newer constructions are assumed to have environmental advantages and are considered to have a future in the European market. It was chosen to focus on climate screen constructions, primarily kinetic wooden façades, representing the most important component of a typical building and the most environmentally damaging part of a building. The building façade is also a central part of the architectural expression. All wooden façades have U-values related to Building European Regulations 2020 and Building Regulations for small houses 1998. All building façades have a U-value of 0.20, while the wooden window and glass façade have a U-value of 1.65. All roofs have a U-value of 0.15. For all constructions, an environmental profile (EP) for 1 m^2 of the building in the analysis is shown. All results shown in the environmental profile are expressed as annual values, i.e., in the calculation of the environmental profile (EP), the environmental impact of each type of wooden material is divided by its projected lifespan in the existing structure and then added together as part of the overall environmental effect of the concept. Besides, a layer cake diagram is shown, which compares the environmental impact in the form of a greenhouse effect, distributed on the parts of the construction in analyzing:

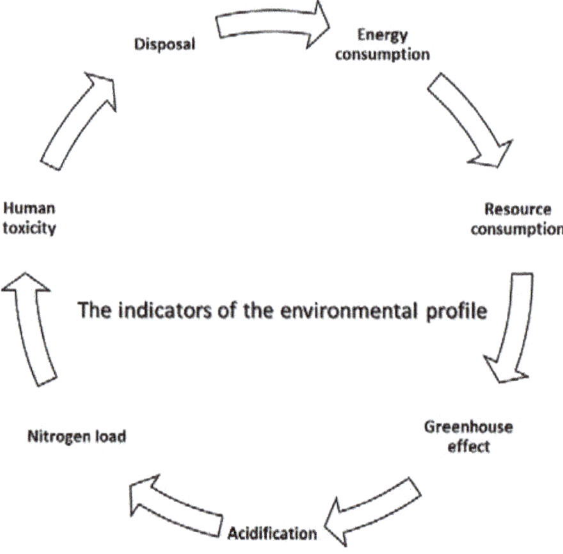

Figure 2. The environmental profile indicators.

5.3. Resource Consumption

The practice of sustainable building refers to various methods in the process of implementing building projects that involve less harm to the environment [35]. It is important to make use of natural resources. They are used as input in production and consumption and form increased safety [66]. The increasing environmental damage require urgent action to reduce environmental degradation [67]. Resource consumption represents an essential factor in the environmental profile (EP). It is about the consumption of non-renewable, or, less often, renewable, resources [68]. It explains the environmental impact of resource consumption (see Figure 3).

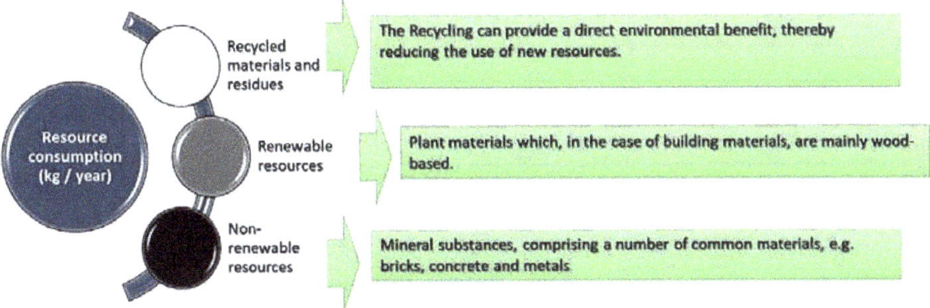

Figure 3. The environmental profile (EP) shows the environmental impact of Resource consumption.

The environmental consequences of resource use may include a lack of energy, increased area consumed, and risks associated with the extraction or cultivation process. The results of applying this indicator to the five chosen wooden façade materials are seen in the diagram below (see Figure 4).

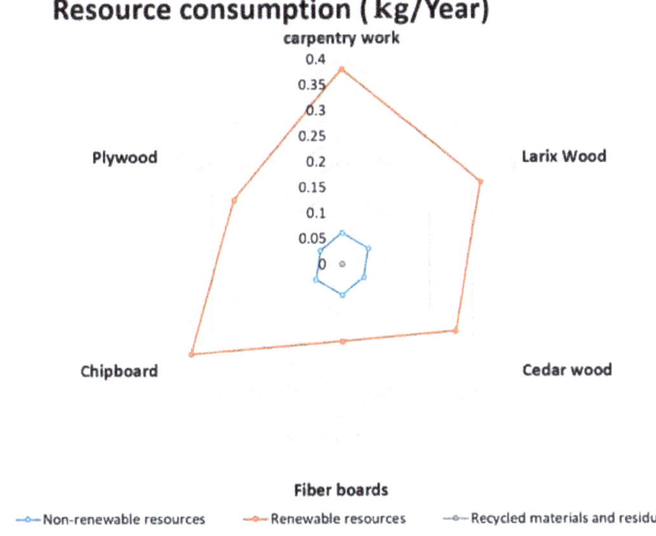

Figure 4. The impact of resource consumption on the selected façade wooden material environmental [68–72].

5.4. Energy Consumption

The environmental indicator shows the environmental impact of resource consumption (see Figure 5).

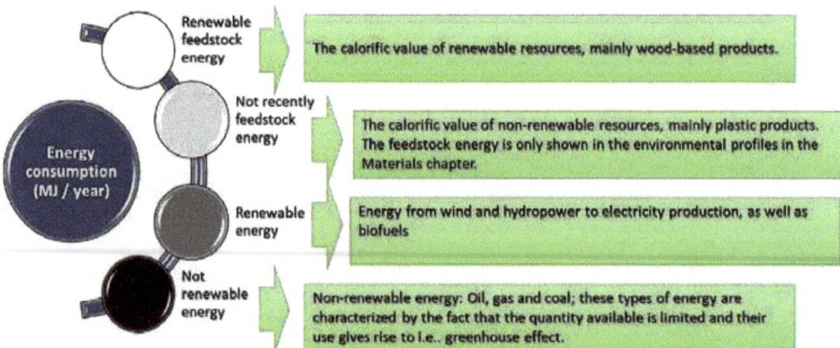

Figure 5. The environmental indicators show the environmental impact of resource consumption.

The environmental consequences of resource use may include a lack of energy, an increased area consumed, and risks associated with the extraction or cultivation process. The results of applying this indicator to the five chosen wooden façade materials are seen in the diagram below (see Figure 6).

Energy consumption (MJ/Year)

carpentry work
Larix Wood
Cedar wood
Fiber boards
Chipboard
Plywood
0 1 2 3 4 5 6
Non-renewable resources — Renewable resources — Recycled materials and residues

Figure 6. The impact of energy consumption on the selected façade wooden material [68–72].

5.5. Greenhouse Effect ($\times$1000/CO_2/Year)

The environmental impact of greenhouse gas pollution is the third predictor of the environmental profile. The human-caused greenhouse effect is caused mainly by the release of fluorinated greenhouse gases (F-GHGs) and other greenhouse gases, including nitrous oxides (NO_X), methane (CH_4), and carbon dioxide (CO_2), that trap heat that would otherwise reflect from the planet to space. They are currently contributing to the warming of the atmosphere. The results of applying this indicator to the five chosen wooden façade materials are seen in the diagram below (see Figure 7).

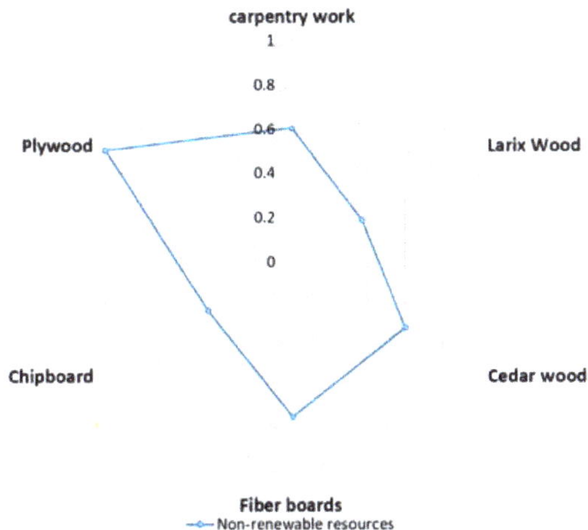

Figure 7. The impact of the greenhouse effect on the selective façade of wooden material [68–72].

5.6. Acidification (gSO_3/Year)

The fourth indicator in the environmental profile (EP) depicts the environmental impact of acidifying compounds (particularly sulfur dioxide and nitrogen oxides), attacking plant leaves and needles and acidifying the soil. The results of applying this indicator to the five chosen wooden façade materials are seen in the diagram below (see Figure 8).

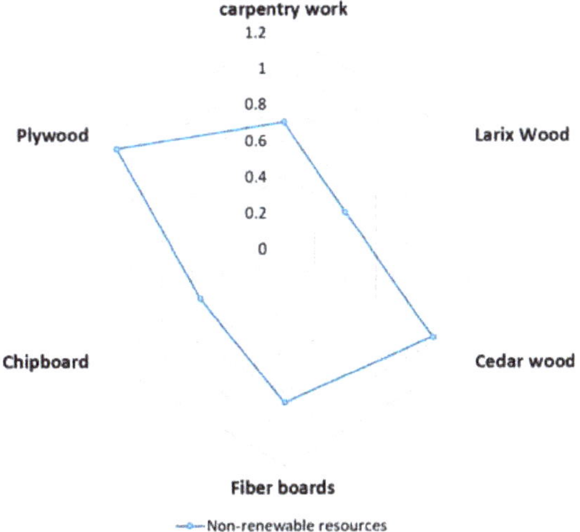

Figure 8. The impact of acidification on the selective façade wooden material [68–72].

5.7. Nitrogen Load (gNO_3/Year)

The fifth measure in the environmental profile (EP) represents the impact of nitrogen- or phosphorus-containing compounds on the environment. They will lead to the expansion of algae or plants to get out of control, which is harmful to the environment. Applying this indicator to the selected five wooden façade materials is shown in the below diagram's results (see Figure 9).

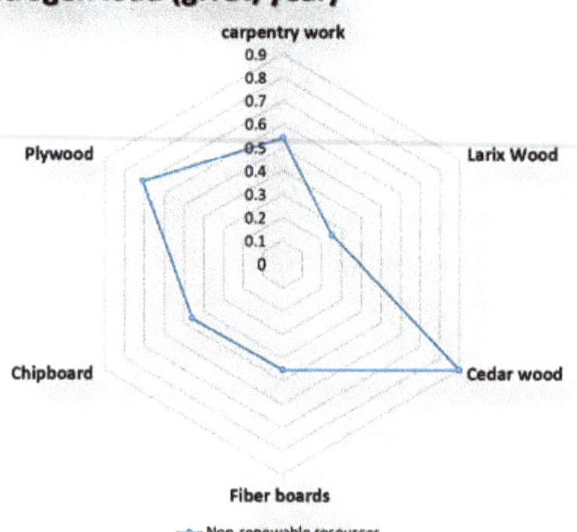

Figure 9. The impact of nitrogen on the selective façade wooden material [67–69].

5.8. Human Toxicity (m^3/Year)

The sixth predictor of the environmental profile (EP) indicates that pollutants with acute and permanent harmful effects on humans have an ecological impact. Contaminants are released into the receiving environment at the life cycle of goods, facilities, and systems, such as air, water, and soil. The human-health toxicity feature is described as DALY per kg of chemicals released into a given environment [72]. Emission inventories of various materials will include hundreds of chemicals, which could cause adverse effects to people and habitats. Applying this indicator to the selected five wooden façade materials is shown in the below diagram's results (see Figure 10).

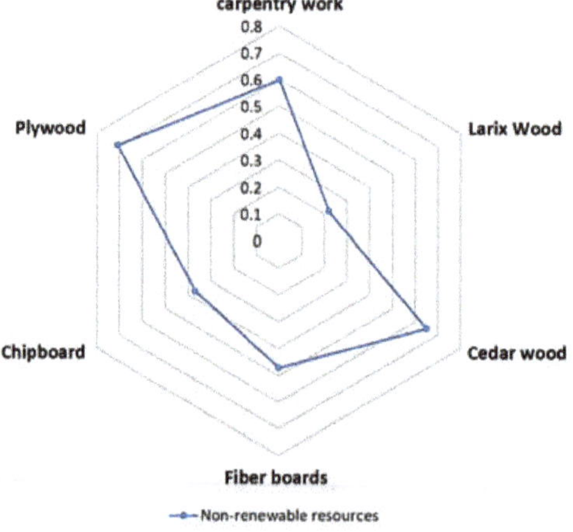

Figure 10. The impact of human toxicity on the selective façade wooden material [68–72].

5.9. Disposal (kg/Year)

The last indicator of the environmental profile (EP) shows the building, construction, or material when its service life is over (see Figure 11).

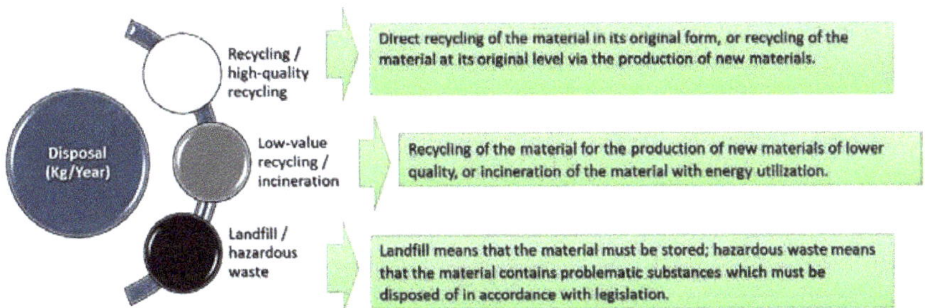

Figure 11. The environmental indicators show the environmental impact of the disposal.

Applying this indicator to the selected five wooden façade materials is shown in the below diagram's results (see Figures 12 and 13).

Disposal (kg/year)

carpentry work
0.3
0.25
0.2
0.15
0.1
0.05
0
Plywood
Larix Wood
Chipboard
Cedar wood
Fiber boards
Non-renewable resources

Figure 12. Shows the impact of disposal on the selective façade wooden material [68–72].

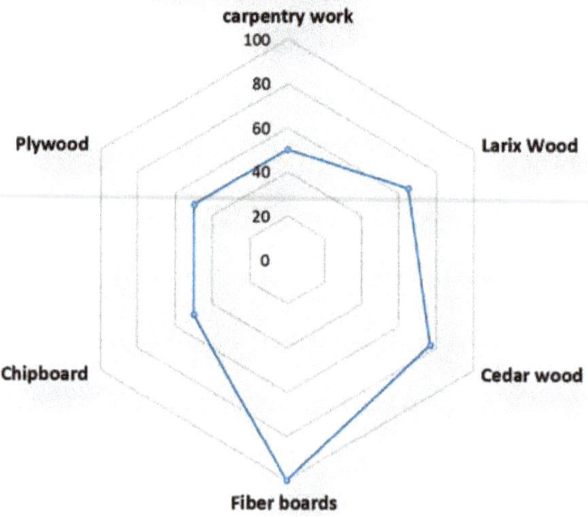

Figure 13. The life cycle façade wooden material [68–70].

6. Discussion

Reserves and renovation speed of materials create limitations in accelerating the useability process. So, there has to be a new creation of it for a specific material to substitute the used one. For example, there are big reserves of natural stone but very low renovation speed; on the contrary, there are limited reserves of wooden materials but very high renovation speed.

The life cycle of a product, applied to the construction area, is all the ways it follows from the acquisition of the raw materials (resources) from nature to make it, passing through its processing, packaging, transportation, installation, use, maintenance, deconstruction, or demolition, obtaining a waste that we can directly reuse, recycle, or dispose of in a controlled landfill [73]. Then, of course, the cycle is closed when we do not waste anymore, but the raw material we had at the beginning, or another material, was equally useful.

Due to the environmental problems experienced today, the trend towards sustainable or recyclable materials in the construction sector has increased, and studies on ecological structure concepts have increased. Wood material is also preferred because it is natural, easy to handle, durable, and easily used with other materials. However, today's comfort conditions have changed; new construction technologies have been developed; the number of floors of buildings has increased; and wooden building materials have been insufficient for these conditions.

The construction sector is lagging when it comes to digitization in comparison to other industry sectors. However, the rapid innovation and change in information technology offer an immense opportunity to implement a circular economy. With the increasing complexity and the substantial majority of materials and products in a building, digitization, process automation, and data standards need to be a prerequisite rather than an exception, besides the revolutionary advantages of digital technology for building and operational processes and materials passports [74].

Digital processes must gather, process, store, and use the massive quantities of data involved. Information stored in materials passports is only useful when the relevant actors can use it at the required time. Materials passports (MPs) need to be integrated into BIM to provide input data for reversible and circular design assessments [75]. BIM, which can be seen as a digital twin, will become a standard tool in the construction industry because it can store referencing and link data of individual components within a building over its life cycle. Materials passports and BIM should be seen in combination as they complement each other.

For effective use of IoT, the development of Artificial Intelligence (AI) plays a vital role. AI can assess information based on patterns (e.g., for information transfer) or when collecting data [76]. For example, an automated building façade scan can be interpreted (e.g., dimensions of windows) and analyzed. Furthermore, within the machine learning process, there is a possibility to identify material composition in an automated way in the future.

The architecture and engineering projects of the Circular Building highlight the transition from the industry-led linear economy paradigm of "take, make, waste" to the circular economy [77]. The project is the third in a series of projects they have worked on with Lewis Blackwell, Chief Strategy Officer of The Building Center in London. Smith constructed the Wikihouse in 2014 and the A House for London in 2015. Both projects were designed to explore the different technologies and construction methods that emerged at that time. The theme of Wikihouse was the open-source design and digital manufacturing, while A House for London was the modular building and housing crisis, and the issues of materials, resources, and waste in the construction industry were addressed [78]. The construction industry provides the necessary infrastructure, offices, and homes for our cities and neighborhoods and undeniably impacts the environment. According to calculations, 10 million tons of the approximately 20 million tons of waste generated in London in 2008 are from construction [79]. As the construction industry practices develop, what to do with the increasing waste is a big problem, as well as simultaneously extracting the mines used in the sector and transforming them into building materials, the environmental damages caused by the energy consumed. The Circular Building idea arises from this point. Cyclic Building is not a circular structure; the adjective "cyclic" describes the journey of materials. Ellen MacArthur, who had been offshore for 71 days, broke the record for traveling worldwide by her sailing boat on February 7, 2005, when she arrived at Ushant in Breton, France [80]. She embarked on another circular journey in September 2010 when she founded the Ellen MacArthur Foundation, promoting the "circular economy." A circular economy reproduces and reuses materials just as in nature, unlike the "take, make, waste" model of our linear economy, which is based on consuming resources. It transforms waste into food thanks to pre-planned design processes. The fundamental philosophy has to maintain goods at their maximum value for as long as possible so that structures and facilities can be used for as long as possible. The "biosphere" and "technosphere", named after Swiss architect Walter Stahel [81] and based on the "cradle to cradle" concept, illuminated by Michael Braungart as a chemist, and William McDonough as an architect, bring the model of material circulation to existence [82].

The idea of a prototype building that is planned to bring together many components from the construction industry, then disassemble the parts and return them to the supply chain, created a new vision for materials engineering, and recycling has been made possible by digital technologies [83]. Suppliers would be responsible for regenerating all of the materials and making them available. The materials with the lowest possible energy use and low carbon impact were chosen. A complete digital model was created and integrated the material passport idea into the building construction. During the dismantling of the building, footprints of every material used were recorded. Thus, the building became a material resource, an archive to be used in the future. Buildings are described as material banks in the context of the material passport. Instead of being dismantled during use, buildings' pieces are properly disassembled and reused or recycled at the same or even higher quality standards than the conventional make-a-waste scheme. Material passport promotes a waste-free economy. Materials passports facilitate circular business models by identifying materials and displaying their circular pathways.

Suppliers, designers, engineers, and end-users are now able to access information from a digital portal. The aim is to create materials with continuous cycles of use and reuse. For example, in the maintenance cycle, products are retained to optimize their usage time, and the value for recovery is assured in the reprocessing and energy production loop. The refurbishment loop allows biosphere materials to be safely cascaded into new products until they ultimately re-enter the biosphere through incineration or composting. Many items are not yet planned to be used circularly. The passports portal offers a feedback loop to refine their content, product design, and facilities for improved circular use materials passports, facilitating adapting to a circular economy. Engineering and design processes in the circular economy require close collaboration, especially with material suppliers. People find this idea very attractive; they are interested in the process itself and the ideas it came up with, and they started using the modular structure we currently use, digital fabrication, the use of materials from cradles to cradles.

Stewart Brand's book entitled "How Buildings Learn" is a fundamental source on this topic [84]. American author and visionary Stewart Brand is also the editor of the famous 1968 work "Whole Earth Catalog" [85], which illuminated the idea of Shearing Layers, developed by architect Frank Duffy [86], in his 1994 book entitled "How Buildings Learn: What Happens After They Are Built" [87]. In this approach, a building is not considered a single entity, but a structure composed of elements that transform according to different timelines. The layers of a building are called Six "s": Site (location); Structure; Skin (surface); Service (electricity, plumbing, heating); Space Plan (layout); and Stuff (items), in other words, other stuff such as furniture that belongs to homes. Different spatial elements have different timelines [84].

The items of the interior can also be designed differently. For example, rug suppliers such as Desso offer the opportunity to rent a rug; the rugs would be returned at the end of the time [88]. They also produce a new rug using the same material. It is possible to adopt the same model for electrical products. Regarding Phillips's "pay as you burn" model for lighting, the user (not the consumer) pays for the light he uses instead of the material. There is no limit to re-evaluating lighting elements. The project also benefits from traditional materials that have undergone chemical applications. A wood material called Accoya is the façade cladding. Accoya has become a game-changer material. Accoya, produced from a fast-growing softwood, is treated with acetylation [86]. It is akin to drying wood pickles. Acetylation similarly hardens the wood and stops the moisture movements in it. It becomes durable and can be used repeatedly. It is a great material that is as hard as wood.

Circular economy practices are spreading, although not yet mainstream. If the construction industry did not support, it would not have made that much impact. It is an important step forward in construction and engineering and a very exciting process; innovators will probably continue to look for ways to disseminate for the next decades. This approach will completely change where the materials are gathered, what they are used for, and how they are utilized. The novelty of this study is to conduct a comprehensive, environmentally comparative analysis of the vast of building materials and define the smart kinetic façade role in modern building design depending on environmental profile (EP) inquiry, regarding the idea of integrating environmental sensitivity into the material passport phenomenon. The results show that considering materials' environmental sensitivity in terms of resource consumption, energy consumption, greenhouse effect, acidification, nitrogen load, human toxicity, disposal, and life cycle, respectively, as a model within the material passport (MP) concepts, will significantly contribute to the use of the right building material at the right cost.

7. Conclusions

Wood is certainly the oldest natural material, renewable, easily recyclable, and it can store carbon dioxide, making high wood buildings a solution capable of meeting key sustainability goals. Moreover, since the early 20th century, thanks to the attention and dissemination of concepts related to the environmental sustainability of processes and production, they are studied and appreciated regarding other values about the ecosystem balance and the perceived environmental comfort in buildings made of wood. Furthermore, while wood is used depending on the size of the tree used in traditional buildings, it can be prepared in industrially desired sizes in today's conditions. This situation has increased the usage area of wood and has made it a material above the standards of other materials we use in today's needs.

Increasing ecological awareness, growing expectations for the health and comfort of home environments, and developing interesting new products from the wood industry are the basis for modern construction designs in the urban context. We take a holistic view of energy consumption and the material cycle in the construction industry, and we see that wood offers many advantages. For example, wood binds carbon dioxide during the growth phase and protects it for years even when it becomes a wood-building material, thus preventing carbon dioxide from entering the atmosphere again.

The study explores the possibilities to generate worth in the human-made facilities by integrating the building material passport and the wooden kinetic façade (WKF), to incorporate circular qualities into field value chains. The importance of data analysis to invent and add value is shown in this tool. Thus, to optimize materials by regeneration and recycling, the notion of waste is revised. The reintegration of materials into the innovative business models is crucial to support LCA, also end-of-life studies. Many barriers in the civil construction field to applying the BMP instrument include systemic consideration of the value chain and flows, resulting in improved cooperation among stakeholders and public assistance, emphasizing regulations and fiscal incentives necessary for a circular economic transition.

Instead of a linear system, a transformation into a circular economy will develop, and new features will appear. Establishing take-back systems is required to provide incentives for participation with the information exchange and innovative business models. To obtain the maximum benefit from materials passports for a circular economy, exchanging relevant, up-to-date information at the right time is key to a functioning value chain.

The availability of material data is a core aspect of a functioning circular economy. As buildings and components have long lifetimes and can have multiple changes of ownership and responsibilities, the data need to be kept up to date and passed on to the relevant actors systematically. A circular supply chain is only as strong as its weakest link, which requires incentives to ensure the participation of all parties. We need to start acting now in implementing the necessary steps in the building industry and its supply chain because establishing a circular economy is a prerequisite for sustainable development towards a sustainable and more circular future.

Author Contributions: Conceptualization, A.A. (Amjad Almusaed) and A.A. (Asaad Almssad); methodology, A.A. (Amjad Almusaed), A.A. (Asaad Almssad); software, A.A. (Amjad Almusaed); validation, M.E.A., İ.A.; analysis, A.A. (Amjad Almusaed); investigation, A.A. (Asaad Almusaed), I.Y., M.E.A., İ.A.; resources, A.A. (Amjad Almusaed); data curation, A.A. (Amjad Almusaed), I.Y.; writing—original draft preparation, A.A. (Amjad Almusaed); writing—review and editing, M.E.A., İ.A. All authors have read and agreed to the published version of the manuscript.

Funding: This research received no external funding.

Institutional Review Board Statement: Not applicable.

Informed Consent Statement: Not applicable.

Data Availability Statement: Not applicable.

Conflicts of Interest: The authors declare no conflict of interest.

References

1. Almusaed, A.; Almssad, A.; Alasadi, A. Analytical interpretation of energy efficiency concepts in the housing design process from hot climate. *J. Build. Eng.* **2019**, *21*, 254–266. [CrossRef]
2. Benachio, G.L.F.; Freitas, M.D.C.D.; Tavares, S.F. Circular economy in the construction industry: A systematic literature review. *J. Clean. Prod.* **2020**, *260*, 121046. [CrossRef]

3. Hopkinson, P.; Chen, H.-M.; Zhou, K.; Wang, Y.; Lam, D.; Wong, Y. Recovery and reuse of structural products from end-of-life buildings. In *Proceedings of the Institution of Civil Engineers—Engineering Sustainability*; Thomas Telford Ltd.: London, UK, 2019; Volume 172, pp. 119–128.
4. Lacy, P.; Rutqvist, J. *Waste to Wealth—The Circular Economy Advantage*; Palgrave Macmillan: London, UK, 2015; pp. 3–18.
5. BIS (Department for Business Innovation & Skills). *Supply Chain Analysis into the Construction Industry: A Report for the Construction Industrial Strategy*; BIS Research Paper, no. 145; Department for Business, Innovation and Skills: London, UK, 2013.
6. Pomponi, F.; Moncaster, A. Circular economy for the built environment: A research framework. *J. Clean. Prod.* **2017**, *143*, 710–718. [CrossRef]
7. Grobman, Y.J.; Yekutiel, T.P. Autonomous Movement of Kinetic Cladding Components in Building Facades. In *Proceedings of the 2nd Annual International Conference on Material, Machines and Methods for Sustainable Development (MMMS2020)*; Springer Science and Business Media LLC: Berlin/Heidelberg, Germany, 2013; pp. 1051–1061.
8. Almusaed, A.; Almssad, A. Efficient daylighting approach by means of light-shelve device adequate for habitat program in Aarhus City. *Int. J. Smart Grid Clean Energy* **2014**, *3*, 441–453. [CrossRef]
9. Wanas, A.; Aly, S.; Farghal, A.; El-Dabaa, R. Use of Kinetic Facades to Enhance Daylight Performance in Office Buildings with Emphasis on Egypt Climate. *J. Eng. Appl. Sci.* **2015**, *62*, 339–361.
10. Barozzi, M.; Lienhard, J.; Zanelli, A.; Monticelli, C. The sustainability of adaptive envelopes: Developments of kinetic archi-tecture. *Procedia Eng.* **2016**, *155*, 275–284. [CrossRef]
11. Panya, D.S.; Kim, T.; Choo, A.S. A Methodology of Interactive Motion Facades Design through Parametric Strategies. *Appl. Sci.* **2020**, *10*, 1218. [CrossRef]
12. De Luca, P.; Carbone, I.; Nagy, J.B. Green building materials: A review of state of the art studies of innovative materials. *J. Green Build.* **2017**, *12*, 141–161. [CrossRef]
13. Bacha, C.B.; Bourbia, F. The effect of kinetic facades on energy efficiency in office buildings-hot dry climates. In Proceedings of the 11th Conference on Advanced Building Skins, Bern, Switzerland, 10–11 October 2016; pp. 458–468.
14. Almusaed, A.; Almssad, A.; Homod, R.Z.; Yitmen, I. Environmental Profile on Building Material Passports for Hot Climates. *Sustainability* **2020**, *12*, 3720. [CrossRef]
15. Sodagar, B.; Gilroy-Scott, B.; Fieldson, R. Design for Sustainable Architecture and Environments. *Int. J. Environ. Cult. Econ. Soc. Sustain. Annu. Rev.* **2008**, *4*, 73–84. [CrossRef]
16. Almusaed, A.; Almssad, A. Improvement of Thermal Insulation by Environmental Means. *Eff. Therm. Insul. Oper. Factor A Passiv. Build. Model* **2012**, *1*. [CrossRef]
17. Ruggiero, G.; Marmo, R.; Nicolella, M. A Methodological Approach for Assessing the Safety of Historic Buildings' Façades. *Sustainability* **2021**, *13*, 2812. [CrossRef]
18. Burnard, M.D.; Kutnar, S. Wood and human stress in the built indoor environment: A review. *Wood Sci. Technol.* **2015**, *49*, 969–986. [CrossRef]
19. Munir, M.T.; Pailhories, H.; Eveillard, M.; Aviat, F.; Lepelletier, D.; Belloncle, C.; Federighi, M. Antimicrobial Characteristics of Untreated Wood: Towards a Hygienic Environment. *Health* **2019**, *11*, 152–170. [CrossRef]
20. Yüksek, I. The Evaluation of Building Materials in Terms of Energy Efficiency. *Period. Polytech. Civ. Eng.* **2015**, *59*, 45–58. [CrossRef]
21. Sekularac, J.I.; Sekularac, N.; Tovarovic, J.C. Wood as element of facade cladding in modern architecture. *Tech. Technollogies Educ. Manag.* **2012**, *7*, 1304–1310.
22. Solovev, V. Reasonable choice of materials for project management in transport construction. *IOP Conf. Series Mater. Sci. Eng.* **2021**, *1103*, 012036. [CrossRef]
23. Zhukov, A.D.; Smirnova, T.V.; Naumova, N.V.; Mustafayev, R.M. Sistemy ekologicheski ustoychivogo stroitel'stva [Environmentally Sustainable Building Systems]. Stroitel'stvo: Nauka i obrazovanie [Construction: Science and Education]. 2013. Available online: http://nso-journal.ru/public/journals/1/issues/2013/03/4.pdf (accessed on 6 April 2021).
24. Khoshnava, S.M.; Rostami, R.; Zin, R.M.; Štreimikienė, D.; Mardani, A.; Ismail, M. The Role of Green Building Materials in Reducing Environmental and Human Health Impacts. *Int. J. Environ. Res. Public Health* **2020**, *17*, 2589. [CrossRef]
25. Lawson, B. *Embodied Energy of Building Materials. The Environmental Design Guide, Pro 2*; Royal Australian Institute of Architects: Canberra, Australia, 1998; pp. 4–5.
26. Haupt, M.; Hellweg, S. Measuring the environmental sustainability of a circular economy. *Environ. Sustain. Indic.* **2019**, *1*, 100005. [CrossRef]
27. Kibert, C. The next generation of sustainable construction. *Build. Res. Inf.* **2007**, *35*, 595–601. [CrossRef]
28. Assefa, Y.; Staggenborg, S.A.; Prasad, V.P.V. Grain Sorghum Water Requirement and Responses to Drought Stress: A Review. *Crop Manag.* **2010**, *9*, 1–11. [CrossRef]
29. Shahda, M.M.M. Vision and Methodology to Support Sustainable Architecture through Building Technology in the Digital Era. *Int. J. Environ. Sci. Sustain. Dev.* **2018**, *2*, 1–14. [CrossRef]
30. Fakourian, F.; Asefi, M. Environmentally responsive kinetic façade for educational buildings. *J. Green Build.* **2019**, *14*, 165–186. [CrossRef]
31. Thayer, R. *Gray World, Green Heart: Technology, Nature, and the Sustainable Landscape*; John Wiley & Sons: New York, NY, USA, 1994; pp. 101–110.

32. Oláh, J.; Aburumman, N.; Popp, J.; Khan, M.A.; Haddad, H.; Kitukutha, N. Impact of Industry 4.0 on Environmental Sustainability. *Sustainability* **2020**, *12*, 4674. [CrossRef]
33. Dwek, M. Integration of Material Circularity in Product Design. Ph.D. Thesis, Université Grenoble Alpes, Grenoble, France, 2017.
34. Munaro, M.R.; Fischer, A.C.; de Azevedo, N.C.; Tavares, S.F. Proposal of a building material passport and its application feasibility to the wood frame constructive system in Brazil. In *Proceedings of the IOP Conference Series: Earth and Environmental Science*; IOP Publishing: Bristol, UK, 2019; Volume 225, p. 012018.
35. Akadiri, P.O.; Chinyio, E.A.; Olomolaiye, P.O. Design of A Sustainable Building: A Conceptual Framework for Implementing Sustainability in the Building Sector. *Buildings* **2012**, *2*, 126–152. [CrossRef]
36. Romano, R.; Aelenei, L.; Aelenei, D.; Mazzucchelli, E.S. What is an adaptive façade? Analysis of Recent Terms and definitions from an international perspective. *J. Facade Des. Eng.* **2018**, *6*, 65–76.
37. Youssef, M.M.; De Temmerman, N.; Brebbia, C.A. Kinetic behavior, the dynamic potential through architecture and design. *Int. J. Comput. Methods Exp. Meas.* **2017**, *5*, 607–618. [CrossRef]
38. Pingale, T.B.; Damugade, S.Y.; Jirge, N.D. Visual Communication in Architecture. *Int. J. Eng. Res. Technol.* **2017**, *10*, 123–126.
39. Shahin, H.S.M. Adaptive building envelopes of multistory buildings as an example of high performance building skins. *Alex. Eng. J.* **2019**, *58*, 345–352. [CrossRef]
40. Salama, W. Design of concrete buildings for disassembly: An explorative review. *Int. J. Sustain. Built Environ.* **2017**, *6*, 617–635. [CrossRef]
41. Harry, S. Dynamic Adaptive Building Envelopes—An Innovative and State-of-The-Art Technology. *Creative Space* **2016**, *3*, 167–184. [CrossRef]
42. Hart, J.; Pomponi, F. More Timber in Construction: Unanswered Questions and Future Challenges. *Sustainability* **2020**, *12*, 3473. [CrossRef]
43. Suchomel, J.; Belanová, K.; Gejdoš, M.; Němec, M.; Danihelová, A.; Mašková, Z. Analysis of fungi in wood chip storage piles. *BioResources* **2014**, *9*, 4410–4420. [CrossRef]
44. Kuru, A.; Oldfield, P.; Bonser, S.; Fiorito, F. A Framework to Achieve Multifunctionality in Biomimetic Adaptive Building Skins. *Buildings* **2020**, *10*, 114. [CrossRef]
45. Omer, A.M. Built Environment: Relating the Benefits of Renewable Energy Technologies. *Int. J. Automot. Mech. Eng.* **2012**, *5*, 561–575. [CrossRef]
46. Zhang, Y.; Yan, K.; Cheng, T.; Zhou, Q.; Qin, L.; Wang, S. Influence of Climate Change in Reliability Analysis of High Rise Building. *Math. Probl. Eng.* **2016**, *2016*, 1–11. [CrossRef]
47. Tiryaki, S.; Bardak, S.; Aydin, A.; Nemli, G. Analysis of volumetric swelling and shrinkage of heat treated woods: Experimental and artificial neural network modeling approach. *Maderas. Ciencia y tecnología* **2016**, *18*. [CrossRef]
48. Almusaed, A. *Intelligent Sustainable Strategies upon Passive Bioclimatic Houses: From Basra (Iraq) to Skanderbeg (Denmark): Postdoc Research*; Arkitektskole i Århus: Aarhus, Denmark, 2004; p. 173.
49. O'Brien, M.; Bringezu, S. What Is a Sustainable Level of Timber Consumption in the EU: Toward Global and EU Benchmarks for Sustainable Forest Use. *Sustainability* **2017**, *9*, 812. [CrossRef]
50. Moezzipour, B.; Ahmadi, M.; Abdolkhani, A.; Doosthoseini, K. Chemical changes of wood fibers after hydrothermal recycling of MDF wastes. *J. Indian Acad. Wood Sci.* **2017**, *14*, 133–138. [CrossRef]
51. Ghazouani, A.; Xia, W.; Ben Jebli, M.; Shahzad, U. Exploring the Role of Carbon Taxation Policies on CO_2 Emissions: Contextual Evidence from Tax Implementation and Non-Implementation European Countries. *Sustainability* **2020**, *12*, 8680. [CrossRef]
52. Risbrudt, C.D.; Ross, R.J.; Blankenburg, J.J.; Nelson, C.A. Forest Products Laboratory: Supporting the nation's armed forces with valuable wood research for 90 years. *For. Prod. J.* **2007**, *57*, 6–14.
53. Ramage, M.H.; Burridge, H.; Busse-Wicher, M.; Fereday, G.; Reynolds, T.; Shah, D.U.; Wu, G.; Yu, L.; Fleming, P.; Densley-Tingley, D.; et al. The wood from the trees: The use of timber in construction. *Renew. Sustain. Energy Rev.* **2017**, *68*, 333–359. [CrossRef]
54. Petersen, E.H.; Krogh, H.; Dinesen, J. *Miljødata for Udvalgte Bygningsdele. SBI Forlag-Rapport 296*; Statens Byggeforskning-Sinstitut: Hørsholm, Denmark, 1998.
55. Vek, V.; Balzano, A.; Poljanšek, I.; Humar, M.; Oven, P. Improving Fungal Decay Resistance of Less Durable Sapwood by Impregnation with Scots Pine Knotwood and Black Locust Heartwood Hydrophilic Extractives with Antifungal or Antioxidant Properties. *Forests* **2020**, *11*, 1024. [CrossRef]
56. Momohara, I.; Ota, Y.; Sotome, K.; Nishimura, T. Assessment of decay risk of airborne wood-decay fungi II: Relation between isolated fungi and decay risk. *J. Wood Sci.* **2011**, *58*, 174–179. [CrossRef]
57. Harlow, B.A.; Duursma, R.A.; Marshall, J.D. Leaf longevity of western red cedar (Thuja plicata) increases with depth in the canopy. *Tree Physiol.* **2005**, *25*, 557–562. [CrossRef] [PubMed]
58. Orlowski, K.A.; Dudek, P.; Chuchala, D.; Blacharski, W.; Przybylinski, T. The Design Development of the Sliding Table Saw Towards Improving its Dynamic Properties. *Appl. Sci.* **2020**, *10*, 7386. [CrossRef]
59. Antov, P.; Krišt'ák, L.; Réh, R.; Savov, V.; Papadopoulos, A.N. Eco-Friendly Fiberboard Panels from Recycled Fibers Bonded with Calcium Lignosulfonate. *Polymers* **2021**, *13*, 639. [CrossRef] [PubMed]
60. Pirayesh, H.; Moradpour, P.; Sepahvand, S. Particleboard from wood particles and sycamore leaves: Physico-mechanical properties. *Eng. Agric. Environ. Food* **2015**, *8*, 38–43. [CrossRef]

61. Dziurka, D.; Mirski, R.; Dukarska, D.; Derkowski, A. Possibility of using the expanded polystyrene and rape straw to the manufacture of lightweight particleboards. *Maderas. Cienc. Tecnol.* **2015**, *17*, 647–656. [CrossRef]
62. Aydin, I. Effects of veneer drying at high temperature and chemical treatments on equilibrium moisture content of plywood. *Maderas. Cienc. Tecnol.* **2014**, *16*, 445–452. [CrossRef]
63. Xiong, X.; Niu, Y.; Zhou, Z.; Ren, J. Development and Application of a New Flame-Retardant Adhesive. *Polymers* **2020**, *12*, 2007. [CrossRef] [PubMed]
64. BRASIL, Ministério das Cidades. *Diretriz SINAT nº 005—Revisão 02—Sistemas Construtivos Estruturados em Peças Leves de Madeira Maciça Serrada, com Fechamentos em Chapas—Sistemas Leves Tipo "Light Wood Framing"*; PBQP-H: Brasília, Brasil, 2017; p. 73.
65. Forsberg, A.; von Malmborg, F. Tools for environmental assessment of the built environment. *Build. Environ.* **2004**, *39*, 223–228. [CrossRef]
66. Liedtke, C.; Bienge, K.; Wiesen, K.; Teubler, J.; Greiff, K.; Lettenmeier, M.; Rohn, H. Resource Use in the Production and Consumption System—The MIPS Approach. *Resources* **2014**, 3, 544–574. [CrossRef]
67. Zhang, Z.; Xue, B.; Pang, J.; Chen, X. The Decoupling of Resource Consumption and Environmental Impact from Economic Growth in China: Spatial Pattern and Temporal Trend. *Sustainability* **2016**, *8*, 222. [CrossRef]
68. Hauschild, M. *Baggrund for Miljøvurdering af Produkter*; Instituttet for Produktudvikling, Danmarks Tekniske Universitet: Lungby, Denmark, 1996; pp. 1–81.
69. Jensen, A.A.; Hoffman, L.; Møller, B.; Schmidt, A.; Christiansen, K.; Elkington, J. Life cycle assessment. A guide to approaches, experiences and information sources. *Environ. Issues Ser.* **1997**, *6*, 237–262.
70. Dinesen, J.; Krogh, H.; Traberg-Borup, S. *Livscyklusbaseret Bygningsprojektering: Opgørelse af Bygningers Energiforbrug og Energirelaterede Miljøpåvirkninger [Life-Cycle-Based Building Design. Calculation of Building's Energy Consumption and Energy Related Environmental Impacts]*; SBI forlag: Hørsholm, Denmark, 1997; pp. 1–57.
71. Skovgaard, M.; Rasmussen, F.N.; Sørensen, C.G.; Birgisdottir, H. *Brugervejledning til LCAbyg: Beregning af Bygningers Miljøprofiler*; SBI, Statens Byggeforskningsin-stitut, Aalborg Universitet: København, Denmark, 2015; pp. 12–15.
72. Abouhamad, M.; Abu-Hamd, M. Life Cycle Assessment Framework for Embodied Environmental Impacts of Building Construction Systems. *Sustainability* **2021**, *13*, 461. [CrossRef]
73. Honic, M.; Kovacic, I.; Sibenik, G.; Rechberger, H. Data- and stakeholder management framework for the implementation of BIM-based Material Passports. *J. Build. Eng.* **2019**, *23*, 341–350. [CrossRef]
74. Ullah, Z.; Al-Turjman, F.; Mostarda, L.; Gagliardi, R. Applications of Artificial Intelligence and Machine learning in smart cities. *Comput. Commun.* **2020**, *154*, 313–323. [CrossRef]
75. Huijbregts, M.A.; Rombouts, L.J.; Ragas, A.M.; van de Meent, D. Human-toxicological effect and damage factors of car-cinogenic and noncarcinogenic chemicals for life cycle impact assessment. *Integr. Environ. Assess. Manag. Int. J.* **2005**, *1*, 181–244. [CrossRef]
76. Andriulaitytė, I.; Vilnius Gediminas Technical University; Valentukeviciene, M. Circular economy in buildings. *Bud. o Zoptymalizowanym Potencjale Energetycznym* **2020**, *10*, 23–29. [CrossRef]
77. Tooze, J.; Baurley, S.; Phillips, R.; Smith, P.; Foote, E.; Silve, S. Open Design: Contributions, Solutions, Processes and Projects. *Des. J.* **2014**, *17*, 538–559. [CrossRef]
78. Oyedele, L.O.; Ajayi, S.O.; Kadiri, K.O. Use of recycled products in UK construction industry: An empirical investigation into critical impediments and strategies for improvement. *Resour. Conserv. Recycl.* **2014**, *93*, 23–31. [CrossRef]
79. Sharp, N. *The First and the Fastest: Comparing Robin Knox-Johnston and Ellen MacArthur's Historic Round-the-World Voyages*; The History Press: Cheltenham, UK, 2018.
80. Stahel, W. *The Performance Economy*; Springer: Berlin/Heidelberg, Germany, 2010.
81. McDonough, W.; Braungart, M. *Cradle to Cradle: Remaking the Way We Make Things*; North Point Press: New York, NY, USA, 2010.
82. Petersen, K.H.; Napp, N.; Stuart-Smith, R.; Rus, D.; Kovac, M. A review of collective robotic construc-tion. *Sci. Robot.* **2019**, *4*. [CrossRef]
83. Brand, S. *How Buildings Learn: What Happens after They're Built*; Penguin Putnam Inc.: New York, NY, USA, 1995.
84. Brand, S. *Whole Earth Catalog*; Point Foundation: New York, NY, USA, 1968.
85. Duffy, F. Measuring Building Performance. *Facilities* **1990**, *8*, 17–20. [CrossRef]
86. Pushkar, S.; Verbitsky, O. LCA of different building lifetime shearing layers for the allocation of green points. *Eco-Archit. V Harmon. Archit. Nat.* **2014**, *142*, 459. [CrossRef]
87. Agrawal, V.V.; Atasu, A.; Van Wassenhove, L.N. OM Forum—New Opportunities for Operations Management Research in Sustainability. *Manuf. Serv. Oper. Manag.* **2019**, *21*, 1–12. [CrossRef]
88. Petrovski, A.; Petrovska-Hristovska, L.; Ivanovic-Sekularac, J.; Šekularac, N. Assessment of the sus-tainability of façade refurbishment. *Therm. Sci.* **2019**, *24*, 991–1006. [CrossRef]

 *materials* 

Article

Optimum Placement of Heating Tubes in a Multi-Tube Latent Heat Thermal Energy Storage

Mohammad Ghalambaz [1,2], Hayder I. Mohammed [3], Ali Naghizadeh [4], Mohammad S. Islam [5], Obai Younis [6,7], Jasim M. Mahdi [8], Ilia Shojaeinasab Chatroudi [9] and Pouyan Talebizadehsardari [1,2,*]

1 Metamaterials for Mechanical, Biomechanical and Multiphysical Applications Research Group, Ton Duc Thang University, Ho Chi Minh City 758307, Vietnam; mohammad.ghalambaz@tdtu.edu.vn
2 Faculty of Applied Sciences, Ton Duc Thang University, Ho Chi Minh City 758307, Vietnam
3 Department of Physics, College of Education, University of Garmian, Kurdistan 46021, Iraq; hayder.i.mohammad@garmian.edu.krd
4 Faculty of Mechanical Engineering, Babol University of Technology, Babol 7116747148, Iran; ali.naghizadeh2412@gmail.com
5 School of Mechanical and Mechatronic Engineering, Faculty of Engineering and Information Technology, University of Technology Sydney, Ultimo, NSW 2007, Australia; MohammadSaidul.Islam@uts.edu.au
6 Department of Mechanical Engineering, College of Engineering at Wadi Addwaser, Prince Sattam Bin Abdulaziz University, Wadi Addwaser 11991, Saudi Arabia; oubeytaha@hotmail.com
7 Department of Mechanical Engineering, Faculty of Engineering, University of Khartoum, Khartoum 11111, Sudan
8 Department of Energy Engineering, University of Baghdad, Baghdad 10071, Iraq; jasim@siu.edu
9 Department of Mechanical Engineering, Sirjan University of Technology, Sirjan 7813733385, Iran; i.s.chatroudi@gmail.com
* Correspondence: ptsardari@tdtu.edu.vn

Citation: Ghalambaz, M.; Mohammed, H.I.; Naghizadeh, A.; Islam, M.S.; Younis, O.; Mahdi, J.M.; Chatroudi, I.S.; Talebizadehsardari, P. Optimum Placement of Heating Tubes in a Multi-Tube Latent Heat Thermal Energy Storage. *Materials* **2021**, *14*, 1232. https://doi.org/10.3390/ma14051232

Academic Editors: Carlos Garcia-Mateo and Antonio Caggiano

Received: 29 January 2021
Accepted: 25 February 2021
Published: 5 March 2021

Abstract: Utilizing phase change materials in thermal energy storage systems is commonly considered as an alternative solution for the effective use of energy. This study presents numerical simulations of the charging process for a multitube latent heat thermal energy storage system. A thermal energy storage model, consisting of five tubes of heat transfer fluids, was investigated using Rubitherm phase change material (RT35) as the. The locations of the tubes were optimized by applying the Taguchi method. The thermal behavior of the unit was evaluated by considering the liquid fraction graphs, streamlines, and isotherm contours. The numerical model was first verified compared with existed experimental data from the literature. The outcomes revealed that based on the Taguchi method, the first row of the heat transfer fluid tubes should be located at the lowest possible area while the other tubes should be spread consistently in the enclosure. The charging rate changed by 76% when varying the locations of the tubes in the enclosure to the optimum point. The development of streamlines and free-convection flow circulation was found to impact the system design significantly. The Taguchi method could efficiently assign the optimum design of the system with few simulations. Accordingly, this approach gives the impression of the future design of energy storage systems.

Keywords: phase change material; melting; latent heat thermal energy storage; Taguchi method; optimization

1. Introduction

Providing facilities to cover all human needs for abundant on-demand energy leads the total power consumption to increase swiftly. Normal life pushes humans to use energy-driven machines and tools such as cooking, cooling, and heating, food storing [1]. Latent heat thermal energy storage systems (LHTES) with phase change materials (PCMs) provide a solution for the mismatches between energy supplies and demands by offering a more compact storage volume compared with conventional hot water storage tanks [2].

Moreover, this method is valuable during both charging and discharging processes, as phase change temperatures are almost constant which makes it more reliable for domestic applications [3]. This technique is included in solar thermal plants, energy management, peak-shaving, water heat recovery, building heating and cooling, and electronic power management [4]. This is relatively promising for applications needing rigid operation temperatures. Because of the weak thermal conductivity of PCMs, the thermal efficiency of thermal energy storage systems (TES) suffers from low heat transfer rates. Different methods are used to enhance the thermal conductivity, such as utilizing different configurations of fins [5], changing the configuration of the geometry [6–8], metal foam [9], and nanotechnology [10,11], using multiple PCMs [12], and using the combinations of different methods [12]. LHTES systems naturally result in a compact thermal energy storage design due to the high energy density of phase change.

Urban areas are one of the main power consumers by utilizing 45% of the total energy usage responsible for huge greenhouse gas emissions [13]. Combining PCMs into energy store units has been extensively studied in the literature [14–18]. The purpose was to shift the required electric power or a portion of it from peak times (high-cost tariff) to off-peak times (low-cost tariff). PCM phase changing delivers an active method to store energy during the night and consume it during the day. In this technique, the relatively low temperature at night is utilized to cool down the storage unit, and then, it could absorb the building's heat during the day. This decreases the condenser operating temperature and consequently higher efficiency.

Mahdi et al. [19] investigated the thermal improvement methods utilizing static structures, nanomaterials, and the combination of both methods (hybrid technique). Generally, it is found that the hybrid improvement procedures utilizing a heat transfer fluid pipe with a fin or metal foam are the most effective methods. Although each method has its challenges and difficulties in operation, there is no desirable technique, considering the effects of conduction and free convection which must be accounted for in the design process. Yamaha et al. [20] studied the effects of various PCM combinations packed in an air channel as a thermal energy storage component for the air-cooling system of a typical building in Nagoya (Japan). They found that a 5.4 kg PCM per square meter is sufficient to maintain a room temperature for 3 h (13:00–15:00).

Zhao et al. [21] integrated a conventional air cooling system with a TES unit to improve the efficiency of the system. The TES unit was made of a shell and a tube and filled with a PCM. The system operated with water and air as heat transfer fluid (HTF) in both melting and solidification loops, respectively. They stated that the coefficient of performance (COP) increases by 25.6% compared with the conventional air conditioning unit.

Many TES units suffer from low heat transfer properties of PCMs and cannot absorb/release the heat on demand. Thus, a proper design of such systems is the key to reach high efficiency. As a result, many researchers have focused on the influences of design parameters on TES units' efficiency. Dolado et al. [22] studied the important design variables of an LHTES unit. The results showed that the geometrical characteristics of the PCM layer and the air channels, as well as the HTF flow rate, could control the unit performance.

Waqas and Kumer [23] investigated a solar unit integrated with a PCM layer to warm a building room. The HTF flow rate, PCM mass, and fusion temperature are the essential design parameters affecting the unit performance.

Diarce et al. [24] utilized numerical simulations and drove a correlation between the solidification rate and a PCM layer's thickness. This correlation was suitable for the calculation of the thickness of the PCM layer. Ren et al. [13] assessed the heat transfer efficiency of TES photovoltaic collectors. The authors reported that the fusion temperature, PCM type, and HTF flow rate are the major parameters manipulating the stored thermal energy. Amin et al. [25] designed an optimized unit of TES through a parametric investigation. In this work, an efficient indicator, integrating the thermal efficiency and the power storage density, was adopted as a goal to optimize the PCM design. Saman et al. [26] investigated a combined TES-roof solar system and found that the HTF temperature and the flow rate are the two main factors affecting the PCM's thermal response of the system.

Developing an optimum technique requires a full study, which covers all cases with different parameters. Therefore, this approach needs a lot of running processes, which takes a long time to study numerical simulation and waste materials for the experimental study. Taguchi method is a technique used to specify the best and optimal cases for running [13,27–32]. Wang et al. [27] combined the extension theory Taguchi method (ETM) to reduce the required running cases and optimized thermal system elements such as full cells, hydrogen tanks, and electrolyzers. They compared their optimum design with the literature works and found the ETM is the most accurate among other methods. Sun et al. [28] utilized liquid fraction as an indicator and tested the energy charging cycles using the Taguchi method. Liu et al. [29] studied the energy efficiency of a hybrid unit collection of a PCMs-ventilated Trombe wall and a photovoltaic/heat panel combined with PCM using the Taguchi method.

The current work presents a design optimization approach for TES units with multiple HTF tubes. Unlike most existing investigations, the optimum design was settled considering the HTF tubes' locations through the PCM domain to achieve the best location of the tubes. Various simulations were designed using the Taguchi method; each design parameter was divided into four levels. The ultimate objective is to increase the signal/noise (S/N) ratio of the corresponding total stored latent heat and categorize the optimal integration of geometric parameters to offer engineers procedural guidance. The proposed optimization strategy offers a technique to optimally place multiple HTF tubes in TES systems.

2. Model and Governing Equations

Figure 1 illustrates the schematic of the proposed model of the LHTES. The actual LHTES is made of many tubes and could be extended by repeating the pattern in the horizontal direction. Due to the repetition of the inner tubes in the left and right sides of the heat exchanger, the symmetrical boundary condition is applied for the left and right boundaries, denoted by dash lines. The upper and lower walls of the heat exchanger are insulated with no-slip boundary conditions. The temperature of the tubes' wall is considered constant equal to 50 °C, and the PCM is initially placed at the temperature of 15 °C. It should be noted that the model can be repeated and extended to the full scale of a wide LHTES unit by employing the symmetry boundary conditions indicated by the dashed lines in Figure 1.

As displayed, each tube's diameter (D) is considered constant equal to 25.4 mm (1 inch) with an outer diameter of 28.575 mm. The height of the shell is considered 11D, equal to 314,325 mm, and the width of each repeated section is considered equal to $3\sqrt{3}D/2$ which is equal to 74.24 mm. Three independent parameters (*HL*1, *HL*2, and *HL*3) are defined variables in this study, which determines the tubes' locations on the left-hand side of the domain. These variables could be optimized to gain the highest melting rate, which is discussed comprehensively in the results and discussion. The locations of the two tubes on the right-hand side of the domain were determined based on the locations of the tubes on the left-hand side of the domain, as illustrated in Figure 1. All the boundary conditions and geometrical variables are included in Figure 1.

For the PCM, RT35 is employed with the mean melting temperature of 32.5 °C, suitable for domestic low-temperature heating applications such as underfloor heating systems. The properties of RT35 are listed in Table 1.

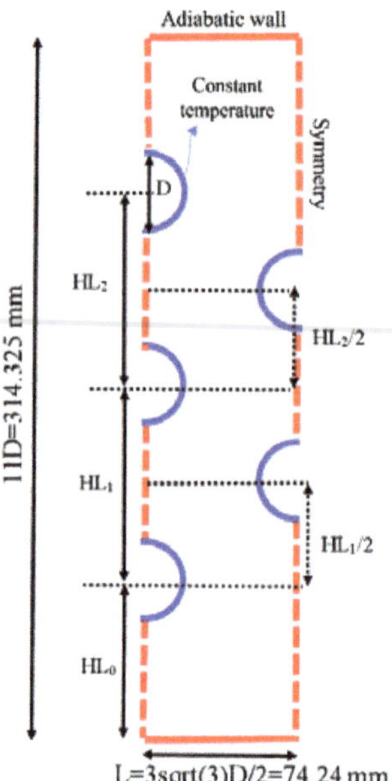

Figure 1. The schematic of the proposed heat exchanger in this study.

Table 1. Thermo-physical properties of RT35 [33].

Property	ρ (kg m^{-3})	L_f (kJ kg^{-1})	C_p (kJ kg^{-1} K)	k (W m^{-1} K)	μ (N s m^{-2})	T_L (°C)	T_S (°C)	β (1 K^{-1})
Values	815	170	2.0	0.2	0.023	35	29	0.0006

According to the enthalpy-porosity method [34,35], the conservation equations of continuity, momentum, and energy are given considering the following assumptions, i.e., laminar and Newtonian fluid flow for the molten PCM as well as neglecting the viscous dissipation and volume expansion of the PCM during the phase change process. The variation of density is negligible except for the buoyancy force, which is modeled considering the Boussinesq approximation. The gravity, which induces the natural convection flow, is directed downward. Thus, considering the abovementioned assumptions, the governing equations are obtained as following Equations (1)–(3):

$$\frac{\partial \rho}{\partial t} + \nabla \cdot \rho \vec{V} = 0, \tag{1}$$

$$\rho \frac{\partial \vec{V}}{\partial t} + \rho\left(\vec{V} \cdot \nabla\right)\vec{V} = -\nabla \mathrm{P} + \mu\left(\nabla^2 \vec{V}\right) - \rho_{ref}\beta\left(T - T_{ref}\right)\vec{g} - A_m \frac{(1-\lambda)^2}{\lambda^3 + 0.001}\vec{V}, \tag{2}$$

$$\frac{\rho C_p \partial T}{\partial t} + \nabla\left(\rho C_p \vec{V} T\right) = \nabla(k \nabla \mathrm{T}) - \left[\frac{\partial \rho \lambda L_f}{\partial t} + \nabla\left(\rho \vec{V} \lambda L_f\right)\right]. \tag{3}$$

The value 10^5 was considered for A_m based on the literature and also validation studies [36–38]. λ was introduced as Equation (4) [39]:

$$\lambda = \frac{\Delta H}{L_f} = \left\{ \begin{array}{ccc} 0 & if & T < T_{Solidus} \\ 1 & if & T > T_{Liquidus} \\ \frac{T - T_{Solidus}}{T_{Liquidus} - T_{Solidus}} & if & T_{Solidus} < T < T_{Liquidus} \end{array} \right\}, \quad (4)$$

where the total enthalpy is the sum of sensitive enthalpy (h) and latent heat (ΔH) is the volumetric enthalpy (H), described as Equations (5) and (6):

$$H = h + \Delta H, \quad (5)$$

where

$$h = h_{ref} + \int_{T_{ref}}^{T} C_p dT, \quad (6)$$

ΔH is calculated based on Equation (4) and the molten liquid fraction (*LF*) is the integration of liquid fraction, λ, over the computational domain. Moreover, the liquid fraction (*LF*) shows the total amount of the melted PCM divided by the total volume of the phase change material, i.e., the normal total amount of the melted PCM.

3. Numerical Method, Mesh Study, and Validation

3.1. Numerical Method

In this study, ANSYS workbench software was employed to perform the simulations and geometry generation and also study the effects of different parameters. In this regard, the QUICK scheme was used for the diffusion fluxes and convection, and the PRESTO scheme was used for the pressure equation. The convergence criteria are also considered 10^{-4} for continuity and momentum equations, while 10^{-6} was chosen for the energy equation.

3.2. Mesh and Time-Step Size Study

The size of the mesh and the time step of computations can affect the accuracy of computations. Employing a very fine mesh size and small time-steps can increase the computations' accuracy, but they also increase the demanded computational resources and time. Thus, the fair trade between the accuracy and computational cost can be performed by a mesh study. Here, three mesh sizes were selected, and the computations were repeated for each mesh size. For the mesh study, case 3 from the L16 table was selected for the mesh independence analysis. The computations were executed with a fine time step size of 0.2. The details of mesh sizes and the number of grid points were as follows: mesh-case 1: 24,500; mesh-case 2: 49,000; mesh-case 3: 73,500. The melting rate over time is plotted in Figure 2a. As can be seen, the results for mesh-cases 2 and 3 were very close. Thus, mesh-case 2 was selected for computations.

The impact of the time-step size on the liquid fraction-melting rate (LF) was studied in Figure 2b over mesh-case 2 for three time-steps of 0.1, 0.2, and 0.4 s. An increase in the time-step enhanced the computational cost significantly, since the governing equations should be solved and converged at each time-step. The results showed that the impact of the time-step on the computed LF was small; however, at an identical time, the liquid fraction varied using 0.4 s as the time-step size compared with that using time-step sizes of 0.2 and 0.1, especially in the middle of the melting process. Thus, the time-step of 0.2 s was selected for all computations of the present research. It should be noted that, after the time-step size analysis, it was concluded that the use of 0.2 s for the time-step size in the mesh independence analysis is correct.

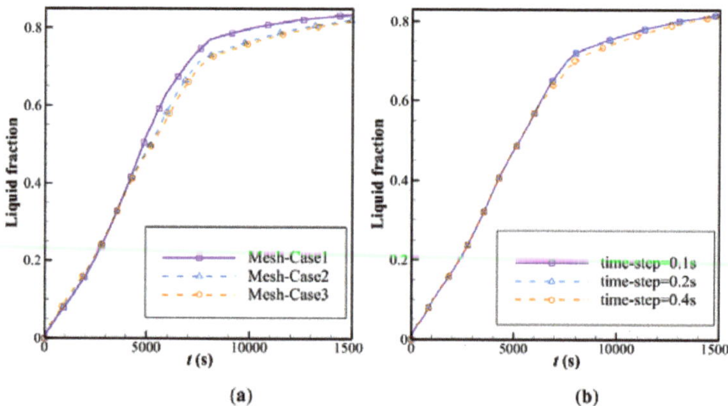

Figure 2. Influence of the mesh size and the time-step on the computed liquid fraction-melting rate (LF) for mesh-case 3 of the L16 table: (**a**) mesh size; (**b**) time-step size for mesh-case 3.

3.3. Validation

To verify the simulations in the present study, the experimental and numerical studies of Mat et al. [36] was employed, and the studied geometry was regenerated for the same geometrical and operational parameters. Mat et al. [36] analyzed a double-tube heat exchanger equipped with longitude fins for storing thermal energy. The TES unit in the study of Mat et al. was filled with RT58 as a PCM, and its wall was kept at a uniform temperature. It should be noted that a double-pipe heat exchanger is a simple form of shell-and-tube heat exchangers with one tube inside the shell. Besides, the boundary condition in the study of Mat et al. is a constant wall temperature, which is similar to what is proposed in this study for melting the PCM. Thus, the study of Mat et al. is suitable to verify the code in the present study. The verification results are shown in Figure 3 for both liquid fraction and average temperature presented in this study and those experimental and numerical data reported by Mat et al. [36]. As it can be seen, there is an excellent agreement for further analysis. It should be noted that, in the study of Mat et al., RT58 was employed as a PCM while RT35 was employed for further analysis in this study, since its melting point is suitable for domestic low-temperature heating applications such as underfloor heating systems working as a more efficient system than conventional radiators.

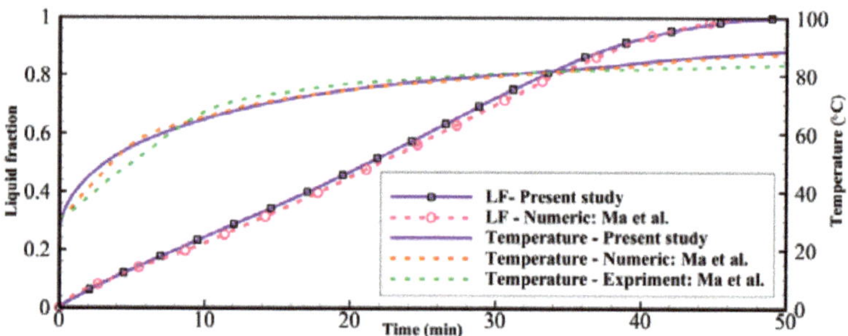

Figure 3. Comparison of the present study with the experimental data of Mat et al. [36].

4. Results and Discussion

The Taguchi method was utilized to find the best design of an LHTES unit in the presence of natural convection effects. The buoyancy forces in the molten PCM induced a free convection circulation flow during the charging process. The free convection improved the charging rate by the increase of heat transfer. The free convection heat transfer typically induces a nonuniform melting in an LHTES unit, since the heated liquid tends to move upward and always the top regions of an enclosure are subject to warm currents. Thus, the proper design of energy storage units to fully benefit from natural convection heat transfer is a critical task. Here, the placement height of HTF tubes was adopted as a design parameter. As depicted in schematic Figure 1, there are three geometrical design variables, which are HL_0/D, HL_1/D, and HL_2/D. Following the Taguchi method, each design parameter should be divided into a few possible levels. Here, four levels were considered for each design parameter, of which the details are summarized in Table 2.

Table 2. The ranges and levels of the control parameters.

Factors	Description	Level 1	Level 2	Level 3	Level 4
A	HL_0/D (height of the first tube)	1.2	1.8	2.4	3.0
B	HL_1/D (height of the second tube)	1.2	1.8	2.4	3.0
C	HL_2/D (height of the fourth tube)	1.2	1.8	2.4	3.0

The design structure shown in Table 2 was used to construct an orthogonal table and explore the design space. Here, the standard L16 table for three design parameters and four levels was selected. It should be noted that the whole design space could be 4^3 designs. The Taguchi method aims to find the best design with a minimum computational cost. Thus, instead of 4^3 simulations, here, only 16 simulations were needed to be executed using the Taguchi method. The computations of PCM melting with convection heat transfer effects were computationally costly. Thus, any reduction in the number of the required simulation is a huge help.

The details of the L16 trial cases are summarized in Table 3. Each row of Table 3 shows an LHTES with a specific placement of the HTF tubes. The target goal for optimization was selected at 75% of the charging process. Thus, a design that could reach a 75% melting volume faster (in a shorter time) is better. Thus, "the smaller, the better approach" was adopted for the Taguchi method.

Table 3. Taguchi L16 orthogonal table for three geometrical design parameters and four levels.

Experiment Number	Design Parameters			Time (s) for $LF = 0.75$	Signal-to-Noise (S/N) Ratio
	HL_0/D	HL_1/D	HL_2/D		
1	1.2	1.2	1.2	13,633	−82.6918
2	1.2	1.8	1.8	13,613	−82.6791
3	1.2	2.4	2.4	9025	−79.1089
4	1.2	3.0	3.0	8205	−78.2816
5	1.8	1.2	1.8	11,977	−81.5670
6	1.8	1.8	1.2	13,850	−82.8290
7	1.8	2.4	3.0	9161	−79.2389
8	1.8	3.0	2.4	9206	−79.2814
9	2.4	1.2	2.4	14,742	−83.3711
10	2.4	1.8	3.0	9374	−79.4385
11	2.4	2.4	1.2	11,096	−80.9033
12	2.4	3.0	1.8	10,332	−80.2837
13	3.0	1.2	3.0	12,603	−82.0095
14	3.0	1.8	2.4	10,710	−80.5958
15	3.0	2.4	1.8	12,379	−81.8537
16	3.0	3.0	1.2	10,104	−80.0899

Each design of Table 3 was simulated, and the required time for 75% melting ($LF = 0.75$) was computed. These data were reported in Table 3. Then, the Taguchi approach was applied to evaluate the value of each design in terms of a characteristic parameter of the S/N ratio.

The relationship for the smaller-is-better S/N ratio was written using log base [40]:

$$S/N = -10 \times log_{10}\left(\frac{\sum(Sa^2)}{n}\right), \tag{7}$$

where Sa denotes the melting time for each test case, and n is the number of observations, which is one in the current research. A design with a large S/N ratio could lead to a shorter charging time. The computed S/N ratios are reported in Table 3. Moreover, the S/N ratios were used to construct Taguchi relations and the ranking table.

Table 4 shows the impact of each design parameter on the charging time of the LHTES unit. Table 4 indicates that the middle right tube (HL_1/D) was the most important design parameter and adopted the first rank of significance. After that, the top tube and the bottom tube were the important design parameters.

Table 4. The rank values of the control parameters based on the S/N ratio.

Level	HL_0/D	HL_1/D	HL_2/D
1	−80.69	−82.41	−81.63
2	−80.73	−81.39	−81.60
3	−81.00	−80.28	−80.59
4	−81.14	−79.48	−79.74
Delta	0.45	2.93	1.89
Rank	3	1	2

The results of Table 4 for each parameter and level were collected and shown in Figure 4. Based on the Taguchi method, a design level with a maximum S/N ratio should be selected as the optimum design level for the parameters. Thus, considering Figure 4, the optimum design corresponded to $HL_0/D = 1.2$, $HL_1/D = 3.0$, and $HL_2/D = 3.0$. The optimum design is depicted in Table 5.

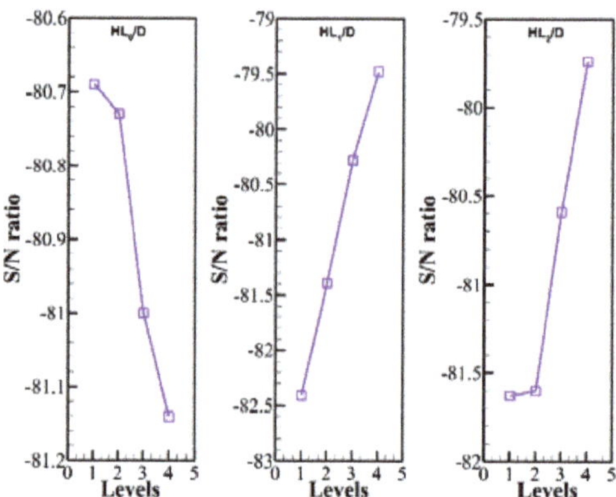

Figure 4. S/N ratios at the levels of the optimum case were as following: $HL_0/D = 1$, $HL_1/D = 4$, and $HL_2/D = 4$.

Table 5. The best design for the minimum thermal charging time based on the Taguchi method.

Optimum Factors			Charging Time of LF = 0.75	
HL_0/D	HL_1/D	HL_2/D	Taguchi Prediction	Tested Case
1.2	3.0	3.0	7915	8205

The Taguchi method estimated a thermal charging time of 7915 s. The actual simulations showed a charging time of 8205 s, which was quite close to the estimated value. In general, the optimum case could be a design available in the L16 table or a design out of this table from 4^3 possible designs. Here, interestingly, the proposed design case was case 4 in the L16 table.

A comparison between the charging times for the optimum case (8205 s) and the worst case of the L16 table (14,472 s) showed that charging time can be changed by 76%. Thus, a simple optimization of the placement of tubes in an LHTES unit could notably improve its performance.

Figure 5 illustrates the melting rate (charging) of the LHTES unit during the time. As can be seen, the optimum case quickly reaches high LF rates. Interestingly, the behavior of all cases at the beginning of the charging process is almost identical. The reason is that initially, the temperature of the LHTES unit is below the solidus temperature, and thus, the PCM is in the solid phase. Thus, the heat transfer mechanism is solely conduction dominant. The PCM around the tubes act as a large domain. Since the thermal conductivity of PCMs is low, in all design cases, the distance between the tubes is sufficiently large, and they cannot influence each other. After the melting process starts and the melting area extends away from the tubes, the natural convection flows appear. In a convection-dominant flow, the tubes' locations can significantly control the convection flow and the heat transfer.

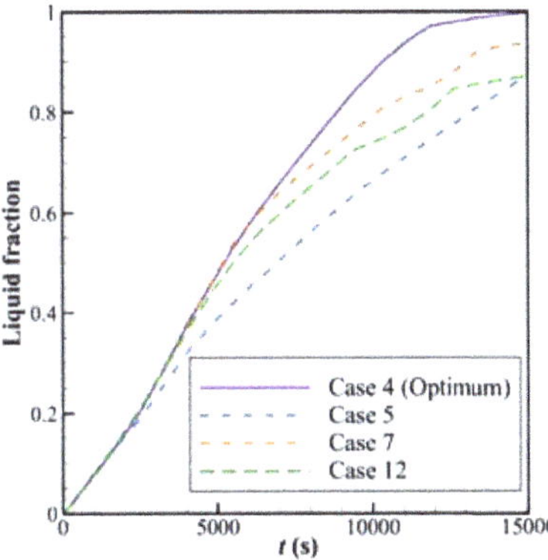

Figure 5. The melting behaviors of the LHTES unit during the time for five selected cases.

A systematic investigation of melting phase change heat transfer can be performed by analyzing streamlines and temperature distribution contours. Figures 6–12 show the streamlines and isotherms for the optimum case (Case 4) and Cases 5, 7, 12, and 13. The results are reported every hour of charging time. The blue region in streamline figures denotes the solid PCM, and the light orange shows the liquid PCM. The thick line between these two regions is the melting interface line.

Figure 6 shows the streamlines just one hour after the commencement of thermal charging, and Figure 7 depicts the corresponding isotherms.

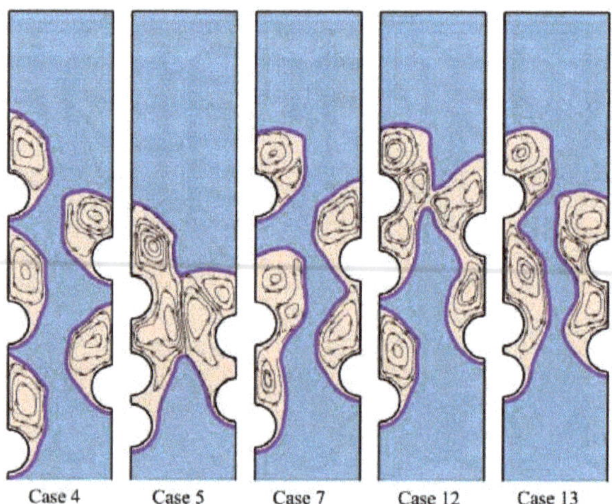

Figure 6. The streamlines and phase change interfaces for five different tubes' arrangement after one hour of charging. Case 4 is the optimum case with the highest melting rate.

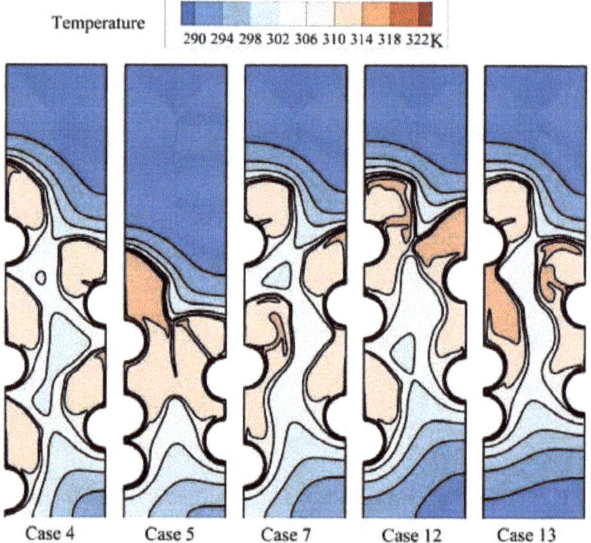

Figure 7. The isotherms for five different arrangements of the tubes after one hour of charging.

As time passes, the heat from HTF tubes accumulates in the solid PCM in the form of sensible heat, and the temperature of PCM increases around the tubes. As can be seen, the melting process starts around the hot tubes, advances to the space between the tubes and inclines toward the top regions. At this stage, it is clear that each tube develops a melting region around it. When the tubes are close to each other, the local melting regions reach each other and merge. Figure 7 displays hot regions around the tubes and cold regions far away from the tubes. The top regions of molten sites are also warmer due to the nature of buoyancy flows.

The melting process after two hours of charging time can be followed in Figures 8 and 9.

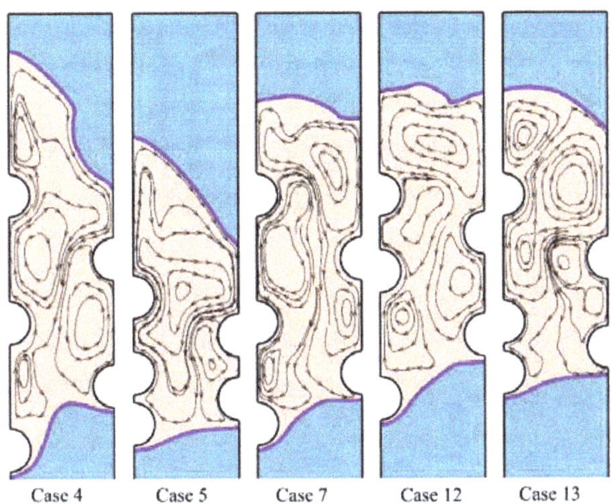

Figure 8. The streamlines and phase change interfaces for five different tubes' arrangement after two hours of charging.

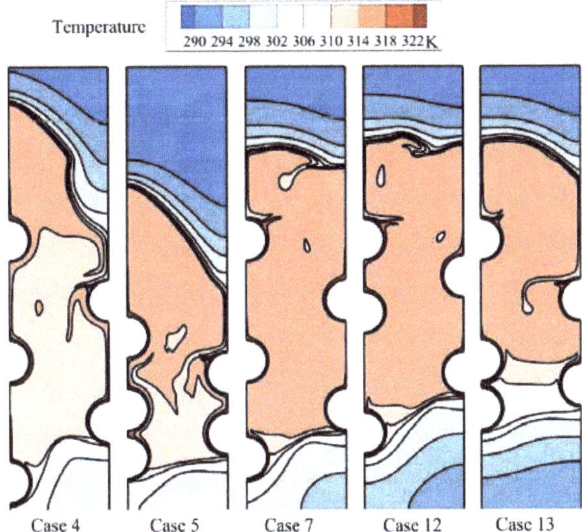

Figure 9. The isotherms for five different arrangements of the tubes after two hours of charging.

The local regions around the HTF tubes merges, and there is a wide molten zone in the center of the enclosure. Still, the top and bottom zones are in a solid-state. As can be seen, the solid region at the bottom is mostly under the influence of the first row of tubes (HL_0/D). This is while the location of other tubes does not show a significant impact on the bottom. In contrast, the second and third tube rows mostly impact the top melting interface. Shifting the tubes toward the top also shifts the melting interface upward. Figure 9 depicts that in all cases except the optimum case, the enclosure's central region is at a high temperature. This is while for the optimum Case 4, only the top region is at a high temperature. This means that the melting rate is strong for the optimum case, and most of the HTF tubes' heat is being consumed by the latent heat.

Figures 10 and 11 show the same trend as Figures 8 and 9.

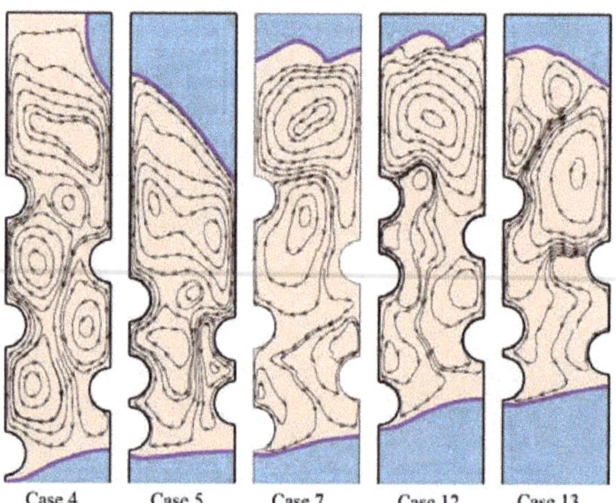

Figure 10. The streamlines and phase change interfaces for five different tubes' arrangement after three hours of charging.

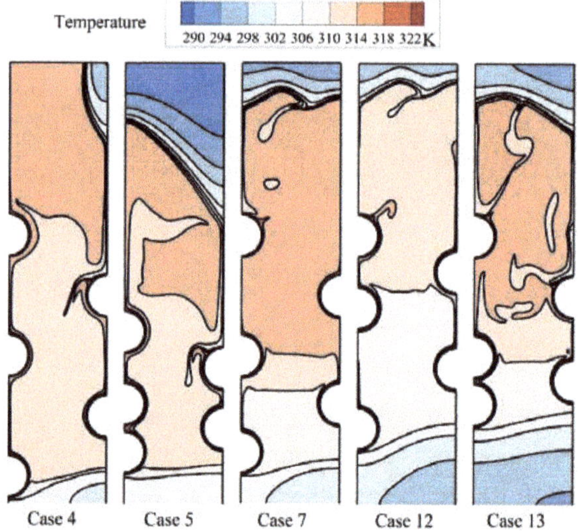

Figure 11. The isotherms for five different tubes' arrangement after three hours of charging.

The difference is that the molten region was further extended. At this stage, the solid regions were shrunk toward the bottom and the top. There is a notable difference between the optimum case and the other cases at this stage. Placing the first row of the HTF tubes at the bottom of the shell helps the shrinking of the solid region in the bottom part of the enclosure. The third row, placed at the highest location, melted most of the top regions. Figure 11 shows that the center regions of all other cases is still very hot while the top region of case 4 is hot. The hot temperature at the top contributes to the melting rate, while a lower temperature at central regions around the HFT tubes improves the heat transfer from the tubes due to large temperature differences.

Figures 12 and 13 show the final stage of melting in the enclosure.

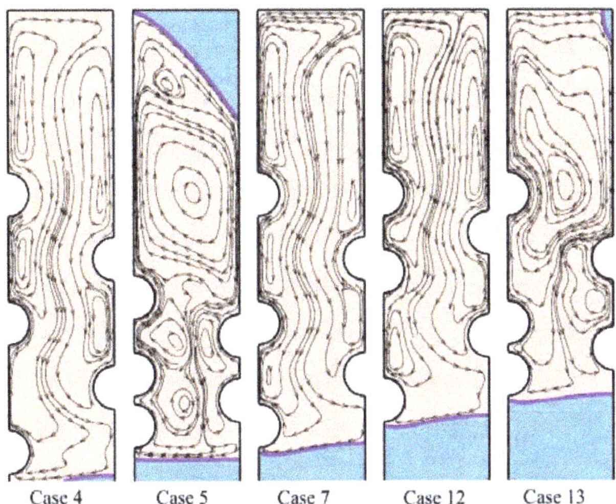

Figure 12. The streamlines and phase change interfaces for five different tubes' arrangement after four hours of charging.

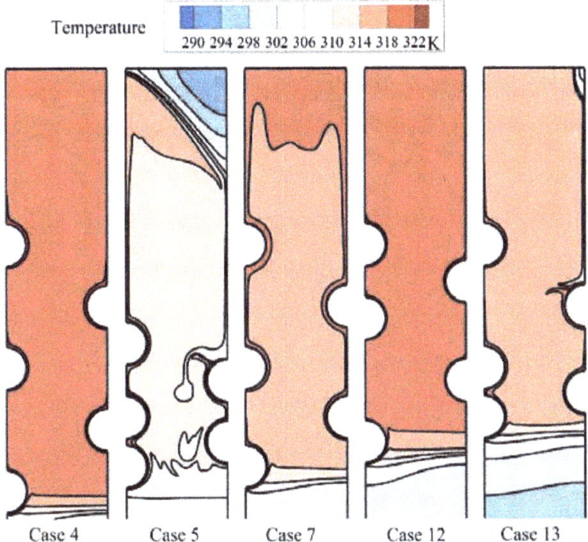

Figure 13. The isotherms for five different tubes' arrangement after four hours of charging.

In the optimum case 4, the top region completely melted down, and there is a negligible amount of the solid PCM at the bottom-right corner. Since there was no significant amount of the solid PCM to absorb the HTF heat in the form of latent heat, the temperature of the molten PCM is reaching the HFT temperature in most parts of the enclosure. The other cases mostly melted the PCM at the top region except case 5, in which the tubes were concentrated at the bottom. The attention to streamlines for this design shows that there are two separate circulation flows in the enclosure. Since the circulation flows are separated, the HTF heat cannot directly reach the top region and slow down the melting of the top zones. The amount of the solid PCM at the bottom is lower than in other cases (cases 7, 12, and 13), since the circulation flow at the bottom is injecting the heat into this region. The analysis on the final stage of melting for all cases showed that the placement of the first row toward the bottom was the most significant approach to melting down the bottom region. Moreover, the even distributions of the second and third rows in the enclosure helped with a general smooth circulation-free convection flow and the rapid melting of the top regions.

5. Conclusions

The melting thermal energy storage of an LHTES unit was addressed theoretically. The LHTES unit was modeled as a symmetric system consisting of five HTF tubes. The geometrical design of tube placement was optimized by employing the Taguchi method. The phase change behavior of the system was systematically investigated using LF graphs, streamlines, and isotherm contours. The main findings of the research can be summarized as follows:

- Based on the Taguchi design, the first row of the HTF tubes should be placed at the lowest possible point while the other tubes should be distributed evenly in the enclosure.
- The charging time of the LHTES unit could be changed by about 76% by just changing the location of tubes in the enclosure.
- From the streamlines and melting interfaces, it can be concluded that the formation of streamlines and free-convection flow circulation in each step of the melting process are the key points in the design of LHTES. Special attention should be paid to the streamline at the final stages of the charging process. A general uniform large circulation flow in the enclosure was much better than several small and week circulation flows.
- The Taguchi method could be used to effectively propose the optimum design of an LHTES unit with few simulations. Thus, this approach seems useful in the future design of energy storage systems.

Author Contributions: Conceptualization, M.G. and P.T.; methodology, M.G. and P.T.; software, M.S.I., O.Y. and P.T.; validation, M.G. and P.T.; formal analysis, M.G., H.I.M., P.T. and J.M.M.; investigation, M.G., H.I.M., A.N., M.S.I., O.Y., J.M.M., I.S.C. and P.T.; resources, M.G. and M.S.I.; writing of the original draft preparation, M.G., H.I.M., A.N., M.S.I., O.Y., J.M.M., I.S.C. and P.T.; writing of review and editing, M.G., H.I.M., A.N., M.S.I., O.Y., J.M.M., I.S.C. and P.T.; visualization, M.G. and P.T.; supervision, M.G. and P.T. All authors have read and agreed to the published version of the manuscript.

Funding: This research received no external funding.

Institutional Review Board Statement: Not applicable.

Informed Consent Statement: Not applicable.

Data Availability Statement: Data is contained within the article.

Conflicts of Interest: The authors declare no conflict of interest.

Nomenclature

A_m (kg m^{-3} s^{-1})	mushy zone constant	t_m (s)	melting/solidification time
C_p (J kg^{-1} K^{-1})	specific heat	T (K)	temperature
g (ms^{-2})	gravity	T_e (K)	mean temperature after the phase change process
k (W m^{-1} K^{-1})	thermal conductivity	T_i (K)	initial temperature
L_f (J kg^{-1})	latent heat of fusion		
LF	the normal total amount of melted phase change mateial		

		Greek symbols	
m (kg)	mass		
P (Pa)	pressure	β (K^{-1})	expansion coefficient
Q (J)	thermal storage/recovery capacity	λ	local liquid fraction
$\dot{Q}$ (W)	thermal storage/recovery rate	μ ($kg\ m^{-1}\ s^{-1}$)	dynamic Viscosity
T_L (K)	liquidus temperature	ρ ($kg\ m^{-3}$)	density
T_s (K)	solidus temperature	ΔH ($J\ kg^{-1}$)	latent heat of fusion
$\vec{V}$ ($m\ s^{-1}$)	velocity		

References

1. Dinker, A.; Agarwal, M.; Agarwal, G. Heat storage materials, geometry and applications: A review. *J. Energy Inst.* **2017**, *90*, 1–11. [CrossRef]
2. Dhaidan, N.S.; Khodadadi, J. Melting and convection of phase change materials in different shape containers: A review. *Renew. Sustain. Energy Rev.* **2015**, *43*, 449–477. [CrossRef]
3. Zhang, T.; Liu, Y.; Gao, Q.; Wang, G.; Yan, Z.; Shen, M. Experimental research on thermal characteristics of PCM thermal energy storage units. *J. Energy Inst.* **2020**, *93*, 76–86. [CrossRef]
4. Farah, S.; Liu, M.; Saman, W. Numerical investigation of phase change material thermal storage for space cooling. *Appl. Energy* **2019**, *239*, 526–535. [CrossRef]
5. Shahsavar, A.; Goodarzi, A.; Mohammed, H.I.; Shirneshan, A.; Talebizadehsardari, P. Thermal performance evaluation of non-uniform fin array in a finned double-pipe latent heat storage system. *Energy* **2020**, *193*, 116800. [CrossRef]
6. Memon, Z.Q.; Pao, W.; Hashim, F.M.; Ali, H.M. Experimental investigation of two-phase separation in T-Junction with combined diameter ratio. *J. Nat. Gas. Sci. Eng.* **2020**, *73*, 103048. [CrossRef]
7. Ali, H.M. Recent advancements in PV cooling and efficiency enhancement integrating phase change materials based systems–A comprehensive review. *Sol. Energy* **2020**, *197*, 163–198. [CrossRef]
8. Sajawal, M.; Rehman, T.-u.; Ali, H.M.; Sajjad, U.; Raza, A.; Bhatti, M.S. Experimental thermal performance analysis of finned tube-phase change material based double pass solar air heater. *Case Stud. Therm. Eng.* **2019**, *15*, 100543. [CrossRef]
9. Sardari, P.T.; Grant, D.; Giddings, D.; Walker, G.S.; Gillott, M. Composite metal foam/PCM energy store design for dwelling space air heating. *Energy Convers. Manag.* **2019**, *201*, 112151. [CrossRef]
10. Singh, R.P.; Kaushik, S.; Rakshit, D. Melting phenomenon in a finned thermal storage system with graphene nano-plates for medium temperature applications. *Energy Convers. Manag.* **2018**, *163*, 86–99. [CrossRef]
11. Hassan, A.; Wahab, A.; Qasim, M.A.; Janjua, M.M.; Ali, M.A.; Ali, H.M.; Jadoon, T.R.; Ali, E.; Raza, A.; Javaid, N. Thermal management and uniform temperature regulation of photovoltaic modules using hybrid phase change materials-nanofluids system. *Renew. Energy* **2020**, *145*, 282–293. [CrossRef]
12. Sadeghi, H.M.; Babayan, M.; Chamkha, A. Investigation of using multi-layer PCMs in the tubular heat exchanger with periodic heat transfer boundary condition. *Int. J. Heat Mass Transf.* **2020**, *147*, 118970. [CrossRef]
13. Ren, H.; Lin, W.; Ma, Z.; Fan, W.; Wang, X. Thermal performance evaluation of an integrated photovoltaic thermal-phase change material system using Taguchi method. *Energy Procedia* **2017**, *121*, 118–125. [CrossRef]
14. Mosaffa, A.; Talati, F.; Tabrizi, H.B.; Rosen, M. Analytical modeling of PCM solidification in a shell and tube finned thermal storage for air conditioning systems. *Energy Build.* **2012**, *49*, 356–361. [CrossRef]
15. Niu, X.; Xiao, F.; Ma, Z. Investigation on capacity matching in liquid desiccant and heat pump hybrid air-conditioning systems. *Int. J. Refrig.* **2012**, *35*, 160–170. [CrossRef]
16. Parameshwaran, R.; Kalaiselvam, S. Energy efficient hybrid nanocomposite-based cool thermal storage air conditioning system for sustainable buildings. *Energy* **2013**, *59*, 194–214. [CrossRef]
17. Rastogi, M.; Chauhan, A.; Vaish, R.; Kishan, A. Selection and performance assessment of Phase Change Materials for heating, ventilation and air-conditioning applications. *Energy Convers. Manag.* **2015**, *89*, 260–269. [CrossRef]
18. Redhwan, A.; Azmi, W.; Najafi, G.; Sharif, M.; Zawawi, N. Application of response surface methodology in optimization of automotive air-conditioning performance operating with SiO 2/PAG nanolubricant. *J. Therm. Anal. Calorim.* **2019**, *135*, 1269–1283. [CrossRef]
19. Mahdi, J.M.; Lohrasbi, S.; Nsofor, E.C. Hybrid heat transfer enhancement for latent-heat thermal energy storage systems: A review. *Int. J. Heat Mass Transf.* **2019**, *137*, 630–649. [CrossRef]
20. Yamaha, M.; Misaki, S. The evaluation of peak shaving by a thermal storage system using phase-change materials in air distribution systems. *Hvac&R Res.* **2006**, *12*, 861–869. [CrossRef]
21. Zhao, D.; Tan, G. Numerical analysis of a shell-and-tube latent heat storage unit with fins for air-conditioning application. *Appl. Energy* **2015**, *138*, 381–392. [CrossRef]
22. Dolado, P.; Lazaro, A.; Marin, J.M.; Zalba, B. Characterization of melting and solidification in a real scale PCM-air heat exchanger: Numerical model and experimental validation. *Energy Convers. Manag.* **2011**, *52*, 1890–1907. [CrossRef]

23. Waqas, A.; Kumar, S. Phase change material (PCM)-based solar air heating system for residential space heating in winter. *Int. J. Green Energy* **2013**, *10*, 402–426. [CrossRef]
24. Diarce, G.; Campos-Celador, Á.; Sala, J.; García-Romero, A. A novel correlation for the direct determination of the discharging time of plate-based latent heat thermal energy storage systems. *Appl. Therm. Eng.* **2018**, *129*, 521–534. [CrossRef]
25. Amin, N.A.M.; Belusko, M.; Bruno, F.; Liu, M. Optimising PCM thermal storage systems for maximum energy storage effectiveness. *Sol. Energy* **2012**, *86*, 2263–2272. [CrossRef]
26. Saman, W.; Bruno, F.; Halawa, E. Thermal performance of PCM thermal storage unit for a roof integrated solar heating system. *Sol. Energy* **2005**, *78*, 341–349. [CrossRef]
27. Wang, M.-H.; Huang, M.-L.; Zhan, Z.-Y.; Huang, C.-J. Application of the extension Taguchi method to optimal capability planning of a stand-alone power system. *Energies* **2016**, *9*, 174. [CrossRef]
28. Sun, X.; Mo, Y.; Li, J.; Chu, Y.; Liu, L.; Liao, S. Study on the energy charging process of a plate-type latent heat thermal energy storage unit and optimization using Taguchi method. *Appl. Therm. Eng.* **2020**, *164*, 114528. [CrossRef]
29. Liu, X.; Zhou, Y.; Li, C.-Q.; Lin, Y.; Yang, W.; Zhang, G. Optimization of a new phase change material integrated photovoltaic/thermal panel with the active cooling technique using taguchi method. *Energies* **2019**, *12*, 1022. [CrossRef]
30. Lin, W.; Ma, Z.; Ren, H.; Gschwander, S.; Wang, S. Multi-objective optimisation of thermal energy storage using phase change materials for solar air systems. *Renew. Energy* **2019**, *130*, 1116–1129. [CrossRef]
31. Harshita, B.; Reddy, K.D.; Venkataramaiah, P. Optimization of Parameters in Thermal Energy Storage System. *Int. J. Appl. Eng. Res.* **2018**, *13*, 15278–15283.
32. Zalba, B.; Sánchez-valverde, B.; Marín, J.M. An experimental study of thermal energy storage with phase change materials by design of experiments. *J. Appl. Stat.* **2005**, *32*, 321–332. [CrossRef]
33. GmbH, R.T. RT35 Data Sheet. Available online: https://www.rubitherm.eu/en/index.php/productcategory/organische-pcm-rt (accessed on 10 November 2020).
34. Talebizadeh Sardari, P.; Walker, G.S.; Gillott, M.; Grant, D.; Giddings, D. Numerical modelling of phase change material melting process embedded in porous media: Effect of heat storage size. *Proc. Inst. Mech. Eng. Part A J. Power Energy* **2020**, *243*, 365–383. [CrossRef]
35. Mahdi, J.M.; Nsofor, E.C. Melting enhancement in triplex-tube latent heat energy storage system using nanoparticles-metal foam combination. *Appl. Energy* **2017**, *191*, 22–34. [CrossRef]
36. Mat, S.; Al-Abidi, A.A.; Sopian, K.; Sulaiman, M.Y.; Mohammad, A.T. Enhance heat transfer for PCM melting in triplex tube with internal–external fins. *Energy Convers. Manag.* **2013**, *74*, 223–236. [CrossRef]
37. Ye, W.-B.; Zhu, D.-S.; Wang, N. Numerical simulation on phase-change thermal storage/release in a plate-fin unit. *Appl. Therm. Eng.* **2011**, *31*, 3871–3884. [CrossRef]
38. Assis, E.; Katsman, L.; Ziskind, G.; Letan, R. Numerical and experimental study of melting in a spherical shell. *Int. J. Heat Mass Transf.* **2007**, *50*, 1790–1804. [CrossRef]
39. Al-Abidi, A.A.; Mat, S.; Sopian, K.; Sulaiman, M.Y.; Mohammad, A.T. Internal and external fin heat transfer enhancement technique for latent heat thermal energy storage in triplex tube heat exchangers. *Appl. Therm. Eng.* **2013**, *53*, 147–156. [CrossRef]
40. Antony, J.; Antony, F.J. Teaching the Taguchi method to industrial engineers. *Work Study* **2001**, *50*, 141–149. [CrossRef]

Article

Effect of Iron (III) Oxide Powder on Thermal Conductivity and Diffusivity of Lime Mortar

Francesc Masdeu *, Cristian Carmona, Gabriel Horrach and Joan Muñoz

Department of Industrial Engineering and Construction, University of Balearic Islands, Ctra. de Valldemossa km 7.5, E07122 Palma de Mallorca, Spain; cristian.carmona@uib.es (C.C.); gabriel.horrach@uib.es (G.H.); joan.munoz@uib.es (J.M.)

* Correspondence: francesc.masdeu@uib.es; Tel.: +34-971-259790

Abstract: One of the challenges in construction is the improvement of energy efficiency of buildings. Development of construction materials of low thermal conductivity is a straightforward way to improve heat isolating capability of an enclosure. Lime mortar has a number of advantageous and peculiar properties and was widely used until the "irruption" of Portland cement. Currently, lime mortar is still used in restoration of traditional buildings or, according to the urban regulations, in catalogued constructions. The goal of the present study is the improvement of the heat isolating capability of lime mortars. The strategy of this work is the addition of iron (III) oxide powder, which is one of the possible components forming the cements, to a base lime mortar. The reason to choose Fe_2O_3 was two-fold. The first reason is low thermal conductivity of Fe_2O_3 compared to lime mortar. The second reason is that the low solubility and small size of iron (III) oxide particles have an effect on the thermal conductivity across the lime particles. The effect of iron (III) oxide powder on the thermal conductivity has been experimentally determined by the hot-box method. It has been found that the insulating capacity and thermal inertia of lime mortar is improved significantly by the addition of Fe_2O_3 powder, increasing the energy saving of the enclosure.

Keywords: energy saving with materials; energy storage; thermal conductivity; thermal diffusivity; lime mortar; iron (III) oxide

Citation: Masdeu, F.; Carmona, C.; Horrach, G.; Muñoz, J. Effect of Iron (III) Oxide Powder on Thermal Conductivity and Diffusivity of Lime Mortar. *Materials* **2021**, *14*, 998. https://doi.org/10.3390/ma14040998

Academic Editor: Antonio Caggiano

Received: 25 January 2021
Accepted: 18 February 2021
Published: 20 February 2021

1. Introduction

In recent years, the increased use of energy from fossil fuels has provoked dramatic climate changes. The greenhouse effect, acid rains, and other phenomena are examples of the consequences of an excessive consumption of this kind of energy. According to the United Nation Environment Program, the energy consumption of buildings represents nearly 40% of the world global energy [1], and around two-thirds of the energy demand in the residential sector is attributed to heating and cooling [2]. The field of construction can assist to mitigate these effects on global warming by improving the performance of construction materials, e.g., increasing heat insulating capability. During recent years, in order to reduce the consumption of energy, great efforts have been devoted in developing low thermal conductivity construction materials and improving the efficiency of materials currently in use [1,3].

One of the potentially favorable materials is the lime mortar. The hardening process of lime is caused by a carbonation reaction [4]. This process requires a long time, especially when compared to Portland cement. However, the use of lime has some advantages, such as strain accommodation (plastic behavior), lower thermal conductivity, or higher breathability, which makes the houses more comfortable [5].

Historically, lime mortar has been widely used around the Mediterranean seaside, mainly as mortar of plaster in vertical walls. The improvement of thermal insulation capacity is especially important for its use as an outer layer of the building enclosure, separating the indoor environment from the outside. From a geological point of view, lime mortar was one of the few binders of high resistance that could be obtained in great abundance in the Mediterranean region. The aim of the present work is to reduce the thermal conductivity or, in other words, to enhance the heat insulating capability of the lime mortar. A common practice to achieve this goal is the addition of filling particles of organic/vegetal origin, such as cork [6], hemp [7,8], olive stone [9], textile waste [10], straw [11], coconut [12], etc. The strategy of the present study is the addition of ceramic submicron particles, whose use and chemical composition is fully compatible with an ecological concept from the point-of-view of generating future, harmless waste for soil and subsoil.

The reasons behind the study of the lime mortar instead of Portland mortar are as follows: (i) better ecological sustainability of lime mortar, since its production requires lower temperatures and less energy [13,14], (ii) lime has a lower thermal conductivity than Portland mortar [15], being a more efficient thermal insulator at the starting point of the study, and (iii) higher indoor comfort provided by lime, since it has a higher breathability [5] and is biocide [16].

The hydraulicity index (HI) that allows one to identify the main chemical components forming the lime cement [17–19] is given by:

$$HI = \frac{[SiO_2] + [Al_2O_3] + [Fe_2O_3]}{[CaO] + [MgO]} \quad (1)$$

where the terms in square brackets are the percentages of the five oxides composing the lime. Taking into account that lime cements are composed mostly of calcium or magnesium oxides, there are three candidates that could be selected as an additive to improve the heat insulating capability: SiO_2, Al_2O_3, and Fe_2O_3. A comparison of the properties of these candidates shows that iron (III) oxide has the lowest thermal conductivity (λ_{Fe2O3} = 0.58 W/(m·K), λ_{SiO2} = 1.1 W/(m·K), λ_{Al2O3} = 25 W/(m·K)), which is also lower than that of the limestone ($\lambda_{limestone}$ = 1.3 W/(m·K)) [20,21]. Moreover, iron (III) oxide is an inexpensive mineral and is the seventh most abundant compound in the Earth's crust [22], which are important factors for a mineral to be used as a construction material.

From the point-of-view of environmental sustainability, iron (III) oxide (hematite) is a component present in farmlands, which is beneficial for plant species. For this reason, the rubbles generated after the stage in the service of buildings would not have a detrimental effect on the environment [23]. Keeping in mind these favorable properties, Fe_2O_3 powder has been selected as an additive to improve the thermal efficiency of lime mortar.

The influence of red and black iron oxides addition on the mechanical and physiochemical properties of a concrete was studied by Kishar et al. [24]. A notable positive effect of both red and black iron oxide particles on slump and compressive strength (up to 22–30%) was reported. Recently, Largeau et al. [25] investigated the effect of Fe_2O_3 on the strength and workability of a Portland cement concrete. They found that fine Fe_2O_3 particles, ca. 200 nm, reduced the porosity and improved compressive strength of concrete for concentrations up to 2.5 wt.%. The present authors are not aware of further research on the effect of iron oxide particles on the thermal properties of lime or Portland mortars.

In this work, the effect of adding small particles of iron (III) oxide on lime mortar has been investigated with the aim of improving the thermal properties of the base lime mortar. It has been found that the insulating capacity and thermal inertia of lime mortar is improved significantly by adding Fe_2O_3 powder, increasing the energy saving of the enclosure [26].

2. Materials and Methods

2.1. Components

Natural hydraulic lime NHL-3.5 Morcem Cal Base 434 CR CSII W0 (Grupo Puma, Malaga, Spain) was selected to use as a base product of the lime mortar. Red iron (III) oxide (Labkem, Barcelona, Spain) of chemical purity higher than 95% was used as the additive.

2.2. Components' Dimensional Characterization

Dimensional characterization of the two components has been carried out using two different techniques. Scanning electronic microscopy (SEM) images, obtained using a Hitachi S-3400N (manufactured by Hitachi Science Systems Ltd., Tokyo, Japan), allowed us to determine characteristic values of particle sizes as well as to visualize the distribution of particles of both Fe_2O_3 and lime.

Precise values of grain size and specific surface, measured respectively by means of Beckman Coulter and a Malvern Mastersizer Micro Plus (Malvern Panalytical, Malvern, UK), are shown in Table 1. The range of grain sizes refers to an interval including more than 80% of particle sizes.

Table 1. Grain size and specific surface of lime mortar and iron (III) oxide powders.

Material	Grain Size Range (μm)	Specific Surface (cm^2/g)
Lime	10–2000	5.7×10^3
Fe_2O_3	0.2–0.5	9.4×10^4

2.3. Mortar Preparation and Curing

In order to analyze the effect of Fe_2O_3 particles on the thermal properties of a lime mortar, five samples with different iron (III) oxide content were prepared. The quantity of water added to the dry mixture was higher as the iron (III) oxide content increases in order to obtain a cement lime mortar of equal workability and elastic consistency, according to the ISO 12439 standard [27]. Table 2 shows the mass and mass fraction of iron (III) oxide powder substituting the lime mortar. The letters LF in the sample notation refer to the mortar components: lime as L and Fe_2O_3 as F, while the number (LF-0, LF-5, . . . , LF-20) denotes the iron (III) oxide mass fraction of each mortar.

Table 2. Samples and compositions.

Sample	Component Mass (g)			Fe_2O_3 Mass Fraction in Mortar (%)
	Lime Mortar	Fe_2O_3	H_2O	
LF-0	1900	0	400	0
LF-5	1805	95	412	5
LF-10	1710	190	436	10
LF-15	1615	285	500	15
LF-20	1520	380	545	20

Cylindrical samples 12 cm in height and 10 cm in diameter (Figure 1) were produced using a plastic mold. Curing time was 60 days. The mass of water needed to obtain the optimal mixing increases with iron (III) oxide content since the addition of Fe_2O_3 small-size particles increases the specific area, demanding an increasing addition of water to surround the surface of the particles.

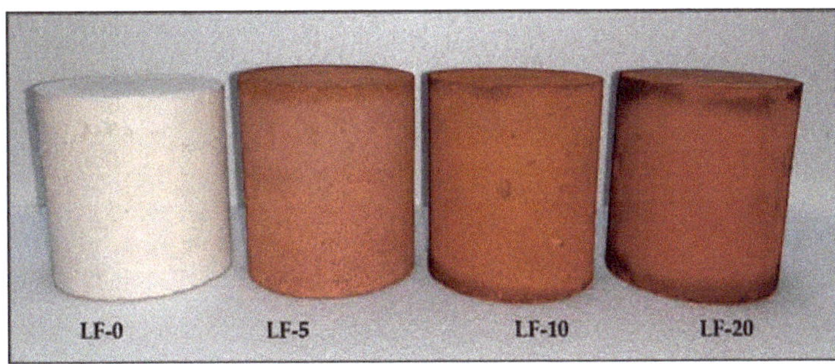

Figure 1. Image of some of the cylindrical samples: LF-0, LF-5, LF-10, and LF-20 (from left to right).

2.4. Experimental Method

2.4.1. Apparent Density

Apparent density of the studied mortars was calculated from their weight and dimensions of the cylindrical samples [9].

2.4.2. Specific Heat Capacity

Specific heat capacity of mortars was measured using a TA Instruments DSC2920 (manufactured by TA Instruments, New Castle, DE, USA) calorimeter in the modulation mode (MDSC), calibrated with a sapphire sample (error lower than 1%). Measurements were performed at 25 °C using mortar samples weighing around 1.0 mg. Due to their low weight, four different samples were taken from each mortar and tested by MDSC in order to balance out the composition heterogeneity of the mortar.

2.4.3. Thermal Conductivity of Mortars

Thermal conductivity was determined by means of a calibrated hot-box method [28] in a home-made device, shown schematically in Figure 2. The case of the device was fabricated from expanded polystyrene (EPS). The hot plate was placed at the bottom of the case and supported the sample. In the steady state conditions, the hot plate maintains a controlled fixed temperature of the hot side of the sample, equal to 61.2 °C. The heat flux on the cold side of the sample is measured using a HFP01 flux sensor (Hukseflux, Delft, The Netherlands). The temperature is measured on both cold and hot sides of samples by means of thermo-couples. The three parameters obtained in this experiment are temperatures on both hot and cold sides, and the heat flux. The electromagnetic protection serves to distribute homogeneously, over the sample section, with the heat generated by the hot plate. Thermal loss sensors permit controlling the isolating efficiency of the box for an optimal measurement of the heat flow through the sample.

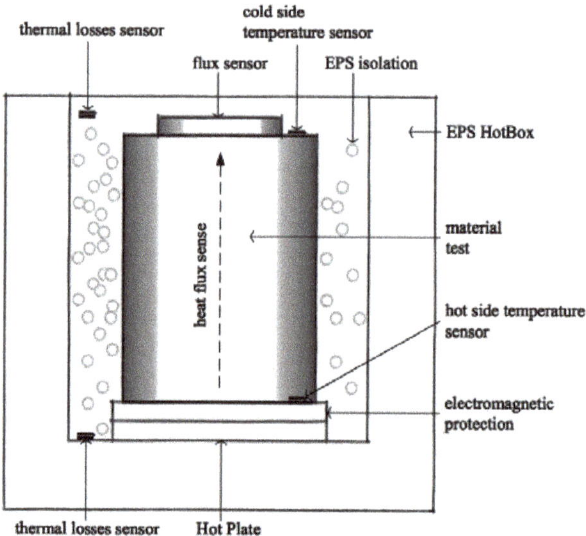

Figure 2. Home-made hot-box design (see text for details).

3. Results and Discussion

3.1. Density and Porosity

The analysis of density and porosity of the sample set provides information crucial for understanding the behavior of thermal conductivity. Theoretical bulk densities were calculated from the bulk density of calcite (2.71 g/cm^3), which is the main component in lime mortar, and iron (III) oxide (5.26 g/cm^3) considering their volume fractions for each sample [29]. Porosity was calculated using the theoretical bulk density and apparent density, as 1-($d_{apparent}/d_{theor}$).

Apparent density and porosity versus iron (III) oxide content from Table 3 are shown in Figure 3. Both parameters show a linear dependence with the iron (III) oxide content. The pores in mortars are created during the process of curing due to evaporation of water. With the increase of the content of Fe_2O_3 submicron particles, the amount of water needed to prepare the mixture increases (see Table 2). Thus, the degree of the porosity generated by releasing water becomes higher.

Table 3. Bulk and apparent densities and porosity of the samples with different iron (III) oxide content.

Sample	Fe_2O_3 Content (mass %)	Theoretical Bulk Density (g/cm^3)	Apparent Density (g/cm^3)	Porosity (%)
LF-0	0	2.71 ± 0.02	1.65 ± 0.03	39.2 ± 0.9
LF-5	5	2.84 ± 0.02	1.67 ± 0.03	41.2 ± 0.9
LF-10	10	2.97 ± 0.02	1.69 ± 0.03	43.1 ± 1.0
LF-15	15	3.09 ± 0.02	1.70 ± 0.03	44.9 ± 1.0
LF-20	20	3.22 ± 0.02	1.71 ± 0.03	46.8 ± 1.1

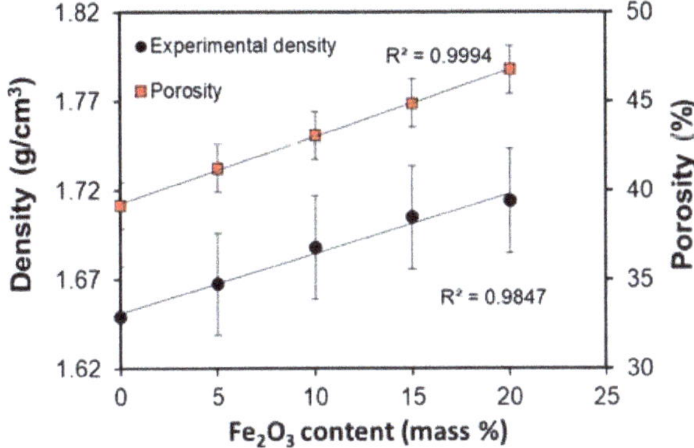

Figure 3. Apparent density and porosity versus Fe_2O_3 content of the studied samples.

Thus, the addition of Fe_2O_3 results in a moderate increase of density (by ca. 4% for LF-20 compared with the base mortar LF-0) and a more substantial increase of porosity (relative increment around 19% for LF-20). The increase of porosity is known to significantly affect thermal conductivity of the material [30–32], which is analyzed in the following section.

3.2. *Thermal Conductivity*

The kinetics of the heat flux for samples with different Fe_2O_3 content is shown in Figure 4. The temperature of the hot plate was set to 61.2 °C at t = 0 and the heating was switched off at t = 23 h. Room temperature was around 15 °C during all tests. According to this experimental protocol, three clearly defined stages are observed in the heat flux kinetics. In the first stage, during the first 8 h, samples heat up from room temperature to an equilibrium value, thus, reaching the steady state regime. During the second stage, between 8 and 23 h, the system is in stationary conditions. After switching off heating of the hot plate at t = 23 h, the samples are cooled down to room temperature.

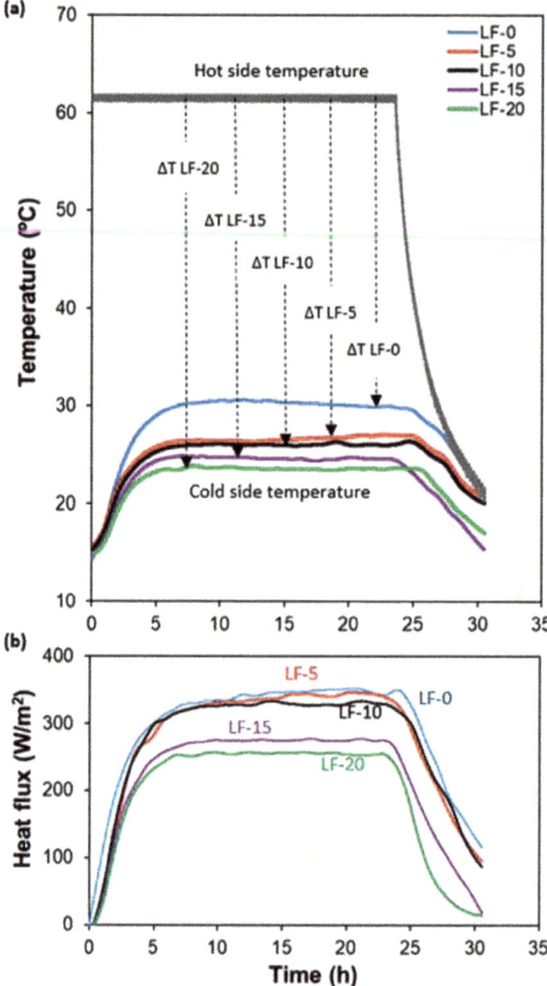

Figure 4. Cold side and hot side temperatures (**a**) and heat flux (**b**) vs. time for samples of lime mortar with a different content of iron (III) oxide.

Figure 4a shows that, in the steady state regime, the temperature difference between the hot and cold sides increases progressively with iron (III) oxide content. Small fluctuations of cold side temperatures observed in this regime could be due to the minor variations of the room temperature during the test. The cold side temperatures of each sample used for the later calculations is an average of measured temperatures in the steady state regime between 8 and 23 h of the test.

More specifically, the values of temperatures at the cold side for the two extreme compositions, LF-0 and LF-20, were 30.5 °C and 23.7 °C, respectively. Taking into account that room temperature during the test was 15 ± 1 °C, these data mean that the increase of temperature for LF-0 and LF-20 are 15.5 °C and 8.7 °C, respectively. Thus, the temperature increase for LF-0 is twice that for LF-20. If the length of all samples was kept similar (less than 2.5% difference), the measured temperatures on the cold side give a first intuitive idea of the different efficiency of thermal isolation of the two materials.

The numerical values of thermal conductivity were calculated using the heat flux through each mortar given by the heat flux-time diagram (Figure 4b), which shows the same tendency as cold side temperature-time dependence. In fact, there exists a direct relation between both parameters, since the heat flux arriving at the cold side of the sample contributes to heating up the material on the cold side. Parameters associated with thermal conductivity are summarized in Table 4. The values of heat flux appearing in Table 4 were determined as an average of results between 8 and 23 h. Experimental thermal conductivity values were obtained from the hot-box measurements at a steady state regime (Table 4). Transmittance, U, was calculated from the measured heat flux, Φ, and the temperature difference between hot and cold sides, ΔT, using the equation obtained from the Fourier's law of heat conduction [33].

$$U = \frac{\Phi}{\Delta T} \tag{2}$$

Table 4. Experimental thermal conductivity and energy saving of the samples of lime mortar with different iron (III) oxide content.

Sample	T_H (°C)	T_C (°C)	ΔT (K)	Φ (W/m^2)	U (W/m^2·K)	L (m)	λ_{exp} (W/m·K)	E.S. (%)
LF-0	61.2	30.5	30.7	166.2	5.42	0.123	0.67 ± 0.01	0.0
LF-5	61.2	26.4	34.8	165.6	4.76	0.120	0.57 ± 0.01	14.3 ± 0.4
LF-10	61.2	26.0	35.2	160.5	4.56	0.121	0.55 ± 0.01	17.2 ± 0.5
LF-15	61.2	24.7	36.5	134.0	3.67	0.122	0.45 ± 0.01	32.8 ± 1.0
LF-20	61.2	23.7	37.5	124.4	3.32	0.120	0.40 ± 0.01	40.1 ± 1.2

T_H: hot side temperature. T_C: cold side temperature. ΔT: difference between hot and cold sides. Φ: heat flux. U: transmittance. L: sample length. λ_{exp}: experimental thermal conductivity. E.S.: energy saving.

The experimental thermal conductivity, λ_{exp}, also shown in Table 4, was calculated from transmittance, U, and sample length, L, using the well-known relationship [33].

$$\lambda = U \cdot L \tag{3}$$

The meaning of energy saving, E.S., is the percentage of reduction of heat losses through a wall of a fixed thickness and area, separating two spaces with a certain temperature difference, made of a mortar containing iron (III) oxide compared to the base mortar.

Another meaning of the E.S. is the percentage of power saved by a heating (or cooling) device to keep a certain temperature difference between two spaces separated by a wall of a fixed thickness and area made of a mortar with Fe_2O_3 addition compared to the base mortar. Correspondingly, in this work, the E.S. was calculated as the percentage of thermal conductivity reduction taking the value for the LF-0 mortar as a reference.

Experimental values of thermal conductivities of different mortars taken from Table 4 are compared in Figure 5. The data show a progressive essentially linear drop of experimental values of thermal conductivity with the increase of Fe_2O_3 content.

Two reasons account for the significant thermal conductivity decrease with adding Fe_2O_3 powder to the limestone. First, the thermal conductivity of the additive (λ_{Fe2O3} = 0.58 W/(m·K)) [21] is half that of the base material ($\lambda_{limestone}$ = 1.2 W/(m·K)) [21]. Second, the porosity of mortar increases with the addition of iron (III) oxide, taking the values from 39.2% in LF-0 to 46.8% in LF-20. An equation, which correlates well with the thermal conductivity of the ceramic bodies with the porosity, was proposed by Aivazov and Domashnev [34].

$$\frac{\lambda}{\lambda_0} = 1 - P + n \cdot P^2 \tag{4}$$

where λ and λ_0 are the thermal conductivities of a porous and pore-free ceramic bodies, respectively, P is the volume fraction of the pores, and n is a constant. According to Equation (3), which can only be used for a constant bulk material, the thermal conductivity drops with the increasing porosity in the mortar as the iron (III) oxide content raises fulfilling a parabolic function. Therefore, as far as the studied mortar is a combination of two different solids, the thermal conductivity of the bulk material, λ_0, diminishes in each mortar as the Fe_2O_3 content rises.

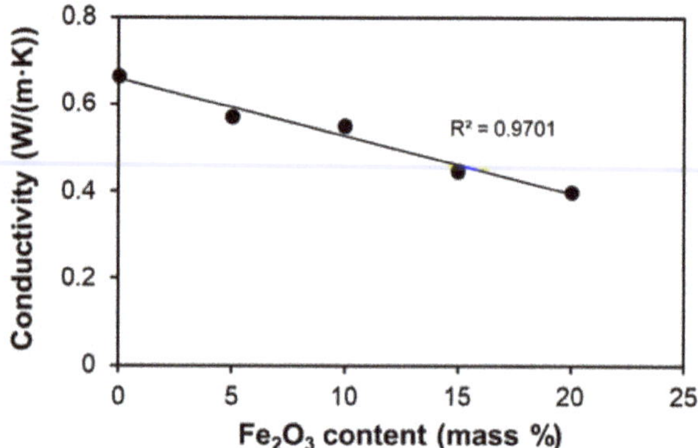

Figure 5. Thermal conductivity dependence with iron (III) oxide content of the lime mortar.

At the microstructural level, two other reasons for a notable decrease of thermal conductivity with Fe_2O_3 additions can be suggested. First, the solubility product constant of the lime cement ($Ca(OH)_2$: $K_{sp} = 4.68 \times 10^{-6}$) [29] is several orders of magnitude higher than those of iron (III) oxide in both the hydrated or ionic forms in alkali media [35], which are between 4.87×10^{-17} and 2.64×10^{-39} [29]. Due to extremely low solubility of Fe forms in aqueous media, the number of Fe^{3+} complex ions available to be transported to the neck between particles is very low. In this way, the area of the neck formed between iron (III) oxide particles is small compared to the necks between lime particles. Taking into account that the heat transfer through lime or the Fe_2O_3 solid phase is more efficient than through air pores (considering their thermal conductivities [20,21,34]), the reduction of the contact area between solid particles forces the thermal conductivity of the mortar to decrease. Second, scanning electron microscopy (SEM) images of LF-5 and LF-20 (Figure 6) shows that relatively small iron (III) oxide particles stick on the lime particles' surfaces. This spatial distribution forces lime particles to keep better separated than in the absence of Fe_2O_3 fine powder, thus, reducing the contact area between better heat conducting lime particles. A comparison of microstructures of LF-5 and LF-20 mortars in Figure 6 indicates that, with the increase of iron (III) oxide content, lime grains are better separated by small Fe_2O_3 particles that stick onto their surfaces.

The overall decrease of the thermal conductivity by the addition of Fe_2O_3 powder is due to the combination of the previously mentioned factors: porosity, thermal conductivity of each mortar component, and microstructure.

Apart from the parameters discussed, absorbed water could modify thermal conductivity in porous ceramics [36]. In order to estimate this contribution, the free water content of samples was tested by thermogravimetry. Mass losses due to free water desorption from base lime mortar, LF-0, and the mortar with 20% of Fe_2O_3 powder, LF-20, were measured during heating from 20 °C up to 200 °C following standard protocols [37]. In both samples, the losses of water at 120 °C are very similar at around 1.5% (weight). As a consequence, the contribution of free water absorbed by the studied samples is not expected to have a significant effect on the variation of thermal conductivity.

Figure 6. SEM images of the microstructure of mortars with 5% and 20% of Fe_2O_3 particles (LF-5 and LF-20, respectively), showing spatial distribution of lime and iron (III) oxide particles.

3.3. *Thermal Diffusivity*

Qualitative property of "thermal inertia" was defined by Ng et al. [38] as the 'property of a material that expresses the degree of slowness with which its temperature reaches that of the environment.' However, the definition that likely best expresses the effects it causes in an enclosure is the 'capacity of a material to store heat and to delay its transmission' due to Ferrari [39]. Then, to keep a constant temperature inside the building when the external temperature changes, the wall's material should have a thermal inertia as high as possible. One of the parameters characterizing thermal inertia is the thermal diffusivity, a, which can be calculated from thermal conductivity, λ, specific heat capacity, C_e, and density, d, [40].

$$a = \frac{\lambda}{d \cdot C_e} \tag{5}$$

The thermal inertia of the material grows if the thermal diffusivity expressed by Equation (4) decreases.

Specific heat capacity was measured using modulated differential scanning calorimeter (MDSC). Applying a sinusoidal heating rate around a linear temperature permits the measurement of the sample's heat capacity [41]. The total heat flow, dH/dt, is equivalent to standard differential scanning calorimeter (DSC) at the same average heating rate, and can be calculated using the following equation [41].

$$\frac{dH}{dt} = C_e \frac{dT}{dt} + f(T, t) \tag{6}$$

where C_e is the specific heat capacity, dT/dt is the measured heating rate, C_e(dT/dt) is the reversing heat flow component of the total heat flow, and f(T,t) is the kinetic component.

Figure 7 shows representative curves of the specific heat capacity, C_e, evolution with time for each mortar, measured around a linear temperature of 273 K. The values of specific heat capacity for each measurement have been calculated as the average of values between 600 and 900 s, when the C_e values of the MDSC measurement is already in a steady state regime.

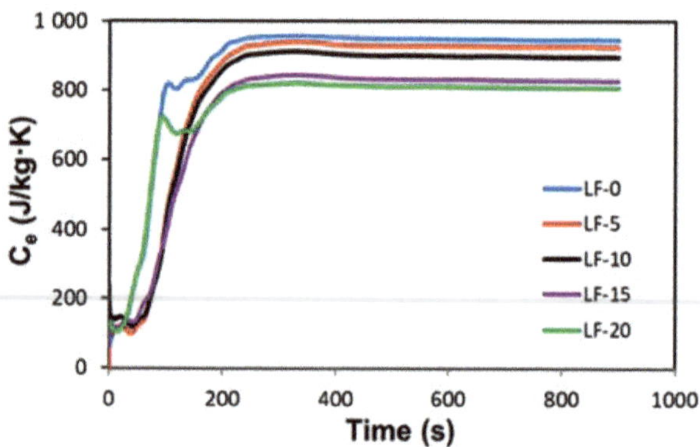

Figure 7. Representative curves of specific heat capacity C_e vs. time measured by MDSC for each mortar composition.

Table 5 shows the values of specific heat capacity for the measurements made on four different samples of each mortar as well as the average values with standard deviations. Figure 7 and Table 5 indicate clearly that the specific heat capacity of the mortar decreases with Fe_2O_3 content. The decrement of specific heat capacity for the LF-20 mortar compared to the base lime mortar LF-0 is 7.6%.

Table 5. Specific heat capacity values, C_e, calculated from MDSC measurements.

Sample	Specific Heat Capacity, C_e (J/(kg·K))							Decrement [1] (%)
	Test 1	Test 2	Test 3	Test 4	Average	Standard Dev.	Rule of Mixture	
LF-0	922	934	926	939	931	8	931 ± 3	0.0
LF-5	933	920	929	925	927	6	913 ± 3	0.43 ± 0.01
LF-10	898	912	900	917	907	9	895 ± 3	2.58 ± 0.03
LF-15	887	896	883	902	892	9	877 ± 3	4.19 ± 0.05
LF-20	866	852	875	847	860	13	859 ± 3	7.63 ± 0.13

[1] Decrement calculated using LF-0 as a reference.

The reason why the specific heat capacity decreases with the addition of Fe_2O_3 powder, is that the specific heat capacity of the additive (Fe_2O_3: C_e = 570 J/(kg·K)) [20] is much lower than that of the base material (lime mortar: C_e = 931 J/(kg·K)). It is worth to note that the specific heat capacity of a mixture of solid materials can be calculated using the rule of mixture as the sum of the mass fraction of each component by its specific heat capacity [42]. The corresponding values, shown in Table 5, are in good agreement with experimental data.

Thermal insulation additives for mortars used in building, as cork [6], expanded clay [43] or expanded polystyrene [44], and have a very low density due to their high content of air (high porosity), which makes it difficult to transfer heat through the material. As a consequence of very low density, those insulation materials have an extremely low heat storage capacity per unit volume, which leads to a high thermal diffusivity, or, in other words, a very poor thermal inertia. Contrary to that, the additive used in the current work is heavier than the base mortar. Therefore, the density of the mortar slightly increases by the addition of Fe_2O_3. The effects of increasing density and decrease of specific heat capacity on thermal diffusivity are opposite (Equation (4)) and nearly compensate each other, as is shown in Table 6. Hence, the decrease of the thermal diffusivity observed is largely due to the variation of the heat conductivity.

Table 6. Thermal diffusivity of the samples of lime mortar with different iron (III) oxide content calculated using Equation (3) from experimental values of thermal conductivity, λ, specific heat capacity, C_e, and density, d.

Sample	Density [1], d (kg/m^3)	Specific Heat Capacity, C_e (kJ/(kg·K))	d·C_e (kJ/m^3·K)	Thermal Conductivity, λ (kW/(m·K))	Diffusivity, a (m^2/s)	Improvement [2] (%)
LF-0	1649	0.931	1535	666	0.434 ± 0.020	0.0
LF-5	1667	0.927	1545	571	0.369 ± 0.016	14.9 ± 1.3
LF-10	1688	0.907	1531	552	0.360 ± 0.017	17.0 ± 1.6
LF-15	1705	0.892	1521	448	0.294 ± 0.014	32.2 ± 3.0
LF-20	1714	0.860	1474	399	0.271 ± 0.014	37.6 ± 3.7

[1] The values of density in Table 6 are those shown in Table 3 as apparent density. [2] Improvement calculated using LF-0 as a reference.

The improvement in efficiency related to thermal inertia of LF-20 is over 35% compared to LF-0. The data from Table 6 yield a linear dependence of thermal diffusivity with iron (III) oxide content in the lime mortar (Figure 8).

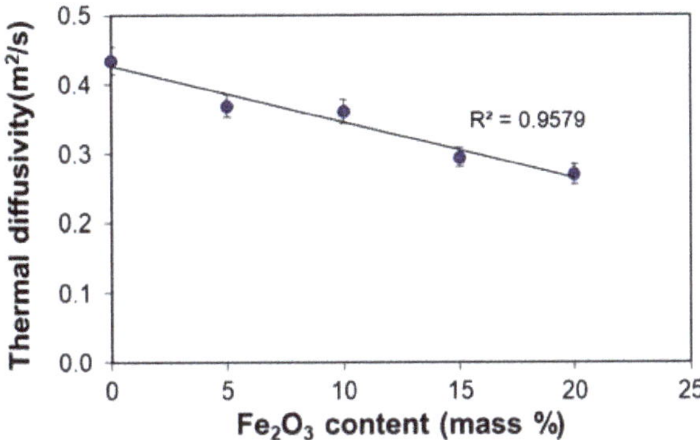

Figure 8. Thermal diffusivity of the lime mortar versus iron (III) oxide content.

4. Conclusions

Addition of iron (III) oxide is an efficient way to enhance the thermal insulating capacity of a lime mortar. Addition of 20% of Fe_2O_3 fine powder to base lime mortar reduces the thermal conductivity by ca. 40% and increases the thermal inertia by ca. 37%, when compared to the base mortar. Thermal conductivity, λ, shows a strong dependence on iron (III) oxide content. The factors improving the thermal properties can be summarized as follows.

1. Thermal conductivity of the used additive, iron (III) oxide, is much lower than that of the lime mortar (base material).
2. The porosity of mortar increases with the addition of Fe_2O_3 fine powder from 39% in the base mortar (0% Fe_2O_3) to 47% in the mortar containing 20% of Fe_2O_3. The thermal conductivity of a porous ceramics drops significantly with the porosity.
3. Due to the extremely low solubility of Fe_2O_3 in aqueous media, the area of the neck formed between iron (III) oxide particles is small compared to the necks between lime particles. Therefore, the effective surface of a solid phase able to transfer the heat by conduction diminishes, thus, improving the thermal insulation capability.
4. The use of iron (III) oxide as an additive, which causes an increase of density and a decrease of thermal conductivity compared to base mortar, leads to a significant improvement of the thermal inertia of the resulting mortar.

The future work should consist in studying the influence of Fe_2O_3 additions on the mechanical properties of lime mortar and their stability. Once the better thermal efficiency of the new material has been demonstrated, the mechanical strength under compression becomes of a prime importance. Previous works on the addition of iron oxide particles to Portland cement provide positive expectations in this sense.

5. Patents

Part of the results of the present article were registered in the following utility model: Masdeu, F.; Muñoz, J.; Carmona, C.; Horrach, G. Mortero de cal termoaislante y su uso en edificación. Spanish Patent ES1222024 U, 2018.

Author Contributions: Conceptualization, F.M., G.H., and J.M. Methodology, F.M. Software, F.M., C.C., and J.M. Validation, F.M., G.H., and J.M. Formal Analysis, F.M. and J.M. Investigation, F.M. and C.C. Resources, G.H. and J.M. Data Curation, F.M. Writing—Original Draft Preparation, F.M. Writing—Review & Editing, F.M. and J.M. Visualization, F.M., C.C., and. J.M. Supervision, F.M. and J.M. Project Administration, G.H. and J.M. All authors have read and agreed to the published version of the manuscript.

Funding: This research received no external funding.

Institutional Review Board Statement: Not applicable.

Informed Consent Statement: Not applicable.

Data Availability Statement: Data Sharing is not applicable.

Acknowledgments: The authors are grateful to J. Cifre, F. Hierro, and R. Sánchez, UIB Scientifico-Technical Services, for their help in experimental work. Kind assistance of Ir. J. Van Humbeeck, Katholieke Universiteit Leuven (Belgium) in particle size determination is gratefully acknowledged.

Conflicts of Interest: The authors declare no conflict of interest.

References

1. Asdrubali, F.; D'Alessandro, F.; Schiavoni, S. A review of unconventional sustainable building insulation materials. *SM&T* **2015**, *4*, 1–17.
2. European Commission. *EU Energy, Transport and GHG Emissions, Trends to 2050 Reference Scenario*; European Commission: Brussels, Belgium, 2013.
3. Brás, A.; Faria, P. Effectiveness of mortars composition on the embodied carbon long-term impact. *Energy Build.* **2017**, *154*, 523–528. [CrossRef]
4. Van Balen, K. Carbonation reaction of lime, kinetics at ambient temperature. *Cem. Concr. Res.* **2005**, *35*, 647–657. [CrossRef]
5. Galván-Ruiz, M.; Velázquez-Castillo, R. Lime, an Ancient Material as a Renewed Option for Construction. *Ing. Investig. Tecnol.* **2011**, *12*, 93–102.
6. Soumia, M.; Youssef, M.; Abou bakr Cherki, A. Thermal properties of the composite material clay/granular cork. *Constr. Build. Mater.* **2014**, *70*, 183–190.
7. Benfratello, S.; Capitano, C.; Peri, G.; Rizzo, G.; Sorrentino, G. Thermal and structural properties of a hemp-lime biocomposite. *Constr. Build. Mater.* **2013**, *48*, 745–754. [CrossRef]
8. Pochwała, S.; Makiola, D.; Anweiler, S.; Böhm, M. The Heat Conductivity Properties of Hemp–Lime Composite Material Used in Single-Family Buildings. *Materials* **2020**, *13*, 1011. [CrossRef]
9. Barreca, F.; Fichera, C.R. Use of olive stone as an additive in cement lime mortar to improve thermal insulation. *Energy Build.* **2013**, *62*, 507–513. [CrossRef]
10. Briga-Sá, A.; Nascimento, D.; Teixeira, N.; Pinto, J.; Caldeira, F.; Varum, H.; Paiva, A. Textile waste as an alternative thermal insulation building material solution. *Constr. Build. Mater.* **2013**, *38*, 155–160. [CrossRef]
11. Ashour, T.; Georg, H.; Wu, W. Performance of straw bale wall: A case of study. *Energy Build.* **2011**, *43*, 1960–1967. [CrossRef]
12. Kochhar, G.S.; Manohar, K. Use of coconut fiber as a low-cost thermal insulator. Insulation materials: Testing and applications. *ASTM* **1997**, *3*, 283–291.
13. Moropoulou, A.; Bakolas, A.; Aggelakopoulou, E. The effects of limestone characteristics and calcination temperature to the reactivity of the quicklime. *Cem. Concr. Res.* **2001**, *31*, 633–639. [CrossRef]
14. Schumacher, G.; Juniper, L. Coal utilisation in the cement and concrete industries. *Coal Handb. Towards Clean. Prod.* **2013**, *2*, 387–426.
15. Liu, K.; Wang, Z.; Jin, C.; Wang, F.; Lu, X. An experimental study on thermal conductivity of iron ore sand cement mortar. *Constr. Build. Mater.* **2015**, *101*, 932–941. [CrossRef]
16. Martínez-Ramírez, S.; Puertas, F.; Blanco-Varela, M.T.; Thompson, G.E.; Almendros, P. Behaviour of Repair Lime Mortars by Wet Deposition Process. *Cem. Concr. Res.* **1998**, *28*, 221–229. [CrossRef]

17. Robador, M.D.; Arroyo, F. Characterization of Roman coatings from the Roman house in Mérida (Spain). *J. Cult. Herit.* **2013**, *14*, s52–s58. [CrossRef]
18. Boyton, R.S. *Chemistry and Technology of Lime and Limestone*; John Wiley & Sons: New York, NY, USA, 1966.
19. Banfill, P.F.G.; Forrester, A.M. A Relationship between Hydraulicity and Permeability of Hydraulic Lime. In Proceedings of the International RILEM Workshop 'Historic Mortars: Characteristics and Tests', RILEM, Paisley, Scotland, 12–14 May 1999; pp. 173–183.
20. Available online: http://www.engineeringtoolbox.com/thermal-conductivity-d_429.html (accessed on 27 November 2020).
21. Shakelford, J.; Alexander, W. *Material Science and Engineering Handbook*, 3rd ed.; CRC Press: Boca Raton, FL, USA, 2001.
22. Krebs, R.E. *The History and Use of Our Earth's Chemical Elements*, 2nd ed.; Greenwood Publishing Group: Westport, CT, USA, 2006; pp. 102–105.
23. Acevedo-Sandoval, O.; Ortiz-Hernández, E.; Cruz-Sánchez, M.; Cruz-Chávez, E. El papel de óxidos de hierro en suelos. *Terra Latinoam.* **2004**, *22*, 485–497.
24. Kishar, E.A.; Alasqalani, M.Y.; Sarraj, Y.R.; Ahmed, D.A. The Effect of Using Commercial Red and Black Iron Oxides as a Concrete Admixtures on its Physiochemical and Mechanical Properties. *Int. J. Sci. Res.* **2015**, *4*, 1389–1393.
25. Largeau, M.A.; Mutuku, R.; Thuo, J. Effect of Iron Powder (Fe2O3) on Strength, Workability, and Porosity of the Binary Blended Concrete. *Open J. Civ. Eng.* **2018**, *8*, 411–425. [CrossRef]
26. Masdeu, F.; Muñoz, J.; Carmona, C.; Horrach, G. Mortero de cal Termoaislante y su uso en Edificación. Spanish Utility Model Patent ES1222024 U, 18 December 2018.
27. ISO 12439. *Mixing Water for Concrete*; International Organization for Standarization: Geneva, Switzerland, 2010.
28. *ASTM C1363-11, ASTM 90 (reapproved), Standard Test Method for Thermal Performance of Building Materials and Envelope Assemblies by Means of a Hot-Box Method Apparatus*; ASTM: West Conshohocken, PA, USA, 2014.
29. Weast, R.; Lide, D. *Handbook of Chemistry and Physics*, 70th ed.; CRC Press: Boca Raton, FL, USA, 1989.
30. Kamseu, E.; Nait-Ali, B.; Bignozzi, M.C.; Leonelli, C.; Rossignol, S.; Smith, D.S. Bulk composition and microstructure dependence of effective thermal conductivity of porous inorganic polymer cements. *J. Eur. Ceram. Soc.* **2012**, *32*, 1593–1603. [CrossRef]
31. Dos Santos, W.N. Effect of moisture and porosity on the thermal properties of a conventional refractory concrete. *J. Eur. Ceram. Soc.* **2003**, *23*, 745–755. [CrossRef]
32. Niubó, M.; Formosa, J.; Maldonado-Alameda, A.; del Valle-Zermeño, R.; Chimenos, J.M. Magnesium phosphate cement formulated with low grade magnesium oxide with controlled porosity and low thermal conductivity as a function of admixture. *Ceram. Int.* **2016**, *42*, 15049–15056. [CrossRef]
33. Bergman, T.L.; Lavine, A.S.; Incropera, F.P.; DeWitt, D.P. *Fundamentals of Heat and Mass Transfer*, 7th ed.; John Wiley & Sons: Jefferson City, MI, USA, 2011; pp. 116–117.
34. Rhee, S.K. Porosity—Thermal conductivity correlations for ceramic materials. *Mat. Sci. Eng.* **1975**, *20*, 89–93. [CrossRef]
35. Housecroft, C.E.; Sharpe, A.G. *Inorganic Chemistry*, 3rd ed.; Pearson: London, UK, 2008; pp. 78–89.
36. Marshall, A.L. Thermal properties of concrete. *Build. Sci.* **1972**, *7*, 167–174. [CrossRef]
37. Gulbea, L.; Vitinab, I.; Setinac, J. The influence of cement on properties of lime mortars. *Procedia Eng.* **2017**, *172*, 325–332. [CrossRef]
38. Ng, S.; Low, K.; Tioh, N. Newspaper sandwiched aerated lightweight concrete wall panels—thermal inertia, transient thermal behavior and surface temperature prediction. *Energy Build.* **2011**, *43*, 1636–1645. [CrossRef]
39. Ferrari, S. Building envelope and heat capacity: Re-discovering the thermal mass for winter energy saving. In Proceedings of the 2nd PALENC Conference and 28th AIVC Conference on Building Low Energy Cooling and Advanced Ventilation Technologies in the 21st Century, Crete, Greece, 27–29 September 2007.
40. Demirel, Y. *Nonequilibrium Thermodynamics. Transport and Rate Processes in Physical, Chemical and Biological Systems*, 3rd ed.; ElSevier: Oxford, UK, 2014; pp. 87–88.
41. Available online: http://www.tainstruments.com/pdf/literature/TP_006_MDSC_num_1_MDSC.pdf (accessed on 27 November 2020).
42. Waples, D.W.; Jacob, S.; Waples, J.S. A Review and Evaluation of Specific Heat Capacities of Rocks, Minerals, and Subsurface Fluids. Part 1: Minerals and Nonporous Rocks. *Nat. Resour. Res.* **2004**, *13*, 97–122. [CrossRef]
43. Bogas, J.A.; Gomes, A.; Pereira, M.F.C. Self-compacting lightweight concrete produced with expanded clay aggregate. *Constr. Build. Mater.* **2012**, *35*, 1013–1022. [CrossRef]
44. Xu, Y.; Jiang, L.; Xu, J.; Li, Y. Mechanical properties of expanded polystyrene lightweight aggregate concrete and brick. *Constr. Build. Mater.* **2012**, *27*, 32–38. [CrossRef]

Article

Fibre-Reinforced Geopolymer Concretes for Sensible Heat Thermal Energy Storage: Simulations and Environmental Impact

Domenico Frattini [1,*], Alessio Occhicone [2], Claudio Ferone [2,*] and Raffaele Cioffi [2]

[1] Graduate School of Energy and Environment, Seoul National University of Science and Technology, Gongneung-ro 232, Nowon-gu, Seoul 01811, Korea
[2] Department of Engineering, University Parthenope of Naples, Centro Direzionale di Napoli Is. C4, 80143 Napoli, Italy; alessio.occhicone@uniparthenope.it (A.O.); raffaele.cioffi@uniparthenope.it (R.C.)
* Correspondence: domenico.frattini@seoultech.ac.kr (D.F.); claudio.ferone@uniparthenope.it (C.F.)

Abstract: Power plants based on solar energy are spreading to accomplish the incoming green energy transition. Besides, affordable high-temperature sensible heat thermal energy storage (SHTES) is required. In this work, the temperature distribution and thermal performance of novel solid media for SHTES are investigated by finite element method (FEM) modelling. A geopolymer, with/without fibre reinforcement, is simulated during a transient charging/discharging cycle. A life cycle assessment (LCA) analysis is also carried out to investigate the environmental impact and sustainability of the proposed materials, analysing the embodied energy, the transport, and the production process. A Multi-Criteria Decision Making (MCDM) with the Analytical Hierarchy Process (AHP) approach, taking into account thermal/environmental performance, is used to select the most suitable material. The results show that the localized reinforcement with fibres increases thermal storage performance, depending on the type of fibre, creating curvatures in the temperature profile and accelerating the charge/discharge. High-strength, high-conductivity carbon fibres performed well, and the simulation approach can be applied to any fibre arrangement/material. On the contrary, the benefit of the fibres is not straightforward according to the three different scenarios developed for the LCA and MCDM analyses, due to the high impact of the fibre production processes. More investigations are needed to balance and optimize the coupling of the fibre material and the solid medium to obtain high thermal performance and low impacts.

Keywords: AHP; carbon fibres; conductivity; geopolymers; LCA; sustainability; thermal storage

Citation: Frattini, D.; Occhicone, A.; Ferone, C.; Cioffi, R. Fibre-Reinforced Geopolymer Concretes for Sensible Heat Thermal Energy Storage: Simulations and Environmental Impact. *Materials* **2021**, *14*, 414. https://doi.org/10.3390/ma14020414

Received: 13 November 2020
Accepted: 12 January 2021
Published: 15 January 2021

1. Introduction

In recent decades, Solar Thermal Power Plants were developed at a large scale to indirectly convert concentrated solar energy into green electricity [1]. These plants work at high temperatures, and as frequently occurs for renewable energy sources, they are highly dependent on the availability of solar radiation, i.e., only during the day [2]. Hence, a reliable and efficient thermal storage system is mandatory to extend the productivity of these energy plants. Sensible heat thermal energy storage (SHTES) is applied for a high-temperature range, i.e., >573 K. In particular, the most feasible and economic storage materials for SHTES are represented by high performance concretes [3]. However, in the long term, they suffer from thermal fracture and failure due to thermal stress and non-uniform temperature distribution during charging/discharging steps. To solve these issues, high-strength-high-conductivity long fibres can be added to increase heat diffusion and mechanical strength. Recently, geopolymer concretes are reported to have a higher thermal stress resistance in repeated heating/cooling cycles at high temperatures [4–6], but improved thermal properties are required to compete with reference benchmarks [7]. Many numerical and simulation approaches have been used in literature to predict, evaluate, and enhance their performance [8–10]. Among them, FEM is the most useful to obtain a reliable image of spatial distribution inside storage material because this is of fundamental importance for module design and performance analysis [11,12]. Fibre spacing, pattern, and thermal properties are the main drivers for SHTES concrete reinforcement and fibre embedding; hence, an efficient simulation of the temperature distribution during the charge/discharge cycle of a thermal storage unit based on fibre-reinforced concretes is necessary. Moreover, considering large scale plants, integrated storage systems, and desired lifetime, a comprehensive design must take into consideration not only the thermal properties of the solid storage medium and its simulated performances but also the environmental impact and sustainability of the materials and processes involved in its production.

Geopolymers are innovative binders that have been extensively studied in recent years consisting of amorphous to semi-crystalline aluminosilicates synthesized using alkaline solutions and solid precursors such as low-Ca fly ash [13], calcined clays [14–16], and other industrial, and natural waste [17–20].

These materials show excellent mechanical properties, low shrinkage (low-Ca precursors), thermal stability, freeze-thaw, acid and fire resistance, long term durability, and recyclability [21,22]; thus, they are a potential alternative to traditional Portland cement in selected applications because in many cases they have also a reduced environmental impact.

On the other hand, owing to their ceramic nature, they have relatively low toughness and low flexural strength and, to improve these properties, geopolymer matrix composite materials have been prepared and studied. Plenty of studies [23,24] have been produced on this topic, and many types of fillers have been tested, such as particulate and various kinds of short and continuous fibres. Novel geopolymer matrix composites and hybrids have also been obtained by the in situ co-reticulation of a geopolymer matrix with an epoxy-based organic resin [25–30].

In addition to the mechanical performance, geopolymer matrix composites can be used to modify geopolymer thermal properties, such as thermal conductivity and fireproofing [31], e.g., sustainable geopolymer concrete with good thermal insulation properties were obtained by incorporating recycled expanded polystyrene spheres in a metakaolin based geopolymer matrix, together with a waste-derived filler [32]. Considering this background, it is demonstrated that geopolymer concretes represent a high potential alternative material for SHTES based on solid media due to their versatile properties, formulations, and feasibility.

In this work, a geopolymer concrete is modelled as a solid medium for SHTES. The enhancement of thermal properties and temperature distribution are obtained by adding fibres with high conductivity in an ordered arrangement around the heat exchanger pipe. Temperature contours, time evolution, and thermal performance of geopolymer concretes are compared with and without fibres to demonstrate the effect of fibres and fibre materials.

In addition to the simulation of the behaviour of the new geopolymer materials, for their use in thermal energy storage, and the comparison with previous literature, an analysis of the environmental impact that the production of 1 m^3 of material has on the ecosystem was performed. This type of approach was useful to understand the technical and environmental limits of the new materials designed for thermal storage units. The environmental impact assessment was carried out following the rules imposed by the life cycle assessment (LCA), following a cradle-to-gate approach under the conditions valid for the assessment of the Environmental Product Declaration (EPD®) for possible future industrialization in Italy.

The International EPD® System is a global program that communicates verified, transparent, and comparable information about the life cycle environmental impact of products. EPD is based on ISO 14025 (ISO 14025:2006 Environmental labels and declarations—Type III environmental declarations—Principles and procedures) and EN 15804 (EN 15804: 2012 Sustainability of construction works, Environmental product declarations, Core rules for the product category of construction products) standards. For this reason, EPD is among the most accepted methods that return, in a standardized and comparable way, the results of the LCA related to a product, a process or an activity [33].

To give support in the decision-making process at the industrial level, the thermal and environmental performances obtained were also evaluated through the Super-Decisions simulation program. This software provides a powerful methodology for combining judgment and data to effectively rank options and predict outcomes. Several authors [34,35] showed that, following the combined LCA-Analytical Hierarchy Process (AHP) Multi-Criteria Decision Making (MCDM) approach, it is possible to obtain the best evaluation of the behaviour of materials from the point of view of performance and environment. This kind of approach has been used by the scientific community, due to its simplicity and robustness for sustainable evaluation [36] and has already been used to choose sustainable materials based on their features [37].

2. Materials and Methods

2.1. Model Design of the Fibre-Reinforced SHTES Unit and Simulation Conditions

The storage module is a complex parallelepiped with an embedded heat exchanger composed by tubes in square arrangement [38]. This structure is often decomposed as a repetition of differential storage elements due to module symmetry. The differential storage element for calculations is usually a hollow cylinder with a cross-section area equivalents to that of a square [9,38,39]. Differently from the cited references, the geometry is not further reduced to a 1D radial problem because the external square represents a real physical limit and, from the thermal point of view, this could be incorrect as the temperature diffusion (e.g., at the corners) could be uneven in certain directions, and could be undetected during the transient simulation. Moreover, a 1D radial modelled element does not allow the local enhancement by inserting discrete objects to create multi-conduction areas for a thermally engineered design.

Therefore, as an extension of previous literature [7,40], in this work, a square-based parallelepiped of unit length (1 m) is the differential storage element for FEM simulations. The geometric mesh was modelled in GAMBIT 2.3.16. The new feature is the arrangement of 16 bunches of fibres in squared pitch around the central heat exchanger pipe as shown in Figure 1.

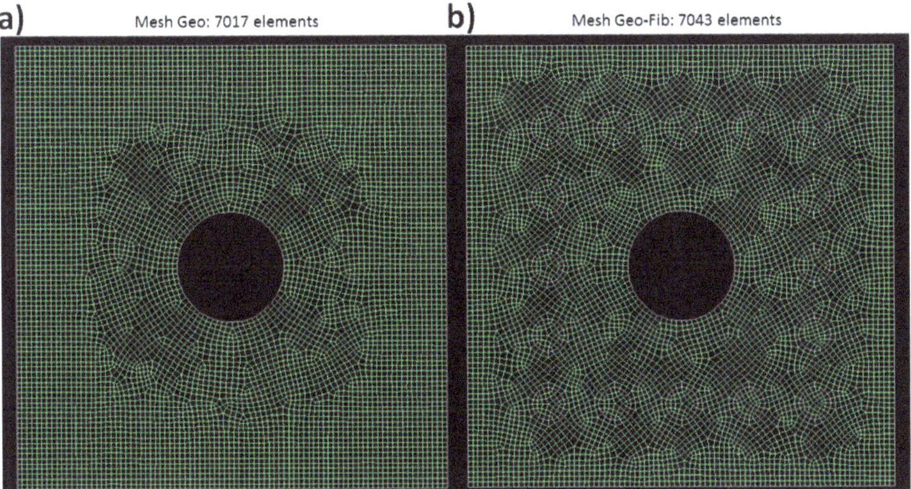

Figure 1. Cross-section mesh of the storage element used for simulations. (**a**) Mesh Geo without fibres. (**b**) Mesh Geo-Fib with fibres.

The storage mesh has an 8×8 cm^2 section area; fibres are centrally spanned on a 5×5 cm^2 square around the tube. The tube diameter is 2 cm, fibre bunch diameter is 0.6 cm, and the interspace between each bunch is 0.65 cm. Further details on the mesh are reported elsewhere [7]. The storage material is a geopolymer concrete (G), while two types of fibres, i.e., Carbon fibre (FibC) or Nickel fibre (FibNi), are considered as reinforcement. Their properties are listed in Table 1. These two types of fibres, i.e., a metal-based and a carbon-based one, were selected because they are two common commercial fibres, not very expensive, and show the same thermal conductivity but different density and specific heat capacity (hence, a different diffusivity). So, it was interesting to compare them for this specific application. Moreover, they are also already used to reinforce concretes and building materials to improve the mechanical resistance and hence the lifetime of the material, aspects that are not directly investigated in this work. Given the improved mechanical resistance and durability, more information about the thermal performance and sustainability of SHTES is needed when considering these fibres.

Table 1. Thermal properties of storage materials used in simulations.

Material	Density ρ (kg/m^3)	Spec. Heat Cap. c (J/kg·K)	Thermal Cond. k (W/m·K)	Thermal Diff. $\alpha \times 10^7$ (m^2/s)	Vol. Thermal Cap. C_{vol} (kWh/m^3·K)
FibC	1810	800	70	483	1448.24
FibNi	8890	456	70	173	4053.84
G	1811	751	1.01	7.43	1360.06

For FibC and FibNi, the thermal properties were retrieved from [41], while those for G are from [7]. Thermal diffusivity α and volumetric thermal capacity C_{vol} were simply calculated from ρ, c, and k as follows:

$$\alpha = \frac{k}{\rho \cdot c} \tag{1}$$

$$C_{vol} = \rho \cdot c \tag{2}$$

Sensible heat storage FEM simulations were run using the computational fluid dynamics CFD software ANSYS Fluent (v. 6.3.26, ANSYS Inc., Canonsburg, PA, USA). Assumptions, governing equations, storage cycle, boundary, and initial conditions are the same as those reported in the literature [7,10,40] and were used without relevant modifications. In brief, the maximum temperature difference for storage was 40 K, i.e., from 623 to 663 K, all the materials were considered isotropic, thermal properties were considered constant within the small temperature range, and the unit was considered perfectly insulated thermally. The imposed thermal cycle was 3600 s of charge/discharge and 3600 s of storage (buffering time or "break"). The time step for simulations was 2 s, convergence criteria were set at 10^{-4}, and the built-in Monitor function of Fluent was used to take the contour profiles of temperature distribution at 5, 20, 40, and 60 min during charge/discharge with a temperature accuracy (colour scale) of 2 K, while one of the external edges of the module was selected to record the average T profile at the wall.

2.2. LCA Analysis

Comparative LCA analysis was carried out on the three geopolymer concretes studied in this work, and on materials studied in previous work for similar applications. In particular, a plain cement concrete (C) [7], a modified concrete with marble sludge (PA0), the same modified concrete with 20% by weight of recycled plastic (PA20) [7], and the material named A4 developed by Guo et al. [42] were compared against the coal fly ash-based geopolymer (G), the same coal fly ash-based geopolymer matrix composite with

Carbon fibres (FibC) or with Nickel fibres (FibNi), as described in the previous section. In Table 2, the compositions of all tested materials are reported.

Table 2. Composition of the investigated materials used in the life cycle assessment (LCA) analysis.

Materials	Unit	C	PA0	PA20	G	FibC	FibNi	A4
CEM II/A-L 42.5R	kg/m^3	280.00	300.00	300.00	-	-	-	-
Sand	kg/m^3	1000.00	-	-	-	-	-	-
Gravel	kg/m^3	400.00	-	-	-	-	-	-
Fine gravel	kg/m^3	200.00	-	-	-	-	-	-
Marble sludge	kg/m^3	-	146.00	171.00	-	-	-	-
Crushed limestone	kg/m^3	-	1648.00	1227.00	1288.83	1234.33	1234.33	-
Plastic aggregate	kg/m^3	-	-	140.00	-	-	-	-
Fly ash	kg/m^3	-	90.00	90.00	313.91	300.63	300.63	-
Alkaline solution *	kg/m^3	-	-	-	208.27	199.46	199.46	-
Superplasticizer **	L/m^3	-	6.86	8.91	-	-	-	-
Fibre	kg/m^3	-	-	-	-	88.69	435.61	-
Calcium aluminate cement	kg/m^3	-	-	-	-	-	-	268.00
Basalt	kg/m^3	-	-	-	-	-	-	991.60
Bauxite	kg/m^3	-	-	-	-	-	-	964.80
Graphite	kg/m^3	-	-	-	-	-	-	268.00
Silica sand	kg/m^3	-	-	-	-	-	-	134.00
Aluminium micropowder	kg/m^3	-	-	-	-	-	-	107.20
Steel	kg/m^3	-	-	-	-	-	-	134.00
Density	kg/m^3	2410.00	2190.86	1936.91	1811.00	1823.11	2170.03	2170.03

* Sodium silicate R = 2 (molar SiO_2/Na_2O ratio) was used as the alkaline solution. ** Superplasticizer is a mix of Lignosulphonate: max. 35%; Naphthalene sulphonate: max. 30%; Melamine sulphonate: max. 45%; Polycarboxylate: max. 35%. Commercial material from the European Federation of Concrete Admixtures Associations Ltd. (EFCA) (EPD EFCA).

The LCA was performed according to ISO 14040:2006 (ISO 14040:2006 Environmental Management—Life Cycle Assessment—Principles and Framework) with the Simapro© software (v. 8.5.2, PRé Sustainability, Amersfoort, The Netherlands).

In this work, the boundary system involved raw materials, transport, and manufacturing and was based on 1 m^3 of material production.

In a specific view:

- The LCA was performed in a way that considered the contribution of the raw materials for all the different mixtures; the production processes have some energy consumption in common, in particular material milling, mixing, and element cutting are similar for each material, so they have the same value for all the products. The difference in the process is the curing necessary for geopolymer products, which need 24 h in a climatic chamber at 60 °C to complete reticulation reaction and hardening. In this case, the electricity consumption linked to this step was estimated using the Italian energetic mix as in [43].
- The impacts of the raw materials were estimated including extraction and all the necessary processes preliminary to their use. For fly ash, just the transportation contributed to the impacts, because it can be used directly in the production process and can be considered as a no-impact material, as demonstrated in previous works [43,44]. In the case of recycled plastic, just the milling and the transport were estimated, with a negative global contribution for different environmental impact items such as global warming potential (all the contributions are shown in Appendix A, Tables A1–A7). This is because the environmental impacts due to traditional management of plastic (disposal mix: in landfills, recycling, composting, burning, etc.) are avoided. For the calcium aluminate cement and the superplasticizer, the environmental impacts were calculated based on the EPD® 830 çimsa RESISTO40 following ISO14040/44 [45] and the European Federation of Concrete Admixtures Associations Ltd. (EFCA) EPD®

according to ISO14025:2011-10 [46], respectively. The estimated impacts of PAN fibre production were also taken from the literature [47].

- All other material impacts were taken from Simapro Ecoinvent 3 database.
- The transportation stage was considered for the delivery of the raw materials to the plant. In particular, based on the average availability of materials in Europe, an average transport distance up to 100 km for all the starting materials was considered. It is worth noting that this estimation had little effect on the overall production impacts of each system, bordering on undetectable.

No treatment of "the end of life" was considered for any of the products, except for the recycled materials used as precursors. In this case, the missed impacts from landfill disposal were evaluated and inserted with a negative value. So, a cradle-to-gate approach was considered, following the EPD.

SimaPro 8.5.2 © is equipped with different methods for assessing impacts. In particular, the ReCiPe (2016) Midpoint method was chosen in the present work. This is a method for Life cycle impact assessment (LCIA) that translates emissions and resource extractions into a limited number of environmental impact scores utilizing the so-called characterization factors [48]. In this work, it was chosen at the midpoint level to better underline the contribution of all precursors in the environment, without the loss of sensitivity linked to the implementation of all the indicators in the three macro areas that constitute the endpoint level of the method [49].

This LCA method includes 18 midpoint impact categories, but, in this work, the midpoint characterization is shown just for some of the items (the remaining ones are reported in Appendix A):

- Climate change: Global Warming Potential (GWP), which quantifies the integrated infrared radiative forcing increase in greenhouse gas (GHG), expressed in kg CO_2-eq (IPCC 2013).
- Stratospheric ozone depletion: The ozone-depleting potential (ODP), expressed in kg CFC-11 equivalents, was used as a characterization factor at the midpoint level. ODP refers to a time-integrated decrease in stratospheric ozone concentration over an infinite time horizon [50].
- Particulate matter: Quantification of the impact of premature death or disability that particulates/respiratory inorganics have on the population, in comparison to $PM_{2.5}$. This includes the assessment of primary (PM_{10} and $PM_{2.5}$) and secondary PM (including the creation of secondary PM due to SO_x, NO_x, and NH_3 emissions) and CO [51].
- Photochemical ozone formation: human health ozone formation potential (HOFP) is expressed in kg NO_{x-eq}. The change in ambient concentration of ozone after the emission of a precursor (nitrogen oxides (NO_x) or non-methane volatile organic compounds (NMVOC)) was predicted with the emission–concentration sensitivity matrices for emitted precursors from the global source-receptor model, TM5-FASST [52].
- Terrestrial acidification: For the midpoint characterization factors of acidifying emissions, the fate of a pollutant in the atmosphere and the soil was calculated as in [53]. Acidification potentials (AP) are expressed in kg SO_2-eq. Changes in acid deposition, following changes in air emission of NO_x, NH_3, and SO_2, were calculated with the GEOS-Chem model [54].
- The midpoint indicator for fossil resource use, determined as the Fossil Fuel Potential (FFP in kg oil-eq), is defined as the ratio between the higher heating value of a fossil resource and the energy content of crude oil [55].

The limitation of these impact factors is for quicker analysis, but all the environmental impact categories are reported in Appendix A.

2.3. *Super-Decisions AHP Analysis*

The six parameters shown in LCA analysis were considered Super-decisions sub-criteria. At the end of environmental analysis, it was possible to investigate the best materials from the point of view of both thermal and environmental performance. To do this, Super-decisions free software (v. 3.2, Creative Decisions Foundation, Pittsburgh, PA, USA) was used. This software provides tools to create and manage AHP and Analytic Network Process (ANP) models, enter judgments, obtain results, and perform sensitivity analysis on the results. In the present work, a hierarchical approach (AHP) was chosen. In this kind of analysis, levels are arranged in descending order of importance. The elements in each level are compared according to dominance or influence for the elements in the level immediately above that level [56].

Figure 2 shows the graph of the hierarchical levels with which the calculations were carried out. Specifically, we have the most suitable material that derives from its thermal performance and from the environmental impact generated by the production of the material itself.

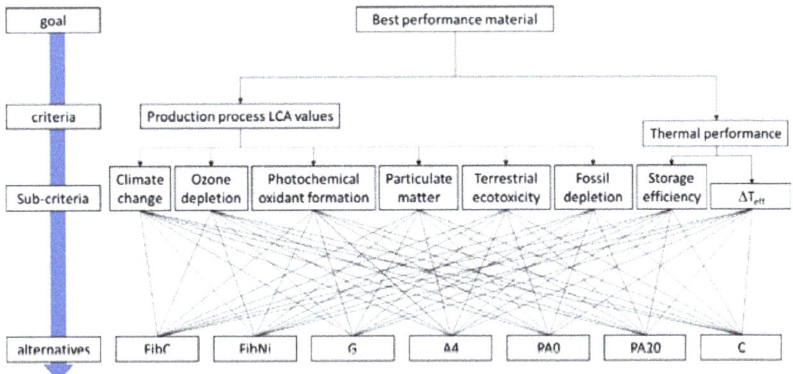

Figure 2. Super-decisions Analytical Hierarchy Process (AHP) model to select the best sensible heat thermal energy storage (SHTES) material.

The chosen criteria consider three different scenarios: (i) the material production process LCA values have the same importance of thermal performance; (ii) the LCA values of materials have a weight four times greater than that of thermal performance; (iii) the weight of thermal performance is four times that of LCA values. The first scenario can represent the case of the short lifetime of the plant; on the other hand, the third scenario can represent the case of a very long lifetime, in which the environmental impact of the production process is spread over a long period in the third scenario, and the durability of the materials becomes fundamental, but this is not the subject of the present work, and this factor could likely see fibre-reinforced materials prevail over the unreinforced ones. In the sub-criteria, a different weight was given to LCA parameters and thermal storage variables. The rank of the different variables was chosen according to the authors' experience [57], based on the LCA results and the thermal performance obtained from the simulations (Appendices A and B).

The scale used to compare different criteria is that suggested by Saaty 2003 [58]: (1) equal, (2) between equal and moderate, (3) moderate, (4) between moderate and strong, (5) strong, (6) between strong and very strong, (7) very strong, (8) between very strong and extreme, (9) extreme. In Table 3, the weights of the 8 variables considered in the sub-criteria are reported. Values obtained for each material (alternative) are reported in Appendix B (Tables A16–A19 and A8–A15).

Table 3. (**a**) Weights of the sub-criteria variables: A1—climate change, A2—ozone depletion, A3—photochemical oxidant formation, A4—particulate matter, A5—terrestrial ecotoxicity, A6—fossil depletion; (**b**) A7—storage efficiency, A8—ΔT_{eff}.

a	A1	A2	A3	A4	A5	A6
A1	A	3	4	4	5	2
A2	1/3	1	2	2	3	1/2
A3	1/4	1/2	1	1	2	1/3
A4	1/4	1/2	1	1	2	1/3
A5	1/5	1/3	1/2	1/2	1	1/4
A6	1/2	2	3	3	4	1
b		**A7**			**A8**	
A7		1			3	
A8		1/3			1	

The alternatives should be compared pairwise as well: as a result, a positive reciprocal matrix for the alternatives should be designed. All the judgements over the criteria and alternatives have to be consistent. This means that the inconsistency index for each element of the reciprocal matrices should not exceed 0.1 [59].

3. Results and Discussion

3.1. Thermal Storage Charge/Discharge Temperature Profiles, Maps and Performance of Fibre-Reinforced Geopolymer Concretes

In Figure 3, the average temperature profile at the wall during a thermal cycle is shown for the three different storage elements investigated. As noticeable, in all cases, the elements could charge and discharge very fast. In detail, the element reinforced with carbon fibres, FibC, attained the highest temperature after the charging step and the deepest discharge state for the given time, i.e., 3600 s for charge/discharge.

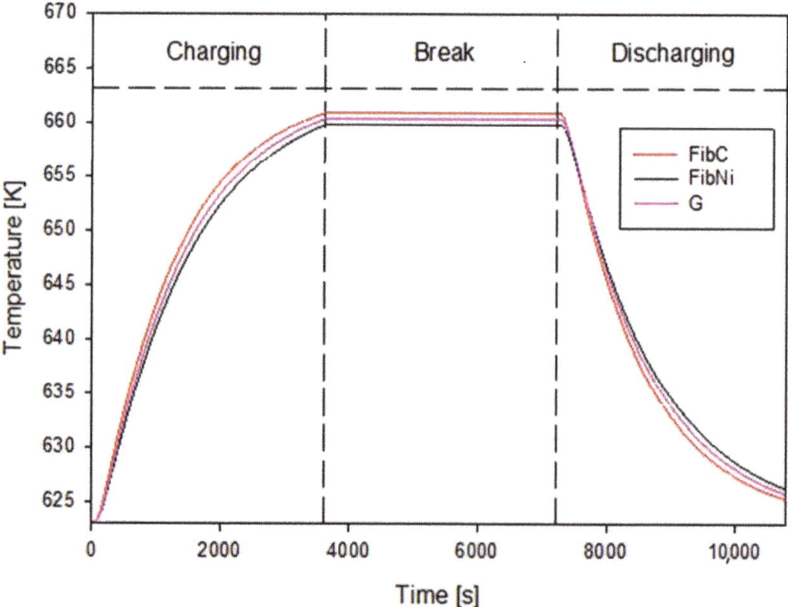

Figure 3. Temperature profile at the wall for the geopolymer concrete with/without fibres.

However, the thermal behaviour of the three elements was slightly different as the best was FibC, due to the presence of the high conductive carbon fibres, but the worst was FibNi and not the plain geopolymer, G. This would mean that the use of Nickel fibres does not improve thermal performance, although is a metal. In reality, the enhancement of inserting the Nickel fibres relies more heavily on the internal temperature distribution and hence the overall storage efficiency because, as discussed later, due to the high heat capacity of these fibres, the external wall temperature is slightly lower than G, but the distribution of the temperature is completely different. These qualitative conclusions are confirmed by quantitative calculations of thermal performance as reported in Table 4.

Table 4. Thermal performance of storage elements.

Material	ΔT_{eff} (K)	Thermal Storage Efficiency (%)	Volume Power Density (kWh/m^3)
FibC	37.71	95.86	14.24
FibNi	36.62	94.63	13.83
G	37.11	92.42	13.28

The FibC storage element has improved storage efficiency and power density due to the higher effective temperature increase (ΔT_{eff}) achieved during charge.

For the local temperature distribution inside the storage element, in Figure 4, the temperature contours at different charging times are compared.

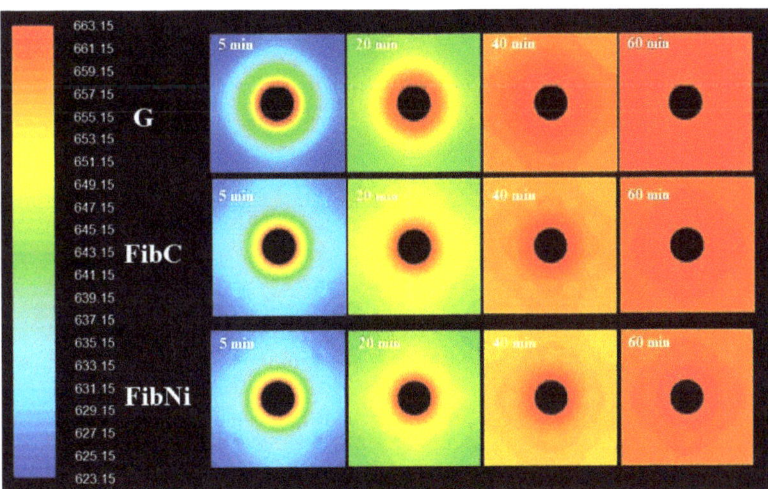

Figure 4. Temperature contours at different charging times for storage elements.

As a consequence of fibre addition, the profile shapes were different. In the plain geopolymer G at the beginning, the temperature contour evolved with a series of concentric waves but, once the wave reached the external wall at the perpendicular direction, the curvature changed and waves slowly propagated to cover corners. At the end of the charge, from the contours, it is evident that temperature distribution was very good but slightly uneven and that the slower propagation at corners was the reason behind this. Differently, when fibres were inserted around the tube, heat propagation was enhanced, and temperature distribution was more uniform at walls as waves proceed smoothly also in corner directions. The two storage elements with fibres both have two well-determined regions, one with uniform temperature all around the walls, and another inner region around the fibres that distribute the heat evenly from the central pipe (that has its heat transfer hot "corona" where the main heat transfer phenomena from the heat transfer fluid take place). The presence of regions with different thermal diffusivity, i.e., fibres and matrix, modified the contour shape and increased the speed of heat propagation near fibres. The lower performances of FibNi, thermal conductivity being the same as for FibC, is ascribed not only to the lower thermal diffusivity but also to the larger volumetric thermal capacity, C_{vol}, i.e., the quantity of thermal energy required to raise the temperature of 1 m^3 of storage material by 1 K. This is because the coupling of the matrix G with the carbon fibres seems quite optimal. After all, they have a similar thermal capacity (i.e., same energy required to charge), but the fibres have far higher conductivity and diffusivity (i.e., faster charge); hence, FibC combines the benefits of both.

The discharging behaviour of the three modules was also simulated, and the contours are reported in the following Figure 5.

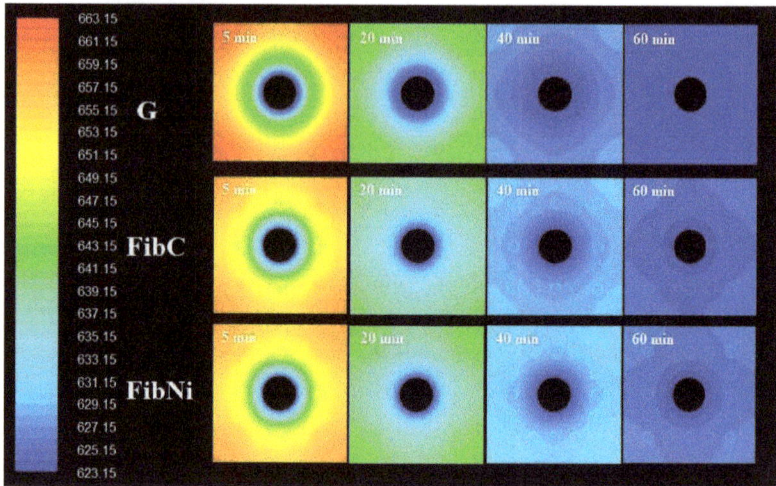

Figure 5. Temperature contours at different discharging times for storage elements.

Discharging behaviour also showed interesting features as it is different for the three materials due to the insertion of the fibres. In G, the discharge was uniform and quite fast, leaving a uniform temperature region all around the walls and another inside the materials. Conversely, the storage elements with fibres both showed a non-circular propagation of the temperature profile in the section due to the fibres. In fact, after discharge, the corners of the elements remained a slightly (i.e., 1–2 K) warmer than the internal region (central pipe corona excluded because it is the "source" of the discharge, similar for all the material). This is probably because the fibres tend to "protect" the element by preventing heat loss, enhancing thermal diffusion because they have their own thermal capacity, i.e., they can store some heat, different from the geopolymer matrix.

However, it should be noted that, although the discharge contours were very different, the average temperature at the wall after discharge (Figure 5) was almost similar for G and FibC, while the FibNi discharged and charged slowly. In fact, at the beginning of the thermal cycle, the three storage modules had the same identical status at the charge, but at the discharge phase, the initial condition of the material was that at the ending of the charging phase; hence, the initial temperature distribution is not the same for all three materials, thus better reflecting the actual behaviour of the modules for real applications. In ideal conditions, the charge and discharge are perfectly reversible; hence, the temperature profiles, e.g., Figure 5, tend to be specular and symmetric, and it is only possible to modulate and tune different times and durations of the charge/discharge phase.

The beneficial effect on thermal performance could be increased by changing the volume loading of the fibres, but an optimal compromise with environmental issues must be carefully considered as it is crucial for the operational life and economic balance of the plant/application, as discussed below.

3.2. LCA Environmental Analysis of the Fibre-Reinforced Geopolymer Concretes for SHTES Units Compared with Other SHTES Unit Materials

The ReCiPe hierarchic midpoint method allowed us to estimate all the impact factors for each material of all the products and the contribution of transport and energy depletion during the production process. The most important impact categories are shown in Figure 6.

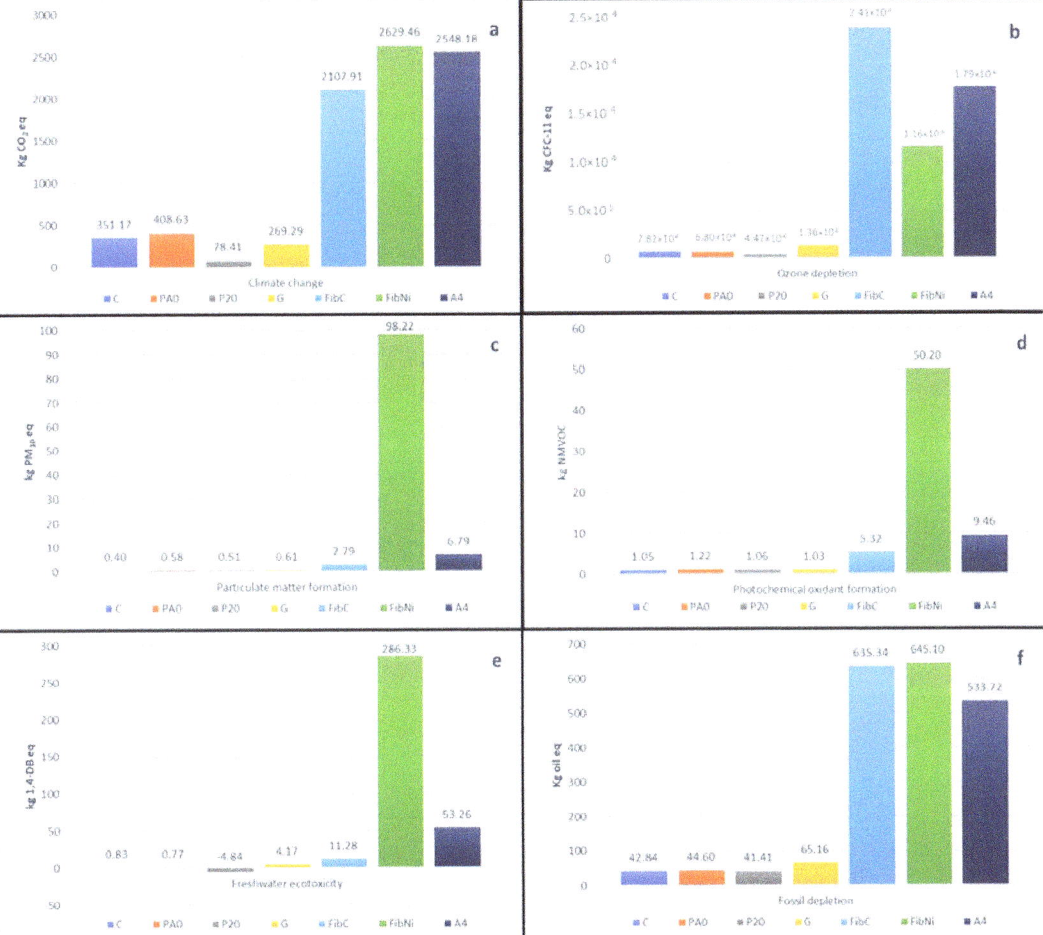

Figure 6. Comparison of environmental impact contribution for the proposed SHTES materials: (**a**) climate change (CO_2 eq. emission); (**b**) ozone depletion (CFC_{11} eq. emission); (**c**) particulate matter ($PM_{2.5}$ emission); (**d**) photochemical oxidant formation (NMVOC eq. formation); (**e**) terrestrial acidification (SO_2 eq. production); (**f**) fossil fuel depletion (oil eq. consumption).

As can be observed, the best performances from the environmental point of view were obtained by recycled plastic concrete (PA20), which shows, in some items, negative values for Freshwater ecotoxicity (all the values shown in Appendix A). In this case, both the limestone sludge and the plastic shavings, used in the concrete mix instead of usual disposal, showed a significant reduction in environmental impacts, making them preferable as a green opportunity.

Good and similar impact values were obtained by the concretes based on Portland cement (C) and concrete with the addition of limestone sludge (PA0). Geopolymer shows better performances from the point of view of global warming potential and acidification, and slightly worse values for fossil and ozone depletion, as detected in other geopolymers/Portland concrete comparisons [33,43].

The worst results from an environmental point of view were obtained by the fibre reinforced geopolymers. These types of materials show very good thermal storage capacity, and thanks to the fibres, they could also acquire excellent mechanical performances. In this case, the environmental impact is considerably increased by the fibre production process, which turns out to be quite energy-intensive due to the high temperatures necessary for melting and extrusion. The use of a small aliquot of fibre (4.9% by volume, adopted in thermal simulation) shows increased environmental impact values even 10 times greater than those obtained with pure geopolymer. In this case, the nature of the fibre has a decisive influence on the environmental impact of the entire product. Results comparable to those of fibrous geopolymer composites are obtained by assessing the impacts associated with the use of A4 concrete, although in this case the environmental impact is slightly lower. This type of material turns out to be quite different from ordinary concretes as it is mainly made up of bauxite and basalt and contains a certain amount of graphite and steel fibre, although in lower weight percentage, compared to those elaborated in the present work.

A4 and composite materials exploit virgin raw materials that have a high energy production cost and, therefore, limit their use if compared to the environmental purposes of this work. However, such solutions should not be discarded based on this analysis because their thermal, mechanical, and volume/weight performances are very interesting, and therefore, an assessment over the entire life cycle would be necessary, for example, with an estimation of the lifetime (which is usually considered greater for geopolymers if compared to cement concretes) and with the possible recovery of raw materials at the end of the lifetime.

To better understand whether these materials can have a real use from an environmental point of view, it was deemed necessary to carry out a weighted comparison of the technical characteristics of the materials and the environmental impact associated with their production process, by using the Super-decisions software. The Super-decisions multi-criteria analysis results are reported in Figure 7. The values are normalized to the best material for each scenario so that the best option has a value of 1 and the others show a lower level of performance proportional to the best material.

In the first scenario, the best performance was obtained by PA0. This material showed good performance due to its elevated thermal properties and relative intermediate values from an LCA point of view. A4, Geopolymer, and PA20 showed global performance very similar to each other, although their characteristics are very different. A4 is the best thermal material among those studied, but its production is not sustainable. The good performance of Geopolymer is related to its intermediate thermal behaviour coupled with a generally low environmental impact, due to the generally low CO_2 emission and the reuse of Fly ash waste material. PA20 instead showed the worst thermal behaviour, but a very low environmental impact due to the recycling of plastics and the use of inert sludge derived from marble processing. In this case, plain concrete C also performed slightly better than FibC, characterized by better thermal performance but worse environmental impacts in the production process. Nickel fibre-reinforced geopolymer material showed the worst global performance.

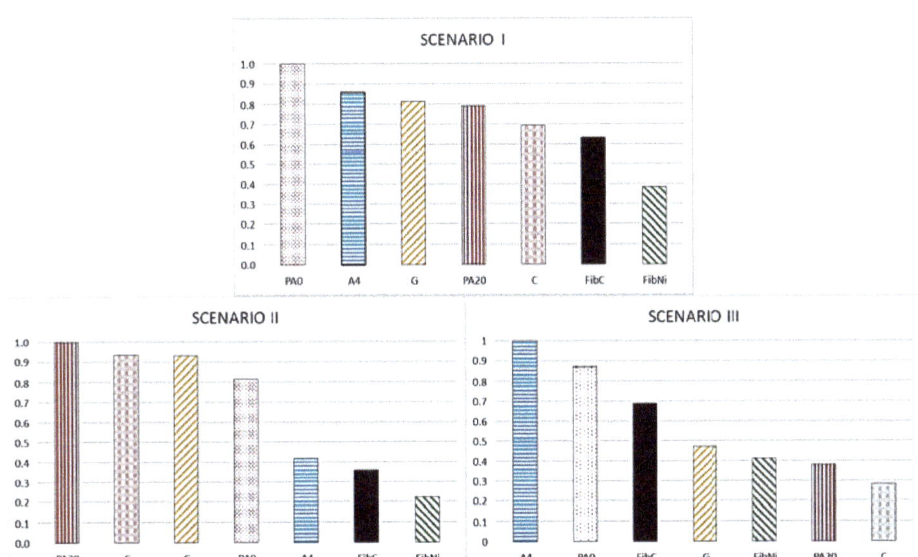

Figure 7. Results of the AHP analysis applied to the proposed SHTES materials, according to the 3 proposed scenarios.

In the second scenario (in which the environmental production impact has the highest importance), the best performance was provided by PA20 concrete, which was better than all the other materials, especially fibre-reinforced ones. In this case, concrete C and Geopolymer had similar performance, better than other materials, starting from PA0 concrete to FibNi material.

In the third scenario, the best material was A4, characterized by the best performances from the storage efficiency and ΔT_{eff} point of view. The performance of PA0 was higher than that of FibC, due to its lower environmental impact and a good value in ΔT_{eff}, although the two materials showed similar storage efficiency. Additionally, in this evaluation, Geopolymer showed good performance, better than FibNi and PA20, with the plain concrete C showing the worst. This underlines that, if the industrial investment has a very small period of use, plain concrete could be a good choice due to its low cost (not analysed in this paper), but it is strongly not recommended for long-term entrepreneurial activities.

It is worth noting that these three different scenarios are the extreme conditions. They were evaluated to simulate the decision-making process of the designers when deciding on thermal storage system materials without any sort of "flexibility" and "sensibility" represented by human factors, and the classification exposed above is a rough list based only on mathematical considerations. For example, since solid data on the durability of singular materials are not available, the weight of the environmental impact obtained during the production phase can have a more or less marked relevance depending on material lifetime, but, in any case, the three scenarios can likely represent a good simulation of the possible real cases.

4. Conclusions

Among the distinguishing features of this work is the use of a square parallelepiped as a differential storage element instead of a hollow cylinder for a more accurate simulation of heat propagation and temperature distribution inside the SHTES media. This approach allows the exact design of the volume of the fibre-reinforced concretes with enhanced thermal properties for performance evaluation. In fact, in this work, the performances of a geopolymer concrete were estimated with and without high-conductive fibres. The results demonstrate that the simulation approach can be useful to design storage modules with different types of fibres and arrangements. For example, the preliminary comparison of two different fibres, i.e., carbon and nickel, shows that metal is not always the best choice, as thermal performances are strongly affected by thermal conductivity, volumetric heat capacity, and thermal diffusivity. The storage efficiency of FibC is around 96% compared to the 94% of FibNi for the identical storage cycle. A similar comparison can be made for power density. This is due to the fact that although FibNi and FibC have the same thermal conductivity, in transient heat transfer, the higher heat capacity of FibNi requires more time and more thermal energy for charging. Further development of this approach can be used to study the optimization of fibre arrangement and properties.

From the environmental point of view, the least impacting product was found to be the benchmark PA20 concrete, obtained with an ordinary Portland cement and recycled materials as aggregates. The benchmark plain concrete and the concrete with marble sludge inert material have a similar impact factor, slightly worse than the geopolymer for CO_2 emission, but generally better in ozone depletion and ecotoxicity. Carbon fibres and Nickel fibres geopolymer matrix composites were found to be the worst, at this stage, due to the high energy demand for fibre production.

On the other hand, these products, with PA0, exhibited the best thermal behaviour from the thermal energy storage point of view, and a careful LCA analysis of the whole TES system, by considering more scenarios and the total lifetime of the system, could suggest ways to reduce the impact factors of the fibre-reinforced geopolymers to exploit their thermal storage benefits, leading to a marked improvement in the environmental performances.

Finally, the main result of this work is that although, theoretically, the addition of long fibres with high thermal properties can improve mechanical resistance and thermal performance, optimizing the charge/discharge times, the combined use of numerical simulation and LCA analysis demonstrates that the addition of fibres can worsen the environmental impact of the overall system, thus reducing the real renewability and sustainability of a storage system.

Thermal and LCA results were used in a multi-criteria decision AHP analysis to simulate the choice of the optimal material. Three different scenarios were simulated, weighing differently the incidence of the environmental impact and the thermal performance. So, depending to the case, the preferred materials are PA20 (showing the lowest environmental impact), PA0 (with a relative intermedia value of environmental impact and very interesting thermal performance), and A4 short-fibred material (the best thermal performing material).

Author Contributions: Conceptualization, D.F., C.F. and R.C.; methodology, D.F. and A.O.; software, D.F. and A.O.; formal analysis, D.F. and A.O.; investigation, D.F. and A.O.; data curation, D.F. and C.F.; writing—original draft preparation, D.F., A.O. and C.F.; writing—review and editing, all authors; supervision, R.C. All authors have read and agreed to the published version of the manuscript.

Funding: This research received no external funding.

Institutional Review Board Statement: Not applicable.

Informed Consent Statement: Not applicable.

Data Availability Statement: All data is contained within the article and referred database and/or sources.

Acknowledgments: The authors would like to acknowledge Mario Russo for his technical support for mesh refinement and simulation curation.

Conflicts of Interest: The authors declare no conflict of interest.

Appendix A

Appendix A is a supplementary section in which all the impacts related to singular species in the analysed materials are shown.

Table A1. Plain Cement Concrete (C).

Item	Unit	Total	Concrete C	Cement, CEM II/A-L 42.5R	Gravel, Crushed	Sand	Gravel, Round
Climate change	kg CO_2 eq	351.17	0.00	323.24	4.83	16.16	6.94
Ozone depletion	kg CFC-11 eq	7.82×10^{-6}	0.00	6.20×10^{-6}	2.62×10^{-7}	9.93×10^{-7}	3.65×10^{-7}
Terrestrial acidification	kg SO_2 eq	1.02	0.00	0.84	0.03	0.10	0.05
Freshwater eutrophication	kg P eq	3.18×10^{-2}	0.00	2.65×10^{-2}	1.28×10^{-3}	3.03×10^{-3}	9.92×10^{-4}
Marine eutrophication	kg N eq	4.34×10^{-2}	0.00	3.53×10^{-2}	1.30×10^{-3}	4.83×10^{-3}	1.97×10^{-3}
Human toxicity	kg 1,4-DB eq	37.90	0.00	31.26	1.42	3.82	1.41
Photochemical oxidant formation	kg NMVOC	1.05	0.00	0.84	0.03	0.13	0.05
Particulate matter formation	kg PM10 eq	0.40	0.00	0.32	0.01	0.04	0.02
Terrestrial ecotoxicity	kg 1,4-DB eq	9.99×10^{-3}	0.00	7.75×10^{-3}	3.98×10^{-4}	1.35×10^{-3}	5.01×10^{-4}
Freshwater ecotoxicity	kg 1,4-DB eq	0.83	0.00	0.62	0.05	0.12	0.04
Marine ecotoxicity	kg 1,4-DB eq	0.90	0.00	0.69	0.04	0.13	0.05
Ionising radiation	kBq U235 eq	18.46	0.00	15.49	0.66	1.79	0.51
Agricultural land occupation	m^2a	2.45	0.00	1.91	0.11	0.31	0.13
Urban land occupation	m^2a	3.27	0.00	1.46	0.24	1.12	0.45
Natural land transformation	m^2	0.04	0.00	0.02	0.00	0.01	0.00
Water depletion	m^3	417.59	0.12	368.09	11.94	27.02	10.42
Metal depletion	kg Fe eq	6.55	0.00	4.07	0.49	1.42	0.57
Fossil depletion	kg oil eq	42.84	0.00	34.23	1.41	5.12	2.08

Table A2. Cement concrete with limestone waste recovery (PA0).

Item	Unit	Total	Concrete PA0	Cement, CEM II/A-L 42.5R	Limestone, Crushed, Washed	Fly Ash	Limestone Residue	Superplasticizer *
Climate change	kg CO_2 eq	408.63	0.00	377.28	19.20	0.06	−0.83	12.93
Ozone depletion	kg CFC-11 eq	6.80×10^{-6}	0.00	5.66×10^{-6}	1.22×10^{-6}	6.76×10^{-9}	-9.66×10^{-8}	4.56×10^{-9}
Terrestrial acidification	kg SO_2 eq	1.25	0.00	1.09	0.17	0.00	−0.01	0.00
Freshwater eutrophication	kg P eq	0.03	0.00	3.04×10^{-2}	2.48×10^{-3}	9.27×10^{-6}	-7.04×10^{-5}	6.25×10^{-6}
Marine eutrophication	kg N eq	0.05	0.00	4.14×10^{-2}	8.45×10^{-3}	7.63×10^{-6}	-2.92×10^{-4}	5.09×10^{-6}
Human toxicity	kg 1,4-DB eq	37.67	0.00	34.23	3.45	0.01	−0.08	0.06
Photochemical oxidant formation	kg NMVOC	1.22	0.00	1.00	2.20×10^{-1}	0.00	−0.01	0.00
Particulate matter formation	kg PM10 eq	0.58	0.00	0.42	1.60×10^{-1}	0.00	0.00	4.43×10^{-5}
Terrestrial ecotoxicity	kg 1,4-DB eq	0.01	0.00	0.01	1.60×10^{-3}	0.00	0.00	2.54×10^{-5}
Freshwater ecotoxicity	kg 1,4-DB eq	0.77	0.00	0.67	1.09×10^{-1}	0.00	0.00	1.65×10^{-4}
Marine ecotoxicity	kg 1,4-DB eq	0.85	0.00	0.73	0.12	0.00	0.00	2.28×10^{-4}
Ionising radiation	kBq U235 eq	11.82	0.00	10.25	1.64	0.01	−0.09	0.01
Agricultural land occupation	m^2a	2.11	0.00	1.90	3.41×10^{-1}	0.00	−0.14	0.00
Urban land occupation	m^2a	2.47	0.00	1.70	9.16×10^{-1}	0.00	−0.15	0.00
Natural land transformation	m^2	0.04	0.00	0.03	5.41×10^{-3}	0.00	5.95×10^{-3}	6.30×10^{-6}
Water depletion	m^3	458.14	0.15	412.78	45.28	0.28	−0.58	0.23
Metal depletion	kg Fe eq	5.34	0.00	4.06	1.32	0.00	−0.04	0.00
Fossil depletion	kg oil eq	44.60	0.00	38.84	6.26	0.02	−0.53	0.01

* Superplasticizer impacts were taken from the European Federation of Concrete Admixtures Associations Ltd. (EFCA) EPD®.

Table A3. Cement concrete with limestone and plastic waste recovery (PA20).

Item	Unit	Total	Concrete PA20	Cement, CEM II/A-L 42.5R	Limestone, Crushed, Washed	Fly Ash	Waste Plastic Mixture	Limestone Residue	Superplasticizer *
Climate change	kg CO_2 eq	78.41	0.00	377.26	14.20	0.06	−328.91	−0.98	16.78
Ozone depletion	kg CFC-11 eq	4.47×10^{-6}	0.00	5.66×10^{-6}	9.07×10^{-7}	6.76×10^{-9}	-2.00×10^{-6}	-1.13×10^{-7}	5.91×10^{-9}
Terrestrial acidification	kg SO_2 eq	1.13	0.00	1.09	0.12	2.04×10^{-4}	-7.51×10^{-2}	-6.70×10^{-3}	1.78×10^{-4}
Freshwater eutrophication	kg P eq	0.03	0.00	0.03	1.80×10^{-3}	9.27×10^{-6}	-2.00×10^{-3}	-8.25×10^{-5}	8.11×10^{-6}
Marine eutrophication	kg N eq	0.03	0.00	0.04	6.27×10^{-3}	7.63×10^{-6}	-1.38×10^{-2}	-3.42×10^{-4}	6.60×10^{-6}
Human toxicity	kg 1,4-DB eq	5.41	0.00	34.23	2.53	8.29×10^{-3}	−31.34	-9.80×10^{-2}	0.07
Photochemical oxidant formation	kg NMVOC	1.06	0.00	1.00	1.64×10^{-1}	1.31×10^{-4}	−0.10	-1.04×10^{-2}	4.84×10^{-3}
Particulate matter formation	kg PM10 eq	0.51	0.00	0.42	1.19×10^{-1}	6.64×10^{-5}	-2.93×10^{-2}	-3.06×10^{-3}	5.75×10^{-5}
Terrestrial ecotoxicity	kg 1,4-DB eq	0.00	0.00	0.01	1.18×10^{-3}	7.64×10^{-6}	-6.12×10^{-3}	-4.53×10^{-5}	3.30×10^{-5}
Freshwater ecotoxicity	kg 1,4-DB eq	−4.84	0.00	0.67	8.02×10^{-2}	2.30×10^{-4}	−5.58	-2.94×10^{-3}	2.14×10^{-4}
Marine ecotoxicity	kg 1,4-DB eq	−3.91	0.00	0.73	8.98×10^{-2}	2.25×10^{-4}	−4.73	-3.73×10^{-3}	2.96×10^{-4}
Ionising radiation	kBq U235 eq	10.46	0.00	10.25	1.20	0.01	−0.90	−0.11	0.01
Agricultural land occupation	m^2a	1.83	0.00	1.90	0.25	0.00	−0.16	−0.16	0.00
Urban land occupation	m^2a	2.13	0.00	1.70	0.68	0.00	−0.08	−0.18	0.00
Natural land transformation	m^2	0.04	0.00	2.61×10^{-2}	4.01×10^{-3}	9.45×10^{-6}	-6.58×10^{-4}	6.97×10^{-3}	8.17×10^{-6}
Water depletion	m^3	420.59	0.15	412.76	27.45	0.28	−19.67	−0.68	0.30
Metal depletion	kg Fe eq	4.43	0.00	4.06	0.98	0.00	−0.56	−0.04	0.00
Fossil depletion	kg oil eq	41.41	0.00	38.84	4.63	0.02	−1.48	−0.62	0.01

* Superplasticizer impacts were taken from European Federation of Concrete Admixtures Associations Ltd. (EFCA) EPD®® [46].

Table A4. Geopolymer concrete (G).

Item	Unit	Total	Limestone, Crushed, Washed	Fly Ash	Sodium Silicate, without Water, in 37% Solution State	Electricity, Medium Voltage
Climate change	kg CO_2 eq	269.29	15.02	0.19	239.47	14.61
Ozone depletion	kg CFC-11 eq	1.36×10^{-5}	9.58×10^{-7}	2.36×10^{-8}	1.08×10^{-5}	1.79×10^{-6}
Terrestrial acidification	kg SO_2 eq	1.57	0.13	7.12×10^{-4}	1.39	5.39×10^{-2}
Freshwater eutrophication	kg P eq	0.09	1.94×10^{-3}	3.24×10^{-5}	8.65×10^{-2}	2.46×10^{-3}
Marine eutrophication	kg N eq	0.07	6.62×10^{-3}	2.66×10^{-5}	5.71×10^{-2}	2.00×10^{-3}
Human toxicity	kg 1,4-DB eq	141.78	2.70	2.89×10^{-2}	1.37×10^{-2}	2.19
Photochemical oxidant formation	kg NMVOC	1.03	0.17	4.58×10^{-4}	0.82	0.03
Particulate matter formation	kg PM10 eq	0.61	0.13	2.32×10^{-4}	0.46	0.02
Terrestrial ecotoxicity	kg 1,4-DB eq	0.02	1.25×10^{-3}	2.67×10^{-5}	0.02	2.03×10^{-3}
Freshwater ecotoxicity	kg 1,4-DB eq	4.17	8.50×10^{-2}	8.03×10^{-4}	4.02	0.06
Marine ecotoxicity	kg 1,4-DB eq	4.14	9.50×10^{-2}	7.85×10^{-4}	3.99	0.06
Ionising radiation	kBq U235 eq	19.03	1.28	3.09×10^{-2}	15.36	2.35
Agricultural land occupation	m^2a	18.45	0.27	3.27×10^{-3}	17.93	0.25
Urban land occupation	m^2a	4.48	0.72	7.65×10^{-4}	3.72	0.05
Natural land transformation	m^2	0.05	4.23×10^{-3}	3.30×10^{-5}	0.04	2.48×10^{-3}
Water depletion	m^3	441.39	35.43	9.82×10^{-1}	329.98	74.99
Metal depletion	kg Fe eq	33.23	1.03	5.10×10^{-3}	31.82	0.38
Fossil depletion	kg oil eq	65.16	4.89	5.95×10^{-2}	55.70	4.50

Table A5. Geopolymer concrete with carbon fibre (FibC).

Item	Unit	Total	Limestone, Crushed, Washed	Fly Ash	Sodium Silicate	Carbon PAN Fibre	Electricity, Medium Voltage (IT)
Climate change	kg CO_2 eq	2107.91	14.38	0.18	229.29	1849.46	14.60
Ozone depletion	kg CFC-11 eq	2.41×10^{-4}	9.17×10^{-7}	2.26×10^{-8}	1.03×10^{-5}	2.28×10^{-4}	1.79×10^{-6}
Terrestrial acidification	kg SO_2 eq	8.32	0.12	6.82×10^{-4}	1.33	6.81	0.05
Freshwater eutrophication	kg P eq	0.39	1.86×10^{-3}	3.10×10^{-5}	0.08	0.30	2.46×10^{-3}
Marine eutrophication	kg N eq	0.32	6.33×10^{-3}	2.55×10^{-5}	0.05	0.25	2.00×10^{-3}
Human toxicity	kg 1,4-DB eq	386.96	2.59	0.03	131.04	251.12	2.19
Photochemical oxidant formation	kg NMVOC	5.32	0.17	4.39×10^{-4}	0.79	4.33	3.41×10^{-2}
Particulate matter formation	kg PM10 eq	2.79	0.12	2.22×10^{-4}	0.44	2.21	1.74×10^{-2}
Terrestrial ecotoxicity	kg 1,4-DB eq	0.28	1.20×10^{-3}	2.55×10^{-5}	1.84×10^{-2}	0.26	2.03×10^{-3}
Freshwater ecotoxicity	kg 1,4-DB eq	11.28	8.14×10^{-2}	7.69×10^{-4}	3.85	7.28	6.04×10^{-2}
Marine ecotoxicity	kg 1,4-DB eq	11.02	9.10×10^{-2}	7.51×10^{-4}	3.82	7.05	5.90×10^{-2}
Ionising radiation	kBq U235 eq	317.69	1.23	2.96×10^{-2}	1.47×10^{1}	299.37	2.35
Agricultural land occupation	m^2a	48.77	0.26	3.13×10^{-3}	1.72×10^{1}	31.09	0.25
Urban land occupation	m^2a	10.15	0.69	7.32×10^{-4}	3.56	5.86	4.64×10^{-2}
Natural land transformation	m^2	0.36	4.05×10^{-3}	3.16×10^{-5}	3.95×10^{-2}	0.32	2.48×10^{-3}
Water depletion	m^3	9954.70	33.92	0.94	3.16×10^{2}	9528.94	74.95
Metal depletion	kg Fe eq	73.97	0.99	4.88×10^{-3}	3.05×10^{1}	42.13	0.38
Fossil depletion	kg oil eq	635.34	4.69	0.06	5.33×10^{1}	572.76	4.50

Table A6. Geopolymer concrete with Nickel fibre (FibNi).

Item	Unit	Total	Limestone, Crushed, Washed	Fly Ash	Sodium Silicate, without Water, in 37% Solution State	Fe-Ni-Cr Alloy	Electricity Medium Voltage
Climate change	kg CO_2 eq	2.64×10^{3}	14.38	0.18	229.29	2371.01	14.60
Ozone depletion	kg CFC-11 eq	1.18×10^{-4}	9.17×10^{-7}	2.26×10^{-8}	1.03×10^{-5}	1.03×10^{-4}	1.79×10^{-6}
Terrestrial acidification	kg SO_2 eq	4.32×10^{2}	0.12	6.82×10^{-4}	1.33	430.63	0.05
Freshwater eutrophication	kg P eq	5.20	1.86×10^{-3}	3.10×10^{-5}	0.08	5.11	2.46×10^{-3}
Marine eutrophication	kg N eq	1.50	6.33×10^{-3}	2.55×10^{-5}	0.05	1.43	2.00×10^{-3}
Human toxicity	kg 1,4-DB eq	9.04×10^{3}	2.59	0.03	131.04	8899.00	2.19
Photochemical oxidant formation	kg NMVOC	5.02×10^{1}	0.17	4.39×10^{-4}	0.79	49.22	0.03
Particulate matter formation	kg PM10 eq	9.82×10^{1}	0.12	2.22×10^{-4}	0.44	97.64	0.02
Terrestrial ecotoxicity	kg 1,4-DB eq	1.50	1.20×10^{-3}	2.55×10^{-5}	0.02	1.48	2.03×10^{-3}
Freshwater ecotoxicity	kg 1,4-DB eq	2.86×10^{2}	8.14×10^{-2}	7.69×10^{-4}	3.85	282.33	6.04×10^{-2}
Marine ecotoxicity	kg 1,4-DB eq	2.93×10^{2}	9.10×10^{-2}	7.51×10^{-4}	3.82	289.46	5.90×10^{-2}
Ionising radiation	kBq U235 eq	3.18×10^{2}	1.23	2.96×10^{-2}	14.71	298.18	2.35
Agricultural land occupation	m^2a	1.40×10^{2}	0.26	3.13×10^{-3}	17.17	122.02	0.25
Urban land occupation	m^2a	6.44×10^{1}	0.69	7.32×10^{-4}	3.56	60.05	0.05
Natural land transformation	m^2	4.47×10^{-1}	0.00	3.16×10^{-5}	0.04	0.40	0.00
Water depletion	m^3	5.10×10^{4}	33.92	0.94	315.96	50,494.71	74.95
Metal depletion	kg Fe eq	9.41×10^{3}	0.99	4.88×10^{-3}	30.46	9379.32	0.38
Fossil depletion	kg oil eq	6.48×10^{2}	4.69	5.70×10^{-2}	53.33	582.52	4.50

Table A7. A4 material.

Item	Unit	Total	Basalt	Bauxite	Graphite	Calcium Aluminate Cement *	Silica Sand	Aluminium Micropowder	Steel, Low-Alloyed, Hot Rolled	Electricity Medium Voltage
Climate change	kg CO_2 eq	2548.18	17.73	152.05	10.83	302.90	3.46	1829.45	188.51	43.24
Ozone depletion	kg CFC-11 eq	0.00	1.10×10^{-6}	9.84×10^{-6}	6.06×10^{-7}	7.18×10^{-9}	2.41×10^{-7}	1.54×10^{-4}	7.44×10^{-6}	5.33×10^{-6}
Terrestrial acidification	kg SO_2 eq	15.17	0.14	1.94	0.08	0.00	0.02	11.97	0.86	0.16
Freshwater eutrophication	kg P eq	1.12	3.42×10^{-3}	1.48×10^{-2}	2.69×10^{-3}	9.84×10^{-6}	6.28×10^{-4}	9.76×10^{-1}	1.11×10^{-1}	7.04×10^{-3}
Marine eutrophication	kg N eq	0.81	0.01	0.06	3.10×10^{-3}	8.02×10^{-6}	7.12×10^{-4}	0.69	0.04	5.90×10^{-3}
Human toxicity	kg 1,4-DB eq	1182.36	3.56	16.77	2.66	0.37	0.74	981.76	170.62	5.87
Photochemical oxidant formation	kg NMVOC	9.46	0.17	1.79	0.07	0.11	0.02	6.29	0.90	0.10
Particulate matter formation	kg PM10 eq	6.79	0.10	0.75	0.04	6.98×10^{-5}	0.01	5.16	0.68	0.05
Terrestrial ecotoxicity	kg 1,4-DB eq	0.13	0.00	1.13×10^{-2}	0.00	1.52×10^{-4}	0.00	0.08	2.05×10^{-3}	5.96×10^{-3}
Freshwater ecotoxicity	kg 1,4-DB eq	53.26	0.11	0.54	0.07	3.24×10^{-4}	0.02	46.29	6.05	0.17
Marine ecotoxicity	kg 1,4-DB eq	49.78	0.12	0.81	0.08	7.94×10^{-4}	0.02	42.63	5.96	0.16
Ionising radiation	kBq U235 eq	176.59	2.35	13.95	1.54	9.41×10^{-3}	0.29	138.51	12.95	7.00
Agricultural land occupation	m^2a	48.19	0.26	1.50	0.28	9.88×10^{-4}	0.13	41.44	3.86	0.73
Urban land occupation	m^2a	47.68	8.41	4.47	0.33	1.86×10^{-4}	0.11	31.30	2.93	0.14
Natural land transformation	m^2	0.83	0.48	0.03	0.00	9.92×10^{-6}	0.00	0.29	0.02	7.37×10^{-3}
Water depletion	m^3	15,296.41	33.07	139.41	23.36	0.90	3.06	13,820.80	1053.02	222.79
Metal depletion	kg Fe eq	356.72	0.87	29.13	0.66	1.53×10^{-3}	0.20	73.91	250.97	0.99
Fossil depletion	kg oil eq	533.72	5.60	50.18	3.18	0.02	0.95	417.10	43.30	13.39

* Calcium aluminate cement impacts were taken from çimsa RESISTO40®® Calcium Aluminate Cement EPD® [45].

Appendix B

Appendix B is the section in which all the criteria ratios for Super-decision analysis are shown.

Table A8. Climate change criteria.

	PA20	PA0	C	A4	G	FibC	FibNi
PA20	1	3	4	8	2	7	9
PA0	1/3	1	2	5	1/3	4	6
C	1/4	1/2	1	4	1/2	4	5
A4	1/8	1/5	1/4	1	1/7	$\frac{1}{2}$	2
G	1/2	3	2	7	1	5	8
FibC	1/9	1/6	1/6	2	1/5	1	3
FibNi	1/7	1/4	1/4	1/2	1/8	1/3	1

Inconsistency 0.0227.

Table A9. Ozone depletion criteria.

	PA20	PA0	C	A4	G	FibC	FibNi
PA20	1	1	1/2	6	3	7	7
PA0	1	1	1/3	6	3	7	7
C	3	3	1	8	2	9	7
A4	1/6	1/6	1/8	1	1/7	2	1/2
G	1/3	1/3	1/2	7	1	8	2
FibC	1/7	1/7	1/9	1/2	1/8	1	1/2
FibNi	1/7	1/5	1/7	2	1/3	2	1

Inconsistency 0.0298.

Table A10. Particulate matter criteria.

	PA20	PA0	C	A4	G	FibC	FibNi
PA20	1	1	1/3	4	1/3	3	7
PA0	1	1	1/3	4	1/3	3	6
C	3	3	1	6	1	3	9
A4	1/4	1/4	1/6	1	1/6	1	2
G	2	3	1	6	1	3	9
FibC	1/3	1/3	1/9	1	1/9	1	2
FibNi	1/7	1/6	1/7	1/2	1/7	1/2	1

Inconsistency 0.00775.

Table A11. Photochemical oxidant formation criteria.

	PA20	PA0	C	A4	G	FibC	FibNi
PA20	1	1	1/2	3	1/2	2	7
PA0	1	1	1/2	3	1/2	2	7
C	2	2	1	4	1	3	9
A4	1/3	1/3	1/4	1	1/4	1/2	2
G	2	2	1	4	1	3	9
FibC	1/2	1/2	1/3	2	1/2	1	5
FibNi	1/7	1/7	1/9	1/2	1/9	1/5	1

Inconsistency 0.0106.

Table A12. Marine ecotoxicity criteria.

	PA20	PA0	C	A4	G	FibC	FibNi
PA20	1	2	1/2	7	1/2	4	9
PA0	1/2	1	1/2	5	1/2	3	7
C	1/2	1	1	5	1	3	7
A4	1/7	1/5	1/4	1	1/4	1/2	2
G	1/3	1/2	1	4	1	2	5
FibC	1/4	1/3	1/3	2	1/3	1	2
FibNi	1/7	1/7	1/9	1/2	1/9	1/5	1

Inconsistency 0.0158.

Table A13. Fossil depletion criteria.

	PA20	PA0	C	A4	G	FibC	FibNi
PA20	1	2	2	7	3	9	9
PA0	1/2	1	1	6	2	7	7
C	1/2	1	1	6	2	7	7
A4	1/7	1/6	1/6	1	1/4	2	2
G	1/3	1/2	1/2	4	1	5	5
FibC	1/9	1/7	1/7	1/2	1/7	1	1
FibNi	1/9	1/7	1/7	1	1/7	1	1

Inconsistency 0.0165.

Table A14. Storage efficiency criteria.

	PA20	PA0	C	A4	G	FibC	FibNi
PA20	1	1/4	3	1/6	1/2	1/4	1/3
PA0	4	1	8	1/2	3	1	2
C	1/3	1/8	1	1/9	1/4	1/8	1/6
A4	6	2	9	1	4	2	3
G	2	1/3	4	1/4	1	1/3	1/2
FibC	4	1	8	1/2	3	1	2
FibNi	3	1/2	6	1/3	2	1/2	1

Inconsistency 0.01567.

Table A15. ΔT_{eff}.

	PA20	PA0	C	A4	G	FibC	FibNi
PA20	1	1/4	2	1/4	1/2	1/3	1
PA0	4	1	7	1	3	2	4
C	1/2	1/7	1	1/7	1/4	1/5	1/2
A4	4	1	7	1	3	2	4
G	2	1/3	4	1/3	1	1/2	2
FibC	3	1/2	5	1/2	2	1	3
FibNi	1	1/4	2	1/4	1/2	1/3	1

Inconsistency 0.0035.

Table A16. LCA criteria.

	Climate Change	Ozone Depletion	Particulate Matter	Photochemical Oxidant Formation	Marine Ecotoxicity	Fossil Depletion
Climate change	1	3	4	4	5	2
ozone depletion	1/3	1	2	2	3	1/2
particulate matter	1/4	1/2	1	1	2	1/3
Photochemical oxidant formation	1/4	1/2	1	1	2	1/3
Marine ecotoxicity	1/5	1/3	1/2	1/2	1	$\frac{1}{4}$
Fossil depletion	1/2	2	3	3	4	1

Inconsistency 0.01151.

Table A17. Thermal performance criteria.

	Storage Efficiency	ΔT_{eff}
Storage efficiency	1	3
ΔT_{eff}	1/3	1

Table A18. Local normalized priority.

	Climate Change	Ozone Depletion	Particulate Matter	Photochemical Oxidant Formation	Marine Ecotoxicity	Fossil Depletion	Storage Efficiency	ΔT_{eff}
A4	0.04	0.03	0.04	0.06	0.03	0.04	0.32	0.29
C	0.16	0.35	0.28	0.26	0.21	0.31	0.02	0.03
FibC	0.05	0.02	0.06	0.09	0.07	0.03	0.20	0.17
FibNi	0.03	0.04	0.02	0.02	0.03	0.03	0.12	0.06
G	0.24	0.24	0.28	0.26	0.13	0.20	0.08	0.10
PA20	0.36	0.15	0.16	0.15	0.33	0.20	0.05	0.06
PA0	0.13	0.15	0.16	0.15	0.21	0.20	0.20	0.30

Table A19. Global results for the three different scenarios from Super-decisions analysis.

LCA: Thermal Performance 50:50			
Name	**Ideals**	**Normals**	**Raw**
A4	0.87	0.17	0.06
C	0.70	0.14	0.05
FibC	0.63	0.12	0.04
FibNi	0.38	0.07	0.02
G	0.82	0.16	0.05
PA20	0.79	0.15	0.05
PA0	1.00	0.19	0.06
LCA: Thermal Performance 80:20			
Name	**Ideals**	**Normals**	**Raw**
A4	0.42	0.09	0.03
C	0.93	0.20	0.07
FibC	0.36	0.08	0.03
FibNi	0.22	0.05	0.02
G	0.93	0.20	0.07
PA20	1.00	0.21	0.07
PA0	0.82	0.17	0.06
LCA: Thermal Performance 20:80			
Name	**Ideals**	**Normals**	**Raw**
A4	1.00	0.24	0.08
C	0.28	0.07	0.02
FibC	0.68	0.17	0.06
FibNi	0.41	0.10	0.03
G	0.47	0.11	0.04
PA20	0.37	0.09	0.03
PA0	0.87	0.21	0.07

References

1. Medrano, M.; Gil, A.; Martorell, I.; Potau, X.; Cabeza, L.F. State of the art on high-temperature thermal energy storage for power generation. Part 2-Case studies. *Renew. Sustain. Energy Rev.* **2010**, *14*, 56–72. [CrossRef]
2. Gil, A.; Medrano, M.; Martorell, I.; Lázaro, A.; Dolado, P.; Zalba, B.; Cabeza, L.F. State of the art on high temperature thermal energy storage for power generation. Part 1-Concepts, materials and modellization. *Renew. Sustain. Energy Rev.* **2010**, *14*, 31–55. [CrossRef]

3. Li, G. Sensible heat thermal storage energy and exergy performance evaluations. *Renew. Sustain. Energy Rev.* **2016**, *53*, 897–923. [CrossRef]
4. Natali, A.; Manzi, S.; Bignozzi, M.C. Novel fiber-reinforced composite materials based on sustainable geopolymer matrix. *Procedia Eng.* **2011**, *21*, 1124–1131. [CrossRef]
5. Kong, D.L.Y.; Sanjayan, J.G. Damage behavior of geopolymer composites exposed to elevated temperatures. *Cem. Concr. Compos.* **2008**, *30*, 986–991. [CrossRef]
6. Celik, A.; Yilmaz, K.; Canpolat, O.; Al-mashhadani, M.M.; Aygörmez, Y.; Uysal, M. High-temperature behavior and mechanical characteristics of boron waste additive metakaolin based geopolymer composites reinforced with synthetic fibers. *Constr. Build. Mater.* **2018**, *187*, 1190–1203. [CrossRef]
7. Ferone, C.; Colangelo, F.; Frattini, D.; Roviello, G.; Cioffi, R.; Maggio, R. Finite Element Method Modeling of Sensible Heat Thermal Energy Storage with Innovative Concretes and Comparative Analysis with Literature Benchmarks. *Energies* **2014**, *7*, 5291–5316. [CrossRef]
8. Prasad, L.; Muthukumar, P. Design and optimization of lab-scale sensible heat storage prototype for solar thermal power plant application. *Sol. Energy* **2013**, *97*, 217–229. [CrossRef]
9. Salomoni, V.A.; Majorana, C.E.; Giannuzzi, G.M.; Miliozzi, A.; Di Maggio, R.; Girardi, F.; Mele, D.; Lucentini, M. Thermal storage of sensible heat using concrete modules in solar power plants. *Sol. Energy* **2014**, *103*, 303–315. [CrossRef]
10. Tamme, R.; Laing, D.; Steinmann, W.D. Advanced thermal energy storage technology for parabolic trough. *J. Sol. Energy Eng. Trans. ASME* **2004**, *126*, 794–800. [CrossRef]
11. Bai, F.; Xu, C. Performance analysis of a two-stage thermal energy storage system using concrete and steam accumulator. *Appl. Therm. Eng.* **2011**, *31*, 2764–2771. [CrossRef]
12. Laing, D.; Bahl, C.; Bauer, T.; Fiss, M.; Breidenbach, N.; Hempel, M. High-temperature solid-media thermal energy storage for solar thermal power plants. *Proc. IEEE* **2012**, *100*, 516–524. [CrossRef]
13. Colangelo, F.; Cioffi, R.; Roviello, G.; Capasso, I.; Caputo, D.; Aprea, P.; Liguori, B.; Ferone, C. Thermal cycling stability of fly ash based geopolymer mortars. *Compos. Part B Eng.* **2017**, *129*, 11–17. [CrossRef]
14. Ferone, C.; Liguori, B.; Capasso, I.; Colangelo, F.; Cioffi, R.; Cappelletto, E.; Di Maggio, R. Thermally treated clay sediments as geopolymer source material. *Appl. Clay Sci.* **2015**, *107*, 195–204. [CrossRef]
15. Pouhet, R.; Cyr, M. Formulation and performance of flash metakaolin geopolymer concretes. *Constr. Build. Mater.* **2016**, *120*, 150–160. [CrossRef]
16. Nenadović, S.S.; Ferone, C.; Nenadović, M.T.; Cioffi, R.; Mirković, M.M.; Vukanac, I.; Kljajević, L.M. Chemical, physical and radiological evaluation of raw materials and geopolymers for building applications. *J. Radioanal. Nucl. Chem.* **2020**, *325*, 435–445. [CrossRef]
17. Liguori, B.; Capasso, I.; De Pertis, M.; Ferone, C.; Cioffi, R. Geopolymerization Ability of Natural and Secondary Raw Materials by Solubility Test in Alkaline Media. *Environments* **2017**, *4*, 56. [CrossRef]
18. Mehta, A.; Siddique, R. An overview of geopolymers derived from industrial by-products. *Constr. Build. Mater.* **2016**, *127*, 183–198. [CrossRef]
19. Ferone, C.; Capasso, I.; Bonati, A.; Roviello, G.; Montagnaro, F.; Santoro, L.; Turco, R.; Cioffi, R. Sustainable management of water potabilization sludge by means of geopolymers production. *J. Clean. Prod.* **2019**, *229*, 1–9. [CrossRef]
20. Capasso, I.; Lirer, S.; Flora, A.; Ferone, C.; Cioffi, R.; Caputo, D.; Liguori, B. Reuse of mining waste as aggregates in fly ash-based geopolymers. *J. Clean. Prod.* **2019**, *220*, 65–73. [CrossRef]
21. Provis, J.L.; van Deventer, J.S.J. *Geopolymers*, 1st ed.; Woodhead Publishing Limited: Cambridge, UK, 2009; ISBN 9781845694494.
22. Zhang, P.; Wang, K.; Li, Q.; Wang, J.; Ling, Y. Fabrication and engineering properties of concretes based on geopolymers/alkali-activated binders—A review. *J. Clean. Prod.* **2020**, *258*, 120896. [CrossRef]
23. Ranjbar, N.; Zhang, M. Fiber-reinforced geopolymer composites: A review. *Cem. Concr. Compos.* **2020**, *107*, 103498. [CrossRef]
24. Silva, G.; Kim, S.; Bertolotti, B.; Nakamatsu, J.; Aguilar, R. Optimization of a reinforced geopolymer composite using natural fibers and construction wastes. *Constr. Build. Mater.* **2020**, *258*, 119697. [CrossRef]
25. Ferone, C.; Roviello, G.; Colangelo, F.; Cioffi, R.; Tarallo, O. Novel hybrid organic-geopolymer materials. *Appl. Clay Sci.* **2013**, *73*, 42–50. [CrossRef]
26. Roviello, G.; Menna, C.; Tarallo, O.; Ricciotti, L.; Ferone, C.; Colangelo, F.; Asprone, D.; di Maggio, R.; Cappelletto, E.; Prota, A.; et al. Preparation, structure and properties of hybrid materials based on geopolymers and polysiloxanes. *Mater. Des.* **2015**, *87*, 82–94. [CrossRef]
27. Roviello, G.; Ricciotti, L.; Ferone, C.; Colangelo, F.; Tarallo, O. Fire resistant melamine based organic-geopolymer hybrid composites. *Cem. Concr. Compos.* **2015**, *59*, 89–99. [CrossRef]
28. Roviello, G.; Ricciotti, L.; Tarallo, O.; Ferone, C.; Colangelo, F.; Roviello, V.; Cioffi, R. Innovative Fly Ash Geopolymer-Epoxy Composites: Preparation, Microstructure and Mechanical Properties. *Materials* **2016**, *9*, 461. [CrossRef]
29. Roviello, G.; Menna, C.; Tarallo, O.; Ricciotti, L.; Messina, F.; Ferone, C.; Asprone, D.; Cioffi, R. Lightweight geopolymer-based hybrid materials. *Compos. Part B Eng.* **2017**, *128*, 225–237. [CrossRef]
30. Roviello, G.; Ricciotti, L.; Molino, A.J.; Menna, C.; Ferone, C.; Cioffi, R.; Tarallo, O. Hybrid Geopolymers from Fly Ash and Polysiloxanes. *Molecules* **2019**, *24*, 3510. [CrossRef]

31. Roviello, G.; Ricciotti, L.; Molino, A.J.; Menna, C.; Ferone, C.; Asprone, D.; Cioffi, R.; Ferrandiz-Mas, V.; Russo, P.; Tarallo, O. Hybrid Fly Ash-Based Geopolymeric Foams: Microstructural, Thermal and Mechanical Properties. *Materials* **2020**, *13*, 2919. [CrossRef]
32. Colangelo, F.; Roviello, G.; Ricciotti, L.; Ferrándiz-Mas, V.; Messina, F.; Ferone, C.; Tarallo, O.; Cioffi, R.; Cheeseman, C.R. Mechanical and thermal properties of lightweight geopolymer composites. *Cem. Concr. Compos.* **2018**, *86*, 266–272. [CrossRef]
33. Del Borghi, A. LCA and communication: Environmental Product Declaration. *Int. J. Life Cycle Assess.* **2013**, *18*, 293–295. [CrossRef]
34. Jeswani, H.K.; Azapagic, A.; Schepelmann, P.; Ritthoff, M. Options for broadening and deepening the LCA approaches. *J. Clean. Prod.* **2010**, *18*, 120–127. [CrossRef]
35. Martín-Gamboa, M.; Iribarren, D.; García-Gusano, D.; Dufour, J. A review of life-cycle approaches coupled with data envelopment analysis within multi-criteria decision analysis for sustainability assessment of energy systems. *J. Clean. Prod.* **2017**, *150*, 164–174. [CrossRef]
36. Campos-Guzmán, V.; García-Cáscales, M.S.; Espinosa, N.; Urbina, A. Life Cycle Analysis with Multi-Criteria Decision Making: A review of approaches for the sustainability evaluation of renewable energy technologies. *Renew. Sustain. Energy Rev.* **2019**, *104*, 343–366. [CrossRef]
37. Khoshnava, S.M.; Rostami, R.; Valipour, A.; Ismail, M.; Rahmat, A.R. Rank of green building material criteria based on the three pillars of sustainability using the hybrid multi criteria decision making method. *J. Clean. Prod.* **2018**, *173*, 82–99. [CrossRef]
38. Laing, D.; Steinmann, W.D.; Tamme, R.; Richter, C. Solid media thermal storage for parabolic trough power plants. *Sol. Energy* **2006**, *80*, 1283–1289. [CrossRef]
39. Jian, Y.; Falcoz, Q.; Neveu, P.; Bai, F.; Wang, Y.; Wang, Z. Design and optimization of solid thermal energy storage modules for solar thermal power plant applications. *Appl. Energy* **2015**, *139*, 30–42. [CrossRef]
40. Frattini, D.; Ferone, C.; Colangelo, F.; de Pertis, M.; Cioffi, R.; Di Maggio, R. Computational evaluation of different construction materials performance in thermal energy storage systems. In Proceedings of the third International Conference on Computational Methods for Thermal Problems, Lake Bled, Slovenia, 2–4 June 2014; pp. 459–462.
41. Callister, W.D., Jr. Appendix B—Selected Properties of Materials. In *Materials Science and Engineering—An Introduction*, 5th ed.; John Wiley & Sons, Ltd: Hoboken, NJ, USA, 1999.
42. Guo, C.; Zhu, J.; Zhou, W.; Chen, W. Fabrication and thermal properties of a new heat storage concrete material. *J. Wuhan Univ. Technol. Sci. Ed.* **2010**, *25*, 628–630. [CrossRef]
43. Ricciotti, L.; Occhicone, A.; Petrillo, A.; Ferone, C.; Cioffi, R.; Roviello, G. Geopolymer-based hybrid foams: Lightweight materials from a sustainable production process. *J. Clean. Prod.* **2020**, *250*, 119588. [CrossRef]
44. Habert, G.; Ouellet-Plamondon, C. Recent update on the environmental impact of geopolymers. *RILEM Tech. Lett.* **2016**, *1*, 17. [CrossRef]
45. Çimsa RESISTO40(R)—Calcium Aluminate Cement. Available online: https://epdturkey.org/wp-content/uploads/cimsa_isidac40_cimento_epdturkey_en.pdf (accessed on 23 December 2020).
46. Concrete Admixtures—Plasticisers and Superplasticisers. Available online: https://epd-online.com/PublishedEpd/Download/7713 (accessed on 23 December 2020).
47. Janssen, M.; Gustafsson, E.; Echardt, L.; Wallinder, J.; Wolf, J. Life cycle assessment of lignin-based carbon fibres. In Proceedings of the 14th Conference on Sustainable Development of Energy, Water and Environment Systems: (SDEWES), Dubrovnik, Croatia, 1–6 October 2019; p. 10.
48. Huijbregts, M.A.J.; Steinmann, Z.J.N.; Elshout, P.M.F.; Stam, G.; Verones, F.; Vieira, M.; Zijp, M.; Hollander, A.; van Zelm, R. ReCiPe2016: A harmonised life cycle impact assessment method at midpoint and endpoint level. *Int. J. Life Cycle Assess.* **2017**, *22*, 138–147. [CrossRef]
49. ReCiPe 2016. Available online: https://www.rivm.nl/en/life-cycle-assessment-lca/recipe (accessed on 23 December 2020).
50. World Meteorological Organization. *Scientific Assessment of Ozone Depletion 2010 Report No. 52*; World Meteorological Organization: Geneva, Switzerland, 2011.
51. Spadaro, J.V.; Rabl, A. Pathway Analysis for Population-Total Health Impacts of Toxic Metal Emissions. *Risk Anal.* **2004**, *24*, 1121–1141. [CrossRef] [PubMed]
52. van Zelm, R.; Preiss, P.; van Goethem, T.; Van Dingenen, R.; Huijbregts, M. Regionalized life cycle impact assessment of air pollution on the global scale: Damage to human health and vegetation. *Atmos. Environ.* **2016**, *134*, 129–137. [CrossRef]
53. Roy, P.-O.; Azevedo, L.B.; Margni, M.; van Zelm, R.; Deschênes, L.; Huijbregts, M.A.J. Characterization factors for terrestrial acidification at the global scale: A systematic analysis of spatial variability and uncertainty. *Sci. Total Environ.* **2014**, *500–501*, 270–276. [CrossRef]
54. Roy, P.-O.; Huijbregts, M.; Deschênes, L.; Margni, M. Spatially-differentiated atmospheric source–receptor relationships for nitrogen oxides, sulfur oxides and ammonia emissions at the global scale for life cycle impact assessment. *Atmos. Environ.* **2012**, *62*, 74–81. [CrossRef]
55. Frischknecht, R.; Jungbluth, N.; Althaus, H.-J.; Bauer, C.; Doka, G.; Dones, R.; Hischier, R.; Hellweg, S.; Humbert, S.; Köllner, T.; et al. *Implementation of Life Cycle Impact Assessment Methods. Ecoinvent Report No. 3, v2.0*; Swiss Centre for Life Cycle Inventories: Dübendorf, Switzerland, 2007.
56. Saaty, T.L. Decision making with the analytic network process (ANP) and its super-decisions software: The National Missile Defense (NMD) example. In Proceedings of the 6th ISAHP Conference, Berne, Switzerland, 2–4 August 2001; pp. 365–382.

57. Rochikashvili, M.; Bongaerts, J.C. Multi-criteria Decision-making for Sustainable Wall Paints and Coatings Using Analytic Hierarchy Process. *Energy Procedia* **2016**, *96*, 923–933. [CrossRef]
58. Saaty, T.L.; Vargas, L.G. *Decision Making with the Analytic Network Process*; International Series in Operations Research & Management Science; Springer: New York, NY, USA, 2006; Volume 95, ISBN 9780387338590.
59. Saaty, T.L. Decision making with the analytic hierarchy process. *Int. J. Serv. Sci.* **2008**, *1*, 83–98. [CrossRef]

Article

Pyrolysis Kinetic Properties of Thermal Insulation Waste Extruded Polystyrene by Multiple Thermal Analysis Methods

Ang Li [1], Wenlong Zhang [2,*], Juan Zhang [2], Yanming Ding [2] and Ru Zhou [3,*]

[1] College of Power Engineering, Naval University of Engineering, 717 Jiefang Ave, Qiaokou District, Wuhan 430032, China; sirius027lai@cug.edu.cn

[2] Faculty of Engineering, China University of Geosciences, 388 Lumo Rd, Hongshan District, Wuhan 430074, China; zhangjuan@cug.edu.cn (J.Z.); dingym@cug.edu.cn (Y.D.)

[3] Jiangsu Key Laboratory of Urban and Industrial Safety, College of Safety Science and Engineering, Nanjing Tech University, 30 Puzhu Rd, Pukou District, Nanjing 211816, China

* Correspondence: zwllhl@cug.edu.cn (W.Z.); maxmuse.zhou@njtech.edu.cn (R.Z.)

Received: 6 November 2020; Accepted: 6 December 2020; Published: 8 December 2020

Abstract: Extruded polystyrene (XPS) is a thermal insulation material extensively applied in building systems. It has attracted much attention because of outstanding thermal insulation performance, obvious flammability shortcoming and potential energy utilization. To establish the reaction mechanism of XPS's pyrolysis, thermogravimetric experiments were performed at different heating rates in nitrogen, and multiple methods were employed to analyze the major kinetics of pyrolysis. More accurate kinetic parameters of XPS were estimated by four common model-free methods. Then, three model-fitting methods (including the Coats-Redfern, the iterative procedure and masterplots method) were used to establish the kinetic model. Since the kinetic models established by the above three model-fitting methods were not completely consistent based on different approximations, considering the effect of different approximates on the model, the reaction mechanism was further established by comparing the conversion rate based on the model-fitting methods corresponding to the possible reaction mechanisms. Finally, the accuracy of the above model-fitting methods and Particle Swarm Optimization (PSO) algorithm were compared. Results showed that the reaction function $g(\alpha) = (1-\alpha)^{-1} - 1$ might be the most suitable to characterize the pyrolysis of XPS. The conversion rate calculated by masterplots and PSO methods could provide the best agreement with the experimental data.

Keywords: extruded polystyrene; pyrolysis; kinetic model; thermal degradation; reaction mechanism

1. Introduction

Energy has played a significant role in promoting economic growth. However, the current global energy problem is already one of the main problems restricting sustainable development [1]. Besides, the issue of energy consumption in buildings is increasingly prominent, and buildings account for more than 30% of global energy consumption [2]. Therefore, many countries are committed to improving energy efficiency, especially in buildings [3]. Therein, the thermal insulation material is one of the most effective approaches to economize energy [4]. Currently, organic polymer foam insulation board [5], such as extruded polystyrene (XPS), is extensively applied as insulation material in building insulation systems due to its outstanding performance, such as low thermal conductivity, lightweight and so on [6].

However, due to the low thermal stability, XPS is easily affected by high temperature and intense solar radiation [7], which damages the characteristics of XPS. In addition, XPS is flammable, more and more building fires can be attributed to XPS, such as Grenfell Tower [8]. The severity of the fire is related to the spread speed, and one of the main reasons for the rapid spread is that the insulation material is easy to ignite, and the degradation products containing gaseous fuel contribute to combustion [9], such as styrene monomer and oligomers [10]. Since pyrolysis is a key component in the combustion process, so it is imperative to study the thermal decomposition characteristic of XPS for predicting the growth of fire [11]. On the other hand, because of the large amount of XPS waste, the issue of disposal of the waste is becoming increasingly urgent [12]. However, improper processing of waste could lead to a series of problems, such as waste of resources [13], environmental contamination [14], fire hazard [15] and so on. Among the commonly-used methods of solid waste treatment, pyrolysis is expected to be a meaningful energy conversion method that can convert solid waste into fuel [16] and recover useful chemicals [17]. What is more, pyrolysis plays an important role in waste plastics for energy recovery [18,19]. Especially, valuable feedstock and fuel are obtained from the pyrolytic process of waste plastics [20,21]. In recent years, the thermal degradation of solid waste has attracted increasing attention owing to the potential to substitute traditional fossil fuels [22].

As discussed above, the knowledge of pyrolysis characteristics of XPS not only is closely related to the fire risk but also facilitates the recycling of XPS waste. There have been many studies on the pyrolysis characteristics of XPS. Jiao et al. [23,24] investigated the pyrolysis of XPS with expanded polystyrene and polyurethane foam and further studied its pyrolysis characteristics in different environments. Jiang et al. [25] studied the pyrolysis behavior of XPS waste to obtain the kinetic model and reconstruct the function of the model. As a result of these studies, the pyrolysis characteristics of XPS can be further understood.

In addition, many researchers pointed out that different approximations used in the calculation of kinetic parameters would affect their accuracy, which brought errors in the reaction mechanism. For example, Farjas et al. [26] noted that the accuracy of the integral isoconversional method was linked to approximations. Vyazovkin et al. [27] also indicated that an error occurred in the calculation of the activation energy because of approximations. However, the reaction mechanism of XPS's pyrolysis is commonly established by coupling the model-free and model-fitting methods [25], which rarely considers the influence of the approximations. Therefore, the purpose of this study is to establish the reaction mechanism of XPS's pyrolysis by multiple methods while considering approximations and find which method can reflect the reaction process with the highest accuracy. Besides, the accurate pyrolysis kinetics of XPS can be used for large-scale fire simulations, such as the Fire Propagation Apparatus [28] and Cone Calorimetry [29]. Furthermore, they contribute to guiding the reactor design [30].

In the current study, thermogravimetric experiments were performed to obtain the pyrolysis characteristics of XPS in nitrogen. More accurate kinetic parameters were estimated by multiple typical model-free methods (such as Flynn–Wall–Ozawa, Starink, Distributed Activation Energy Model and Tang method). Then, model-fitting methods (including Coats–Redfern, the iterative procedure, and masterplots method) were used to establish the kinetic model of XPS. Considering the effect of the approximations on the model, the pyrolysis reaction mechanism was further established by comparing the conversion rate based on the model-fitting methods corresponding to the possible reaction mechanisms. Finally, the accuracy of the above model-fitting methods and Particle Swarm Optimization (PSO) algorithm were compared.

2. Materials and Methods

2.1. Materials

The XPS employed in this study was milled to powder and then put into an oven to lower the water content before testing. The element analysis was performed by Vario EL cube. The results showed that elements C, H, N and S on a dry basis were 71.60%, 6.43%, 1.24% and 0.918%, respectively.

2.2. Thermogravimetric Measurements

Thermogravimetry experiments were conducted by TA Instruments on SDT Q600 (New Castle, DE, USA). The 6 mg sample was evenly placed in an aluminum oxide crucible during the experimental temperature 300–1000 K. Nitrogen was a purge gas, and its flow rate was 100 mL/min. In order to be close to the heating rates of real fires, the heating rates of 5 K/min, 20 K/min, 40 K/min, 60 K/min and 80 K/min were selected.

2.3. Pyrolysis Kinetics

Thermogravimetry provides an ideal environment for the degradation of the small solid sample in which the atmosphere and heating rates can be well controlled [31]. The solid reaction rate during the decomposition can be written as

$$\frac{d\alpha}{dt} = k(T)f(\alpha) \tag{1}$$

where $f(\alpha)$ denotes the differential function. t is time, α represents conversion rate, and $k(T)$ denotes a constant with temperature T. α and $k(T)$ can be defined as follows:

$$\alpha = \frac{m_0 - m_t}{m_0 - m_\infty} \tag{2}$$

$$k(T) = A\exp\left(\frac{-E_a}{RT}\right) \tag{3}$$

Three types of m (m_0, m_t and m_∞) stand for initial, transient and final mass, respectively. E_a represents activation energy, A refers to the pre-exponential factor, and R means the universal gas constant.

Considering the linear relationship between temperature and the heating rate (β), $\beta = dT/dt$, the reaction rate can be substituted as

$$\frac{d\alpha}{dT} = \frac{A}{\beta}f(\alpha)\exp\left(\frac{-E_a}{RT}\right) \tag{4}$$

Then the integral function $g(\alpha)$ is expressed as

$$g(\alpha) = \int_0^\alpha \frac{d\alpha}{f(\alpha)} = \frac{A}{\beta}\int_{T_0}^T \exp^{-\left(\frac{E_a}{RT}\right)}dT \approx \frac{AE_a}{\beta R}P(x) \tag{5}$$

where $x = E_a/RT$. $P(x)$ indicates the temperature integral. There are many approximations of $P(x)$ introduced in the literature [26], and they can be represented as

$$P(x) \approx \exp(-1.0518x - 5.330) \tag{6}$$

$$P(x) \cong \frac{\exp(-1.0008x - 0.312)}{x^{1.92}} \tag{7}$$

$$-\ln(P(x)) \approx 0.377739 + 1.894661\ln x + 1.00145x \tag{8}$$

$$P(x) \approx \frac{\exp(-x)}{x^2}\left(\frac{x^5 + 40x^4 + 552x^3 + 3168x^2 + 7092x + 4320}{x^6 + 42x^5 + 630x^4 + 4200x^3 + 12{,}600x^2 + 15{,}120x + 5040}\right) \tag{9}$$

$$P(x) = \frac{\exp(-x)}{x^2} \times \left(1 + \frac{2!}{-x}\right) \tag{10}$$

2.4. Methods

Model-free and model-fitting methods are common methods for analyzing kinetics. For the model-free methods, the advantage is that the activation energy can still be calculated when the reaction mechanism is not known [32], while the model-fitting methods can determine the reaction mechanism and obtain a set of corresponding kinetic parameters based on the reaction mechanism [25]. Common solid reaction mechanisms are listed in Table 1.

Table 1. Common solid reaction mechanisms [33,34].

No.	$g(\alpha)$	$f(\alpha)$	Rate-Determining Model
1	$1-(1-\alpha)^{2/3}$	$3/2(1-\alpha)^{1/3}$	Chemical Reaction
2	$1-(1-\alpha)^{1/4}$	$4(1-\alpha)^{3/4}$	Chemical Reaction
3	$(1-\alpha)^{-1/2}-1$	$2(1-\alpha)^{3/2}$	Chemical Reaction
4	$(1-\alpha)^{-1}-1$	$(1-\alpha)^{2}$	Chemical Reaction
5	$(1-\alpha)^{-2}-1$	$1/2(1-\alpha)^{3}$	Chemical Reaction
6	$\alpha^{3/2}$	$2/3\alpha^{-1/2}$	Nucleation
7	$-\ln(1-\alpha)$	$1-\alpha$	First Order, $n=1$
8	$[-\ln(1-\alpha)]^{2/3}$	$3/2(1-\alpha)[-\ln(1-\alpha)]^{1/3}$	Avrami–Erofeev
9	$[-\ln(1-\alpha)]^{1/2}$	$2(1-\alpha)[-\ln(1-\alpha)]^{1/2}$	Avrami–Erofeev
10	α	1	Contracting Disk
11	$1-(1-\alpha)^{1/2}$	$2(1-\alpha)^{1/2}$	Contracting Cylinder
12	$1-(1-\alpha)^{1/3}$	$3(1-\alpha)^{2/3}$	Contracting Sphere
13	α^{2}	$1/2\alpha$	1-*D* Diffusion
14	$\alpha+(1-\alpha)\ln(1-\alpha)$	$[-\ln(1-\alpha)]^{-1}$	2-*D* Diffusion
15	$[1-(1-\alpha)^{1/3}]^{2}$	$(3/2)(1-\alpha)^{2/3}[1-(1-\alpha)^{-1/3}]^{-1}$	3-*D* Diffusion
16	$1-2\alpha/3-(1-\alpha)^{2/3}$	$(3/2)[(1-\alpha)^{-1/3}-1]^{-1}$	3-*D* Diffusion
17	$[(1+\alpha)^{1/3}-1]^{2}$	$(3/2)(1+\alpha)^{2/3}[(1+\alpha)^{1/3}-1]^{-1}$	3-*D* Diffusion
18	$1+2\alpha/3-(1+\alpha)^{2/3}$	$(3/2)[(1+\alpha)^{-1/3}-1]^{-1}$	3-*D* Diffusion
19	$[(1+\alpha)^{-1/3}-1]^{2}$	$(3/2)(1+\alpha)^{4/3}[(1+\alpha)^{-1/3}-1]^{-1}$	3-*D* Diffusion

2.4.1. Model-Free Methods

Two forms of model-free methods, namely differential and integral conversion methods, are widely employed [31]. However, Vyazovkin et al. [31] noted that the differential methods were not more accurate than the integral methods. Therefore, in the current study, the integral isoconversional methods are applied. These integral isoconversional methods are different depending on the approximations.

Flynn–Wall–Ozawa Method (FWO)

The FWO method [35,36] estimate E_a by the slope of the linear plot of lnβ versus 1/T. The equation can be written as Equation (11) based on the approximation of Equation (6).

$$\ln\beta=\ln\left(\frac{AE_a}{Rg(\alpha)}\right)-5.331-1.052\left(\frac{E_a}{RT}\right) \tag{11}$$

Starink Method

Similar to the FWO method, the Starink [37] method is also employed to calculate the E_a by the slope (ln$\beta/T^{1.92}$ versus 1/T). The equation based on Equation (7) can be expressed as

$$\ln\frac{\beta}{T^{1.92}}=\ln\left(\frac{AE_a}{Rg(\alpha)}\right)-0.312-1.0008\left(\frac{E_a}{RT}\right) \tag{12}$$

Tang Method

Besides, the Tang [38] method adopts Equation (8), which can be expressed as Equation (13) to estimate the E_a by the slope of ln$\beta/T^{1.894661}$ versus 1/T.

$$\ln\frac{\beta}{T^{1.894661}}=\ln\left(\frac{AE_a}{Rg(\alpha)}\right)+3.635041-1.89466\ln E_a-1.00145\left(\frac{E_a}{RT}\right) \tag{13}$$

Distributed Activation Energy Model Method (DAEM)

The DAEM method is an extensively accepted method to calculate the pyrolysis kinetics of complex materials [39]. Its simplified function is presented in Equation (14) based on Equation (9) [40].

$$\ln\frac{\beta}{T^2} = \ln\left(\frac{AR}{E_a}\right) + 0.6075 - \frac{E_a}{RT} \tag{14}$$

As shown in Equation (14), both E_a and ln$A$ can be obtained from the slope and intercept by plotting lnβ/T^2 versus $1/T$.

2.4.2. Model-Fitting Methods

The model-fitting methods match the theoretical kinetic models according to the thermogravimetric experimental data, and the corresponding model is determined as the kinetic model of solid when the theoretical value of the kinetic parameters is best fitted with the experimental value [41]. The common model-fitting methods contain the Coats–Redfern method (CR), the iterative procedure and masterplots method and so on. Especially, Ding et al. obtained the woody biomass pyrolysis kinetic model through the optimization algorithms, such as Shuffled Complex Evolution (SCE) [42], PSO [43] and Genetic Algorithm method [44]. Therefore, from the perspective of establishing the kinetic model, optimization algorithms can also be considered as a model-fitting method [43].

Coats–Redfern Method

The CR method [45] is one of the most commonly-used model-fitting methods, and the equation can be expressed as Equation (15) using the approximation of Equation (10).

$$\ln\frac{g(\alpha)}{T^2} = \ln\left(\frac{AR}{\beta E_a}\right) - \frac{E_a}{RT} \tag{15}$$

Kinetic parameters (E_a and ln$A$) corresponding to each reaction function $g(\alpha)$ are obtained by the plot of ln$(g(\alpha)/T^2)$ versus $1/T$.

The Iterative Procedure

In addition, the iterative procedure [46] is also applied to determine the solid kinetic model. The expression of the iterative procedure method, namely $g(\alpha)$ function is written as

$$\ln(g(\alpha)) = \left(\ln\left(\frac{AE_a}{R}\right) + \ln(P(x))\right) - \ln\beta \tag{16}$$

If the kinetic model can reflect the solid pyrolysis process appropriately, there is a linear relationship between ln$(g(\alpha))$ versus lnβ, and the slope should be close to −1, and the linear correlation coefficient R^2 is higher [47]. The $P(x)$ applies to the approximation of Equation (9).

Masterplots Method

Masterplots method [48] is obtained by taking α = 0.5 into Equation (5), and it is expressed by the following equation:

$$\frac{g(\alpha)}{g(0.5)} = \frac{P(x)}{P(x_{0.5})} \tag{17}$$

where $x_{0.5} = E_a/RT_{0.5}$. To quantify the application of Equation (17), statistics number F for estimating the fitness of each model is applied, as shown in Equations (18) and (19) [25].

$$S_j^2 = \frac{1}{n-1}\sum_{i=1}^{n}\left(\frac{P_i}{P_{0.5}} - \frac{g_j(\alpha_i)}{g_{j0.5}}\right)^2 \tag{18}$$

$$F_j = \frac{S_j^2}{S_{\min}^2} \tag{19}$$

where i and j are conversion rate and heating rate, respectively. If $F = 1$ for each heating rate of the model, the model is regarded as a kinetic model of solid pyrolysis.

2.4.3. Particle Swarm Optimization Method

The optimization algorithms have been employed to optimize kinetic parameters due to high efficiency and good accuracy, especially the reaction mechanism established reflects the process of solid pyrolysis when the kinetic parameters are globally optimal [29,49]. The fitness value of PSO can be obtained from the following formulas:

$$\phi = \phi_\alpha + \phi_{d\alpha/dt} \tag{20}$$

$$\phi_\alpha = \sum_{j=1}^{N}\left[w_{\alpha,j}\frac{\sum_{k=1}^{n}\left(\alpha_{\text{mod},k} - \alpha_{\text{exp},k}\right)^2}{\sum_{k=1}^{n}\left(\alpha_{\text{exp},k} - \frac{1}{n}\sum_{p=1}^{n}\alpha_{\text{exp},p}\right)^2}\right] \tag{21}$$

$$\phi_{d\alpha/dt} = \sum_{j=1}^{N}\left[w_{d\alpha/dt,j}\frac{\sum_{k=1}^{n}\left(d\alpha/dt_{\text{mod},k} - d\alpha/dt_{\text{exp},k}\right)^2}{\sum_{k=1}^{n}\left(d\alpha/dt_{\text{exp},k} - \frac{1}{n}\sum_{p=1}^{n}d\alpha/dt_{\text{exp},p}\right)^2}\right] \tag{22}$$

where Φ refers to the objective value. α and $d\alpha/dt$ denote the cumulative values of conversion rate and reaction rate, respectively. N and n indicate the number of experiments and experimental data points, respectively. w presents the weighted value. Subscript mod and exp are calculated values from simulations and experiments.

Suppose to search in the D-dimensional space of n particles, the position and velocity vectors of the ith particle are expressed as $x_i = (x_{i1}, x_{i2}, \ldots, x_{iN})$ and $v_i = (v_{i1}, v_{i2}, \ldots, v_{iN})$, respectively. The particle update can be obtained by the following equations:

$$v_{id}^{k+1} = wv_{id}^{k} + c_1r_1\left(p_{id} - x_{id}^{k}\right) + c_2r_2\left(p_{gd} - x_{id}^{k}\right) \tag{23}$$

$$x_{id}^{k+1} = x_{id}^{k} + v_{id}^{k+1} \tag{24}$$

where i and k are the number of particle and iteration, respectively. d indicates the search direction. p_{id} and p_{gd} are the optimal personal position and the global position, respectively. c_1 and c_2 are constants of positive acceleration that represent the individual and global properties of the swarm. r_1 and r_2 are random numbers from 0 to 1.

3. Results and Discussion

3.1. Thermogravimetric Analysis

Figure 1 illustrates the derivative mass loss (DTG) and conversion rate profiles of degradation processes of XPS at different heating rates.

As shown in Figure 1, the movement of the DTG and conversion rate curves is related to the heating rates. As the heating rate increases, the reaction range is gradually delayed to a higher temperature to complete the reaction. For example, the peak temperatures T_P of the DTG curves at five heating rates are 681 K, 707 K, 721 K, 731 K and 737 K. In addition, the initial decomposition temperature of XPS is 575–625 K, and the final temperature is 750–825 K, and the reaction temperature range is about 175 K. Moreover, Jun et al. [50] introduced a classic method called Coats–Redfern to calculate the kinetic parameters of expandable polystyrene and suggested that if there was just one peak in the DTG curve, it indicated that one kind of reaction occurred. Since each DTG curve has only one peak, the pyrolysis of XPS in nitrogen is a one-step reaction.

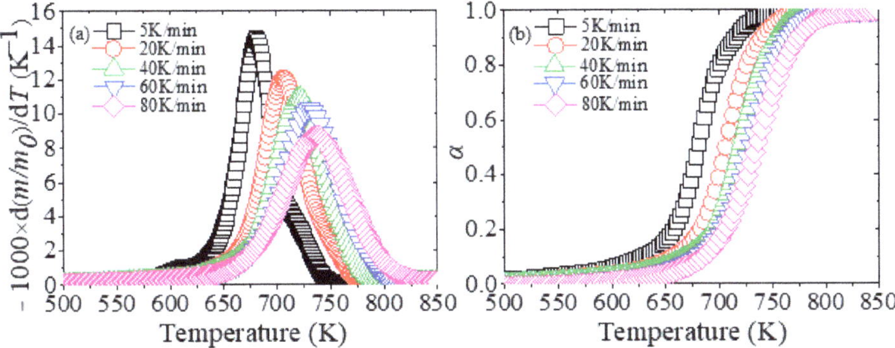

Figure 1. The (**a**) derivative mass loss (DTG) and (**b**) conversion rate profiles of degradation processes of extruded polystyrene (XPS) at different heating rates.

3.2. Kinetic Analysis by the Model-Free Methods

The activation energy E_a calculated based on the FWO, DAEM, Starink and Tang methods is shown in Table 2. It is noted that the E_a of Jiang et al. [12] and Jiao et al. [24] is obtained by the Kissinger–Akahira–Sunose (KAS) method. There are some reasons why the KAS method is not chosen to calculate the E_a in this study, but the E_a obtained by KAS in References [12,24] is compared. The E_a can be obtained by the slope of the linear relationship between the heating rate β and temperature T. For the KAS method, E_a is obtained through the slop of $\ln(\beta/T^2)$ and $1/T$. However, it is the same as the value of the DAEM method. Since the DAEM method has some advantages [38], the DAEM rather than KAS is used to estimate the E_a in the current study. For Reference [12], FWO and KAS methods were applied to estimate the kinetics of XPS. However, the FWO method is slightly inaccurate compared with other model-free methods [31]. Furthermore, the E_a calculated from the FWO method is larger than that calculated by KAS and Starink [13]. In this study, the E_a is also a little larger than that of DAEM, Starink and Tang methods. Besides, the E_a was only calculated by the KAS method in Reference [24]. Therefore, the calculated results obtained by the KAS method in the literature are compared.

Table 2 shows that by comparing References [12,24], there will be a difference in E_a. There are many factors that affect the calculated values of kinetics, such as raw material source, heating rates, temperature, gas flow and so on [51,52]. Jiang et al. [12] selected four heating rates (5 °C/min, 10 °C/min, 15 °C/min and 20 °C/min) and conducted the thermogravimetric analysis with a gas flow of 20 mL/min in nitrogen. Furthermore, the sample weighed about 6 mg, and it was cut to powder and heated up to 800 °C. In the test of Jiao et al. [24], 4 mg particulate sample was heated to 700 °C with four heating rates (5 K/min, 10 K/min, 20 K/min and 30 K/min), and the flow rate of nitrogen was 75 mL/min. In this study, the 6 mg powdered sample was tested at heating rates (5 K/min, 20 K/min, 40 K/min, 60 K/min and 80 K/min), and the temperature was 300–1000 K, and the flow rate of nitrogen was 100 mL/min. It is noted that the calculated values of the literature are only used to compare with calculated results of this study, and they are not the boundaries of the range.

It can also be seen from Table 2 that the E_a maintains constant, and the average E_a is 200.4 kJ/mol (average of four methods). Furthermore, researchers [53,54] noted that if the deviation between the maximum and minimum E_a was less than 20–30% of the average E_a, then the E_a was independent of α. Table 2 shows that the calculated values by four methods are less than 20% of the average E_a, so the pyrolysis of XPS is a one-step reaction in nitrogen, which is also proved by Figure 1.

Table 2. The E_a is calculated by four methods based upon thermogravimetric data.

α	E_a (kJ/mol)					
	FWO	**DAEM**	**Starink**	**Tang**	**Jiao et al. [24]**	**Jiang et al. [12]**
0.1	180.9	179.4	176.7	178.5	147.4	368
0.2	212.1	211.7	212.0	210.7	164.6	298
0.3	211.5	211.0	211.3	208.4	165.0	277
0.4	205.4	204.4	204.7	203.8	164.1	270
0.5	201.2	199.9	200.2	199.4	163.3	270
0.6	197.3	195.6	196.0	196.0	161.8	263
0.7	195.6	193.7	194.0	194.0	161.1	256
0.8	208.1	206.7	207.0	207.0	164.1	253
Average	201.5	200.3	200.2	199.7	161.4	281.9
Value [a]	15.5%	16.1%	17.6%	16.1%	10.9%	40.8%

[a] The value is the deviation between the maximum and minimum E_a and the average E_a percentage [53,54].

3.3. Establishment of Reaction Mechanisms

In this study, the calculated E_a of the CR method at different heating rates is illustrated in Table 3. Then, it compares with that previously obtained E_a using the four model-free methods. The pyrolysis reaction mechanism of XPS should be established when the average E_a of the kinetic model based on the CR method is the closest to that of model-free methods [32].

Table 3. Calculation values of E_a for the CR method.

No.	$g(\alpha)$	E_a (kJ/mol)					
		5 K/min	**20 K/min**	**40 K/min**	**60 K/min**	**80 K/min**	**Average**
1	$1-(1-\alpha)^{2/3}$	111.2	121.5	120.1	122.4	114.5	117.9
2	$1-(1-\alpha)^{1/4}$	125.3	136.9	135.4	138.2	129.5	133.1
3	$(1-\alpha)^{-1/2}-1$	155.1	169.2	167.7	171.4	161.0	164.9
4	$(1-\alpha)^{-1}-1$	178.0	194.1	192.5	196.9	185.2	189.3
5	$(1-\alpha)^{-2}-1$	230.1	250.7	249.1	255.2	240.5	245.1
6	$\alpha^{3/2}$	157.2	171.5	169.5	172.6	161.8	166.5
7	$-\ln(1-\alpha)$	134.6	147.0	145.5	148.5	139.4	143.0
8	$[-\ln(1-\alpha)]^{2/3}$	86.0	94.1	93.1	95.0	88.9	91.4
9	$[-\ln(1-\alpha)]^{1/2}$	61.7	67.7	66.8	68.3	63.63	65.6
10	α	101.1	110.5	109.1	111.1	103.8	107.1
11	$1-(1-\alpha)^{1/2}$	116.6	127.4	126.0	128.5	120.8	123.9
12	$1-(1-\alpha)^{1/3}$	122.4	133.7	132.2	134.8	126.4	129.9
13	α^2	213.3	232.6	229.9	234.1	219.7	225.9
14	$\alpha+(1-\alpha)\ln(1-\alpha)$	232.7	253.7	251.0	255.7	240.2	246.7
15	$[1-(1-\alpha)^{1/3}]^2$	255.9	278.9	276.2	281.7	264.8	271.5
16	$1-2\alpha/3-(1-\alpha)^{2/3}$	240.3	262.0	259.3	264.3	248.4	254.9
17	$[(1+\alpha)^{1/3}-1]^2$	193.9	211.4	208.9	212.6	199.3	205.2
18	$1+2\alpha/3-(1+\alpha)^{2/3}$	200.1	218.2	215.6	219.5	205.9	210.0
19	$[(1+\alpha)^{-1/3}-1]^2$	175.8	191.8	189.3	192.5	180.4	186.0

Table 3 shows that the values of E_a of four reaction models are closest to 200.4 kJ/mol, and their models are No. 4 (189.3 kJ/mol), No. 17 (205.2 kJ/mol), No. 18 (210.0 kJ/mol) and No. 19 (186.0 kJ/mol), respectively. Jiang et al. [25] established a pyrolysis reaction function $g(\alpha) = -\ln(1-\alpha)$ of XPS in nitrogen. However, the E_a of this reaction function is 143.0 kJ/mol by the CR method in this study, and the difference is large, which indicates reaction function $g(\alpha) = -\ln(1-\alpha)$ is not applicable to the current study.

Since the E_a is estimated through the model-free methods and the CR method in different approximations [55], and the difference corresponding to these models is very small (11.1 kJ/mol, 4.8 kJ/mol, 9.6 kJ/mol and 14.4 kJ/mol), so it cannot 100 percent determine the kinetic model of XPS in nitrogen. To improve accuracy, masterplots and the iterative procedure methods are also applied to determine possible kinetic models. The calculation results of the two methods are listed in Table 4.

As presented in Table 4, the model of No. 18 is the best by masterplots method. However, the model of No. 4 is the best by the iterative procedure method.

Table 4. Reaction mechanisms are determined by masterplots and the iterative procedure methods.

No.	$g(\alpha)$	Masterplots					$\ln(g(\alpha))$ vs. $\ln\beta$	
		5 K/min	20 K/min	40 K/min	60 K/min	80 K/min	Slope	R^2
1	$1-(1-\alpha)^{2/3}$	3.96	5.14	1.51	4.11	9.46	0.639	0.992
2	$1-(1-\alpha)^{1/4}$	1.54	2.78	1.24	2.29	4.25	0.717	0.995
3	$(1-\alpha)^{-1/2}-1$	11.64	0.19	0.61	0.04	1.13	0.877	0.995
4	$(1-\alpha)^{-1}-1$	102.66	4.37	0.18	2.61	17.48	0.999	0.992
5	$(1-\alpha)^{-2}-1$	1335.39	87.80	0.98	61.10	271.06	1.277	0.981
6	$\alpha^{3/2}$	11.61	2.78	1.26	2.38	4.33	0.876	0.988
7	$-\ln(1-\alpha)$	4.18	1.47	1.05	1.24	1.70	0.767	0.996
8	$[-\ln(1-\alpha)]^{2/3}$	40.12	5.94	1.57	4.63	11.30	0.511	0.996
9	$[-\ln(1-\alpha)]^{1/2}$	69.58	8.91	1.82	6.82	18.38	0.384	0.996
10	α	48.30	6.93	1.69	5.47	13.64	0.584	0.988
11	$1-(1-\alpha)^{1/2}$	23.50	4.19	1.41	3.39	7.32	0.669	0.994
12	$1-(1-\alpha)^{1/3}$	15.82	3.24	1.30	2.65	5.24	0.701	0.995
13	α^2	0.63	0.299	0.83	0.40	0.15	1.167	0.988
14	$\alpha+(1-\alpha)\ln(1-\alpha)$	21.53	0.31	0.50	0.17	2.78	1.274	0.992
15	$[1-(1-\alpha)^{1/3}]^2$	121.99	5.37	0.14	3.41	21.38	1.401	0.995
16	$1-2\alpha/3-(1-\alpha)^{2/3}$	43.19	1.17	0.37	0.67	6.53	1.316	0.993
17	$[(1+\alpha)^{1/3}-1]^2$	2.92	1.44	1.08	1.35	1.75	1.059	0.984
18	$1+2\alpha/3-(1+\alpha)^{2/3}$	1	1	1	1	1	1.094	0.985
19	$[(1+\alpha)^{-1/3}-1]^2$	13.17	3.00	1.30	2.58	4.84	0.959	0.979

To establish a suitable reaction mechanism, "kinetic compensation effects (KCE)" is generally accepted [56]. If the model is proper, good linear relation occurs between E_a and $\ln A$, as expressed in Equation (25)

$$\ln A = a + bE_a \tag{25}$$

The KCE of the four models is shown in Figure 2. It shows that the linear relationship for $g(\alpha) = 1 + 2\alpha/3 - (1+\alpha)^{2/3}$ is not suitable. It also shows that other models are suitable, and the linear relationships are expressed as $\ln A = 0.184E_a - 6.94$ (No. 4 model, $R^2 = 0.995$), $\ln A = 0.148E_a - 3.69$ (No. 17 model, $R^2 = 0.971$), $\ln A = 0.153E_a - 5.24$ (No. 19 model, $R^2 = 0.989$). Jiang et al. [25] noted that the kinetic model corresponding to the highest R^2 did not mean that it was the real reaction model. Therefore, the three models mentioned above are most possibly the pyrolysis kinetic models of EPS in nitrogen.

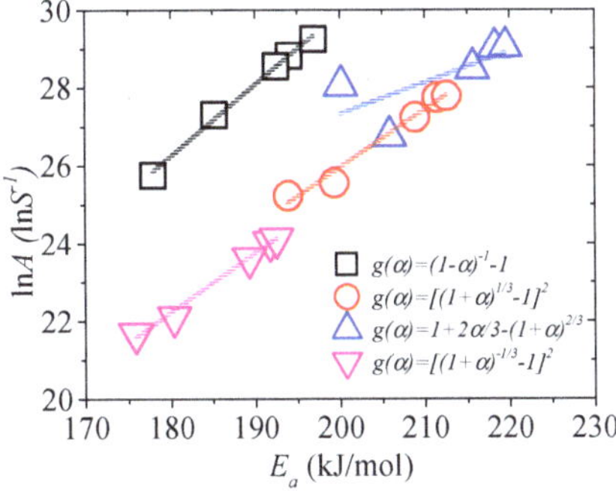

Figure 2. The plot of $\ln A$ versus E_a.

If the reaction mechanism of XPS's pyrolysis is selected correctly, the reaction parameters can be in good agreement with experimental data throughout the pyrolysis process [57]. Thus, the conversion rate α of the theoretical value is fitted to the experimental data. The theoretical α of three reaction models can be estimated by model-fitting methods. The comparison of experimental and theoretical α at 40 and 80 K/min is shown in Figure 3.

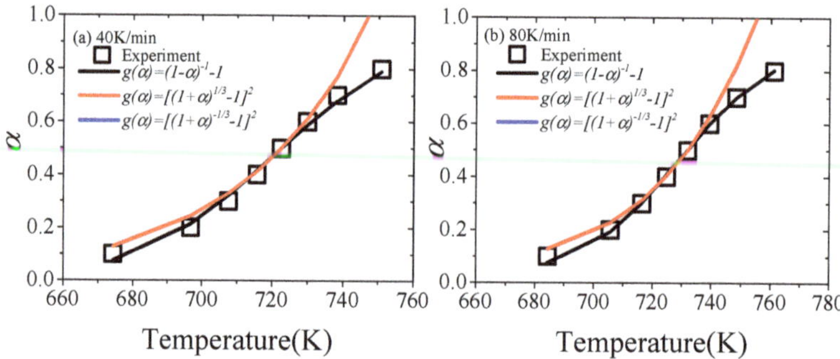

Figure 3. The experimental and theoretical α in the cases of (**a**) 40 K/min and (**b**) 80 K/min.

Figure 3 shows that among the three possible reaction mechanisms, the α corresponding to the No. 4 model obtained by the masterplots method has a good consistency with the experimental value throughout the experiment. The α of the No. 17 model partially fits the experimental value in the masterplots method, but the deviation is larger. As for the No. 19 model, the deviation is the largest, and the value of α is negative, so it is not shown in Figure 3. Therefore, the reaction function of XPS in nitrogen is $g(\alpha) = (1 - \alpha)^{-1} - 1$.

It is noted that the XPS products are various, and the properties may be different. Therefore, the reaction mechanism determined as the most suitable may not completely use on the results from research shown in the literature [12,24]. For example, although this paper and Jiang et al. [25] have both studied the pyrolysis characteristics of XPS, the reaction function of Jiang et al. [25] was not suitable for this study.

3.4. Comparison of Multiple Kinetic Methods

There are many methods to obtain the kinetic model of solid state. To obtain a more accurate analysis, it is necessary to compare multiple kinetic methods. As shown in Figure 4, the α of CR, the iterative procedure, masterplots and PSO methods is compared with the experimental value in the cases of 20, 40, 60 and 80 K/min.

Vyazovkin et al. [31] noted that the kinetic model mainly consisted of three forms by reaction profiles (α vs. T), including sigmoidal form, decelerating form and accelerating form. As presented in Figure 4, the reaction temperature range corresponding to the heating rate is different, but the trend of change is consistent. The α of the masterplots and PSO methods is basically consistent with the experimental value in the process of pyrolysis. The model is a decelerating model [58]. Besides, through the comparison between the masterplots method and the PSO method, it is found that their agreement with the experimental value varies slightly with the heating rates. For 20 and 40 K/min, the α calculated by the PSO method matches the experimental value better than that calculated by the masterplots method. However, for 60 and 80 K/min, the accuracy of the calculated value of the masterplots method is better than the PSO method. Although the CR method and the iterative procedure method are widely applied, the real pyrolysis process of XPS is not shown by them. The trend of the calculation results of the CR method is mostly accelerating, but Figure 4c is decelerating. Besides, the iterative procedure method has a larger deviation.

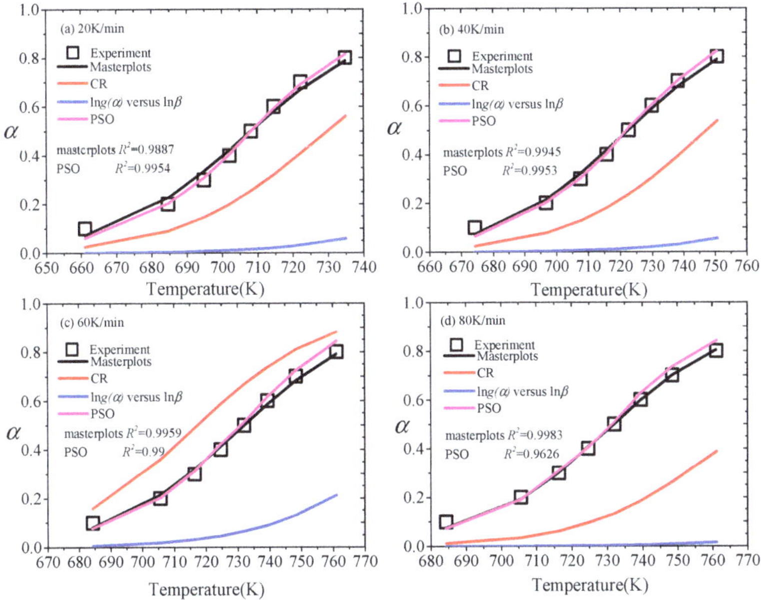

Figure 4. The α of CR, the iterative procedure, masterplots and PSO methods are compared with experimental data, (**a**) 20 K/min, (**b**) 40 K/min, (**c**) 60 K/min, (**d**) 80 K/min.

4. Conclusions

To study whether approximations affect the accuracy when establishing the reaction mechanism of XPS's pyrolysis, which method can reflect the reaction process and have the highest accuracy among the multiple methods, the kinetic model of XPS pyrolysis was investigated from 5 K/min to 80 K/min in this study. Four model-free methods (such as FWO, DAEM, Starink and Tang method) were employed to calculate the more accurate kinetic parameters, and four kinetic methods (including CR, the iterative procedure, masterplots and PSO) were applied to estimate the conversion rate with the comparison of experimental data. The results showed that four reaction mechanisms were close if only the activation energy between model-free methods and the CR method is compared. What is more, the reaction mechanisms of XPS's pyrolysis established via multiple kinetic methods were different. Therein, the reaction function $g(\alpha) = (1 - \alpha)^{-1} - 1$ might be the most suitable to characterize the pyrolysis of XPS in nitrogen. Furthermore, masterplots and PSO methods were more accurate than the CR and the iterative procedure methods. The pyrolysis kinetics of XPS can be used for large-scale fire simulations, such as the Fire Propagation Apparatus and Cone Calorimetry. Furthermore, they are important guidance for reactor design.

Author Contributions: Conceptualization, A.L. and Y.D.; methodology, W.Z. and Y.D.; software, J.Z., W.Z. and Y.D.; validation, Y.D. and R.Z.; formal analysis, A.L.; investigation, W.Z.; resources, A.L. and Y.D.; data curation, R.Z.; writing—original draft preparation, A.L. and W.Z.; writing—review and editing, W.Z. and Y.D.; visualization, J.Z.; supervision, J.Z.; project administration, A.L.; funding acquisition, Y.D. and R.Z. All authors have read and agreed to the published version of the manuscript.

Funding: This study is sponsored by the National Natural Science Foundation of China (No. 51806202) and Jiangsu Key Laboratory of Hazardous Chemicals Safety and Control (No. HCSC201901).

Conflicts of Interest: The authors declare no conflict of interest.

References

1. Ding, Y.; Zhang, J.; He, Q.; Huang, B.; Mao, S. The application and validity of various reaction kinetic models on woody biomass pyrolysis. *Energy* **2019**, *179*, 784–791. [CrossRef]
2. Mavromatidis, G.; Orehounig, K.; Bollinger, L.; Hohmann, M.; Marquant, J.; Miglani, S.; Morvaj, B.; Murray, P.; Waibel, C.; Wang, D. Ten questions concerning modeling of distributed multi-energy systems. *Build. Environ.* **2019**, *165*, 106372. [CrossRef]
3. Ibrahim, M.; Nocentini, K.; Stipetic, M.; Dantz, S.; Caiazzo, F.; Sayegh, H.; Bianco, L. Multi-field and multi-scale characterization of novel super insulating panels/systems based on silica aerogels: Thermal, hydric, mechanical, acoustic, and fire performance. *Build. Environ.* **2019**, *151*, 30–42. [CrossRef]
4. Ni, X.; Wu, Z.; Zhang, W.; Lu, K.; Ding, Y.; Mao, S. Energy utilization of building insulation waste expanded polystyrene: Pyrolysis kinetic estimation by a new comprehensive method. *Polymers* **2020**, *12*, 1744. [CrossRef] [PubMed]
5. Huang, X.; Sun, J.; Ji, J.; Zhang, Y.; Wang, Q.; Zhang, Y. Flame spread over the surface of thermal insulation materials in different environments. *Chin. Sci. Bull.* **2011**, *56*, 1617–1622. [CrossRef]
6. Si, J.; Tawiah, B.; Sun, W.; Lin, B.; Wang, C.; Yuen, A.C.Y.; Yu, B.; Li, A.; Yang, W.; Lu, H. Functionalization of mxene nanosheets for polystyrene towards high thermal stability and flame retardant properties. *Polymers* **2019**, *11*, 976. [CrossRef]
7. Krause, P.; Nowoświat, A. Experimental studies involving the impact of solar radiation on the properties of expanded graphite polystyrene. *Energies* **2019**, *13*, 75. [CrossRef]
8. An, W.; Yin, X.; Cai, M.; Gao, Y.; Wang, H. Influence of vertical channel on downward flame spread over extruded polystyrene foam. *Int. J. Therm. Sci.* **2019**, *145*, 105991. [CrossRef]
9. Ding, Y.; Swann, J.; Sun, Q.; Stoliarov, S.; Kraemer, R. Development of a pyrolysis model for glass fiber reinforced polyamide 66 blended with red phosphorus: Relationship between flammability behavior and material composition. *Compos. B Eng.* **2019**, *176*, 107263. [CrossRef]
10. Mehta, S.; Biederman, S.; Shivkumar, S. Thermal degradation of foamed polystyrene. *J. Mater. Sci.* **1995**, *30*, 2944–2949. [CrossRef]
11. Li, K.; Huang, X.; Fleischmann, C.; Rein, G.; Ji, J. Pyrolysis of medium-density fiberboard: Optimized search for kinetics scheme and parameters via a genetic algorithm driven by Kissinger's method. *Energy Fuels* **2014**, *28*, 6130–6139. [CrossRef]
12. Jiang, L.; Xiao, H.; He, J.; Sun, Q.; Gong, L.; Sun, J. Application of genetic algorithm to pyrolysis of typical polymers. *Fuel Process. Technol.* **2015**, *138*, 48–55. [CrossRef]
13. Zhang, W.; Zhang, J.; Ding, Y.; He, Q.; Lu, K.; Chen, H. Pyrolysis kinetics and reaction mechanism of expandable polystyrene by multiple kinetics methods. *J. Clean. Prod.* **2020**, 125042. [CrossRef]
14. Zhang, Q.; Cai, H.; Yi, W.; Lei, H.; Liu, H.; Wang, W.; Ruan, R. Biocomposites from organic solid wastes derived biochars: A review. *Materials* **2020**, *13*, 3923. [CrossRef]
15. Sun, Y.; Wang, Y.; Liu, L.; Xiao, T. The preparation, thermal properties, and fire property of a phosphorus-containing flame-retardant styrene copolymer. *Materials* **2020**, *13*, 127. [CrossRef]
16. Pan, L.; Jiang, Y. Evaluating the effects of KCl on thermal behavior and reaction kinetics of medium density fiberboard pyrolysis. *Materials* **2019**, *12*, 1826. [CrossRef]
17. Faravelli, T.; Pinciroli, M.; Pisano, F.; Bozzano, G.; Dente, M.; Ranzi, E. Thermal degradation of polystyrene. *J. Anal. Appl. Pyrolysis* **2001**, *60*, 103–121. [CrossRef]
18. Aguado, J.; Serrano, D.; San, M.; Castro, M.; Madrid, S. Feedstock recycling of polyethylene in a two-step thermo-catalytic reaction system. *J. Anal. Appl. Pyrolysis* **2007**, *79*, 415–423. [CrossRef]
19. Cafiero, L.; Fabbri, D.; Trinca, E.; Tuffi, R.; Ciprioti, S. Thermal and spectroscopic (TG/DSC–FTIR) characterization of mixed plastics for materials and energy recovery under pyrolytic conditions. *J. Therm. Anal. Calorim.* **2015**, *121*, 1111–1119. [CrossRef]
20. Benedetti, M.; Cafiero, L.; Angelis, D.; Dell'Era, A.; Pasquali, M.; Stendardo, S.; Tuffi, R.; Ciprioti, S. Pyrolysis of WEEE plastics using catalysts produced from fly ash of coal gasification. *Front. Environ. Sci. Eng.* **2017**, *11*, 11. [CrossRef]
21. Panda, A.; Singh, R.; Mishra, D. Thermolysis of waste plastics to liquid fuel: A suitable method for plastic waste management and manufacture of value added products-A world prospective. *Renew. Sustain. Energy Rev.* **2010**, *14*, 233–248. [CrossRef]

22. Januszewicz, K.; Kazimierski, P.; Suchocki, T.; Kardas, D.; Lewandowski, W.; Klugmann-Radziemska, E.; Luczak, J. Waste rubber pyrolysis: Product yields and limonene concentration. *Materials* **2020**, *13*, 4435. [CrossRef] [PubMed]
23. Jiao, L.; Xu, G.; Wang, Q.; Xu, Q.; Sun, J. Kinetics and volatile products of thermal degradation of building insulation materials. *Thermochim. Acta* **2012**, *547*, 120–125. [CrossRef]
24. Jiao, L.; Sun, J. A thermal degradation study of insulation materials extruded polystyrene. *Procedia Eng.* **2014**, *71*, 622–628. [CrossRef]
25. Jiang, L.; Zhang, D.; Li, M.; He, J.; Gao, Z.; Zhou, Y.; Sun, J. Pyrolytic behavior of waste extruded polystyrene and rigid polyurethane by multi kinetics methods and Py-GC/MS. *Fuel* **2018**, *222*, 11–20. [CrossRef]
26. Farjas, J.; Roura, P. Isoconversional analysis of solid state transformations. *J. Therm. Anal. Calorim.* **2011**, *105*, 757–766. [CrossRef]
27. Vyazovkin, S.; Dollimore, D. Linear and nonlinear procedures in isoconversional computations of the activation energy of nonisothermal reactions in solids. *J. Chem. Inf. Comput. Sci.* **1996**, *36*, 42–45. [CrossRef]
28. Ding, Y.; Zhou, R.; Wang, C.; Lu, K.; Lu, S. Modeling and analysis of bench-scale pyrolysis of lignocellulosic biomass based on merge thickness. *Bioresour. Technol.* **2018**, *268*, 77–80. [CrossRef]
29. Ding, Y.; Fukumoto, K.; Ezekoye, O.; Lu, S.; Wang, C.; Li, C. Experimental and numerical simulation of multi-component combustion of typical charring material. *Combust. Flame* **2020**, *211*, 417–429. [CrossRef]
30. Cepeliogullar, O.; Haykiriacma, H.; Yaman, S. Kinetic modelling of RDF pyrolysis: Model-fitting and model-free approaches. *Waste Manag.* **2016**, *48*, 275–284. [CrossRef]
31. Vyazovkin, S.; Burnham, A.; Criado, J.; Pérez-Maqueda, L.; Popescu, C.; Sbirrazzuoli, N. ICTAC Kinetics Committee recommendations for performing kinetic computations on thermal analysis data. *Thermochim. Acta* **2011**, *520*, 1–19. [CrossRef]
32. Ding, Y.; Ezekoye, O.; Lu, S.; Wang, C.; Zhou, R. Comparative pyrolysis behaviors and reaction mechanisms of hardwood and softwood. *Energy Convers. Manag.* **2017**, *132*, 102–109. [CrossRef]
33. Jiang, H.; Wang, J.; Wu, S.; Wang, B.; Wang, Z. Pyrolysis kinetics of phenol–formaldehyde resin by non-isothermal thermogravimetry. *Carbon* **2010**, *48*, 352–358. [CrossRef]
34. Vlaev, L.; Nedelchev, N.; Gyurova, K.; Zagorcheva, M. A comparative study of non-isothermal kinetics of decomposition of calcium oxalate monohydrate. *J. Anal. Appl. Pyrolysis* **2008**, *81*, 253–262. [CrossRef]
35. Flynn, J.; Wall, L. A quick, direct method for the determination of activation energy from thermogravimetric data. *J. Polym. Sci. Part C Polym. Lett.* **1966**, *4*, 323–328. [CrossRef]
36. Ozawa, T. A new method of analyzing thermogravimetric data. *Bull. Chem. Soc. Jpn.* **1965**, *38*, 1881–1886. [CrossRef]
37. Starink, M. The determination of activation energy from linear heating rate experiments: A comparison of the accuracy of isoconversion methods. *Thermochim. Acta* **2003**, *404*, 163–176. [CrossRef]
38. Tang, W.; Liu, Y.; Zhang, H.; Wang, C. New approximate formula for Arrhenius temperature integral. *Thermochim. Acta* **2003**, *408*, 39–43. [CrossRef]
39. Sonobe, T.; Worasuwannarak, N. Kinetic analyses of biomass pyrolysis using the distributed activation energy model. *Fuel* **2008**, *87*, 414–421. [CrossRef]
40. Miura, K. A new and simple method to estimate $f(E)$ and $k_0(E)$ in the distributed activation energy model from three sets of experimental data. *Energy Fuels* **1995**, *9*, 302–307. [CrossRef]
41. Slopiecka, K.; Bartocci, P.; Fantozzi, F. Thermogravimetric analysis and kinetic study of poplar wood pyrolysis. *Appl. Energy* **2012**, *97*, 491–497. [CrossRef]
42. Ding, Y.; Wang, C.; Chaos, M.; Chen, R.; Lu, S. Estimation of beech pyrolysis kinetic parameters by Shuffled Complex Evolution. *Bioresour. Technol.* **2016**, *200*, 658–665. [CrossRef] [PubMed]
43. Ding, Y.; Zhang, Y.; Zhang, J.; Zhou, R.; Ren, Z.; Guo, H. Kinetic parameters estimation of pinus sylvestris pyrolysis by Kissinger-Kai method coupled with particle swarm optimization and global sensitivity analysis. *Bioresour. Technol.* **2019**, *293*, 122079. [CrossRef] [PubMed]
44. Ding, Y.; Zhang, W.; Yu, L.; Lu, K. The accuracy and efficiency of GA and PSO optimization schemes on estimating reaction kinetic parameters of biomass pyrolysis. *Energy* **2019**, *176*, 582–588. [CrossRef]
45. Coats, A.; Redfern, J. Kinetic parameters from thermogravimetric data. *Nature* **1964**, *201*, 68–69. [CrossRef]
46. Li, L.; Chen, D. Application of iso-temperature method of multiple rate to kinetic analysis. *J. Therm. Anal. Calorim.* **2004**, *78*, 283–293.

47. Shahcheraghi, S.; Khayati, G.; Ranjbar, M. An advanced reaction model determination methodology in solid-state kinetics based on Arrhenius parameters variation. *J. Therm. Anal. Calorim.* **2015**, *122*, 175–188. [CrossRef]
48. Criado, J.; Ortega, A.; Gotor, F. Correlation between the shape of controlled-rate thermal analysis curves and the kinetics of solid-state reactions. *Thermochim. Acta* **1990**, *157*, 171–179. [CrossRef]
49. Ding, Y.; Huang, B.; Wu, C.; He, Q.; Lu, K. Kinetic model and parameters study of lignocellulosic biomass oxidative pyrolysis. *Energy* **2019**, *181*, 11–17. [CrossRef]
50. Jun, H.; Oh, S.; Lee, H.; Kim, H. A kinetic analysis of the thermal-oxidative decomposition of expandable polystyrene. *Korean J. Chem. Eng.* **2006**, *23*, 761–766. [CrossRef]
51. Liu, J.; Zhong, F.; Niu, W.; Su, J.; Gao, Z.; Zhang, K. Effects of heating rate and gas atmosphere on the pyrolysis and combustion characteristics of different crop residues and the kinetics analysis. *Energy* **2019**, *175*, 320–332. [CrossRef]
52. Xu, F.; Wang, B.; Yang, D.; Hao, J.; Qiao, Y.; Tian, Y. Thermal degradation of typical plastics under high heating rate conditions by TG-FTIR: Pyrolysis behaviors and kinetic analysis. *Energy Convers. Manag.* **2018**, *171*, 1106–1115. [CrossRef]
53. Sbirrazzuoli, N. Is the Friedman method applicable to transformations with temperature dependent reaction heat? *Macromol. Chem. Phys.* **2007**, *208*, 1592–1597. [CrossRef]
54. Boonchom, B. Kinetic and thermodynamic studies of $MgHPO_4 \cdot 3H_2O$ by non-isothermal decomposition data. *J. Therm. Anal. Calorim.* **2009**, *98*, 863–871. [CrossRef]
55. Cao, H.; Jiang, L.; Duan, Q.; Zhang, D.; Chen, H.; Sun, J. An experimental and theoretical study of optimized selection and model reconstruction for ammonium nitrate pyrolysis. *J. Hazard. Mater.* **2019**, *364*, 539–547. [CrossRef]
56. Chen, R.; Li, Q.; Xu, X.; Zhang, D. Pyrolysis kinetics and reaction mechanism of representative non-charring polymer waste with micron particle size. *Energy Convers. Manag.* **2019**, *198*, 111923. [CrossRef]
57. Ding, Y.; Huang, B.; Li, K.; Du, W.; Lu, K.; Zhang, Y. Thermal interaction analysis of isolated hemicellulose and cellulose by kinetic parameters during biomass pyrolysis. *Energy* **2020**, *195*, 117010. [CrossRef]
58. Song, Z.; Li, M.; Pan, Y.; Shu, C. A generalized differential method to calculate lumped kinetic triplet of the nth order model for the global one-step heterogeneous reaction using TG data. *J. Loss Prev. Process Indust.* **2020**, 104094. [CrossRef]

Article

Highly Insulated Wall Systems with Exterior Insulation of Polyisocyanurate under Different Facer Materials: Material Characterization and Long-Term Hygrothermal Performance Assessment

Emishaw Iffa [1,*], Fitsum Tariku [2] and Wendy Ying Simpson [2]

1 Oak Ridge National Laboratory, Building Envelope & Urban Systems Research, Oak Ridge, TN 37831, USA
2 British Columbia Institute of Technology, BCIT Building Science Centre of Excellence, Burnaby, BC V5G3H2, Canada; fitsum_tariku@bcit.ca (F.T.); wendy_simpson@bcit.ca (W.Y.S.)
* Correspondence: iffaed@ornl.gov; Tel.: +865-3410-470

Received: 28 June 2020; Accepted: 24 July 2020; Published: 30 July 2020

Abstract: The application of exterior insulation in both new construction and retrofits is a common practice to enhance the energy efficiency of buildings. In addition to increased thermal performance, the rigid insulation can serve to keep the sheathing board warm and serve as a water-resistive barrier to keep moisture-related problems due to condensation and wind-driven rain. Polyisocyanurate (PIR) rigid boards have a higher thermal resistance in comparison to other commonly used exterior insulation boards. However, because of its perceived lower permeance, its use as exterior insulation is not very common. In this study, the hygrothermal property of PIR boards with different facer types and thicknesses is characterized. The material data obtained through experimental test and extrapolation is used in a long term hygrothermal performance assessment of a wood frame wall with PIR boards as exterior insulation. Results show that PIR with no facer has the smallest accumulated moisture on the sheathing board in comparison to other insulation boards. Walls with a bigger thickness of exterior insulation perform better when no vapor barrier is used. The PIR exterior insulation supports the moisture control strategy well in colder climates in perfect wall scenarios, where there is no air leakage and moisture intrusion. In cases where there is trapped moisture, the sheathing board has a higher moisture content with PIR boards with both aluminum or fiberglass type facers. An innovative facer material development for PIR boards can help efforts targeting improved energy-efficient and durable wall systems.

Keywords: drying and wetting; hygrothermal performance; Polyisocyanurate board; moisture content; thermal performance; vapor permeability

---

This manuscript has been authored in part by UT-Battelle, LLC, under contract DE-AC05–00OR22725 with the US Department of Energy (DOE). The US government retains and the publisher, by accepting the article for publication, acknowledges that the US government retains a nonexclusive, paid-up, irrevocable, worldwide license to publish or reproduce the published form of this manuscript, or allow others to do so, for US government purposes. DOE will provide public access to these results of federally sponsored research in accordance with the DOE Public Access Plan (http://energy.gov/downloads/doe-public-access-plan).

1. Introduction

The ever-increasing demand from building codes for improved energy efficiency and society's increasing awareness of environmental sustainability is driving the building construction and manufacturing industries to develop innovative solutions for durable, high-performance buildings. It is well documented that the application of exterior insulation increases the overall thermal performance of new construction and retrofits [1–5]. While studying the thermal performance of a building envelope, the moisture durability needs proper consideration as well [6,7] The moisture control performance requires additional investigation and researchers are examining the effects of rigid insulation on the durability of wall systems incorporating innovative materials and construction practices [8–14].

Among the commonly used insulation materials, Polyisocyanurate (PIR) is known for its higher R-value and consequently providing an increased energy efficiency when it is compared with most of the other foam-based insulation materials [15]. However, the building industry considers PIR as impermeable or semi-permeable, and its application as exterior insulation is limited. Regarding material characterization and hygrothermal performance assessment, relatively smaller research studies are reported in comparison to the thermal performance of exterior insulation.

Burch and Desjarlais conducted a water vapor measurement test for PIR core and facer [16]. Their measurement shows that the glass-mat facers permeance varies from 600 ng/sm2.Pa (10 perm) to 2800 ng/sm2.Pa (49 perm). A study on advanced material preparation and characterization of PIR foams is an active research topic. Kosmela et al. have found that the addition of up to 30% by weight of bio-polyol, instead of foams prepared solely with a petrochemical polyol, have increased the reactivity of the polyol mixture in rigid Polyurethane-Polyisocyanurate(PUR-PIR) foams, which in turn has enhanced the thermal performance of the rigid foam [17]. Borowiciz et al. have also found that PIR foams modified with bio-polyol based on mustard seed oil have lowered the thermal conductivity and water absorption [18]. Berardi and Madzarevic analyzed the aging behavior and tested the blowing agent concentration of a PIR foam. A decrease in 11% and 85% of a blowing agent is measured from the aged PIR foams [19].

The closed cells created during PIR manufacturing are filled with the vaporized blowing agent during the foaming reaction [20]. To keep the blowing agent from migrating out and in return affecting the R-value of the PIR foam, different types of facers are used during PIR production [20–23]. Mackaveckas et al. reported on the influence of different PIR facings on thermal performance. Their findings show a significant heat loss in PIR insulation boards with aluminum facing at wall corners [24].

The main purpose of this study is to investigate the long-term hygrothermal performance of wall systems with PIR exterior insulation. To study the optimal use of the PIR insulation board, the thickness and type of facer materials were varied and their performance under different scenarios: varying moisture loads, vapor barrier applications, and different climates were investigated. Their performance was compared with another exterior rigid insulation board, extruded polystyrene (XPS).

The material characterization for the PIR took place at BCIT's Building Science Centre of excellence (BSCE). The measured thermal and hygrothermal material properties include density, thermal conductivity, heat capacity, sorption isotherm, vapor permeability, water absorption coefficient, and porosity. The next sections of this paper discuss the experimental test, the simulation setup, results obtained, and the conclusions based on experiment and simulation results.

2. Hygrothermal Property Characterization of PIR Foam with Different Facers

To study the long-term performance of hygrothermal simulation performance of PIR as an exterior insulation under different climates, a hygrothermal material property characterization of three types of PIR products is conducted: fiberglass-faced PIR, PIR with aluminum facer and an unfaced PIR insulation. A series of laboratory measurements are carried out to characterize the hygrothermal properties of a rigid PIR insulation board with different facing materials. The hygrothermal properties measured are density, thermal conductivity, specific heat capacity, sorption isotherm, water vapor permeability, water absorption coefficient, and porosity. Measurements are conducted in accordance with the ASTM Standards [25–28]. The measurement procedures, standards used and material characterization results are shown in the section below.

The PIR hygrothermal property characterization under different facer materials for PIR of thickness 12.7 mm (0.5 in.), 25.4 mm (1 in.) and 38.1 mm (1.5 in.) are measured through a laboratory test.

The obtained properties are extrapolated for other thickness sizes. The measured hygrothermal properties of PIR core insulation, PIR with fiberglass facer, and PIR with foil-faced PIR insulation are presented below.

The density is determined from physical dimensions and oven-dry mass measurements of six specimens for all three PIR samples, as shown in Table 1.

Table 1. Density measurement of Polyisocyanurate (PIR) board.

PIR Facer Type	Nominal Thickness mm (in.)	Actual Thickness mm (in.)	Density Kg/m^3
Fiberglass facer	25.4 (1)	23.9 mm (15/16)	55.14 ± 0.42
	12.7 (0.5)	12.2 (0.48)	104.07 ± 6.33
Aluminum facer	38.1 (1.5)	36.7 mm (1–7/16)	34.3 ± 0.51
No facer core board	38.1 (1.5)	36.7 mm (1–7/16)	27.86 ± 0.48

The thermal conductivity of the PIR core board and with foil and fiberglass facers is carried out according to ASTM Standard C518: "Standard Test Method for Steady-State Thermal Transmission Properties by Means of the Heat Flow Meter Apparatus "using four – 30 cm × 30 cm specimens (average thickness 24.86 mm). The measurements are done at a temperature difference of 20 °C (68 °F) and a mean temperature of 24 °C (75.2 °F). Table 2 shows the average thermal conductivity and resistance values of the four PIR board samples for all types of facers. In addition, the maximum and the minimum measured values are also presented in the table.

Table 2. Thermal conductivity of PIR board.

Facer Type	Specimen Thickness mm	Mean Temperature °C	Hot Plate Temperature °C	Cold Plate Temperature °C	Thermal Conductivity W/(m·K)	Thermal Resistance m^2·K/W
Fiberglass	24.12 ± 0.04	24.00 ± 0.03	34.01 ± 0.04	14.01 ± 0.04	0.026 ± 0.0007	0.944 ± 0.031
Aluminum facer	24.98 ± 0.02	24.02 ± 0.00	34.02 ± 0.00	14.02 ± 0.00	0.024 ± 0.0008	1.072 ± 0.038
No facer	24.86 ± 0.02	23.95 ± 0.05	33.96 ± 0.05	13.95 ± 0.05	0.023 ± 0.0008	1.076 ± 0.047

The heat capacity of the PIR for the three samples is determined using a LaserComp Fox heat flow meter (TA instruments, New Castle, DE, USA) and WinTherm32 software (TA instruments, New Castle, DE, USA) for analysis and determination of transient heat transfer through the specimens. Based on four samples measurements, both volumetric heat capacity and specific heat capacity are measured for all three types of PIR with different facers. Table 3 shows the average volumetric and heat capacity values of PIR boards.

Table 3. Heat capacity of PIR board.

PIR Facer	Volumetric Heat Capacity J/(m^3·K)	Specific Heat Capacity J/(kg·K)
Fiberglass facer	62943 ± 1083	1144 ± 23
Aluminum facer	49357 ± 1058	1439 ± 31
No facer	38090 ± 944	1257 ± 14

Sorption Isotherm: PIR boards with fiberglass facer and no facer are used under this study. Because of its low permeability, the aluminum-faced PIR board is not investigated for its hygrothermal property characterization. The equilibrium moisture contents at 50%, 70%, 80%, 90% and 95% relative humidity (RH) conditions and saturation moisture content in 100% RH are determined according to ASTM Standard C1498: "Standard Test Method for Hygroscopic Sorption Isotherms of Building Materials". For each test point, three specimens with dimensions of 100 mm × 100 mm with a thickness of 24 mm are used. The measurements are carried out in controlled climatic chambers that are maintained at a constant temperature of 23 °C and the desired relative humidity set point. A water immersion test is used to determine the capillary saturation moisture contents of the samples. Table 4 shows the measured equilibrium moisture contents of PIR board specimens at different relative humidity.

Table 4. Equilibrium moisture contents of PIR board samples.

RH, %	Moisture Content, kg/m^3	
	Fiberglass Facer PIR	**No Facer PIR**
50	0.48 ± 0.02	0.36 ± 0.008
70	0.63 ± 0.03	0.49 ± 0.009
80	0.72 ± 0.02	0.55 ± 0.009
90	0.94 ± 0.01	0.68 ± 0.004
95	1.05 ± 0.02	0.77 ± 0.006
100	11.30 ± 0.262	8.79 ± 0.468

The water vapor permeability of PIR insulation is determined according to ASTM Standard E-96: "Standard Test Methods for Water Vapor Transmission of Materials". Using climatic chambers with wet cups and dry cups methods, the water vapor transmission rates of PIR samples are determined. The climatic chambers are set at 50%, 70% and 90% relative humidity and 23 °C temperature. The dry (0%) and wet (100%) relative humidity conditions in the test caps are provided by calcium chloride ($CaCl_2$) desiccant and distilled water, respectively. For each test, three circular specimens of 11.94 cm in diameter are used as a replica. The average water vapor permeability values of PIR boards at different mean sample relative humidity are shown in Table 5.

Table 5. Vapor permeability of PIR board.

Sample Mean RH (%)	Chamber RH (%)	Cup RH (%)	Permeability kg/(s·m·Pa)	
			PIR with Fiberglass Facer	**No Facer PIR**
25	50	0	$9.813 \times 10^{-13} \pm 8.897 \times 10^{-13}$	$2.732 \times 10^{-12} \pm 1.789 \times 10^{-13}$
35	70	0	$1.110 \times 10^{-12} \pm 5.045 \times 10^{-13}$	$2.849 \times 10^{-12} \pm 1.711 \times 10^{-13}$
45	90	0	$1.28 \times 10^{-12} \pm 1.009 \times 10^{-13}$	$3.022 \times 10^{-12} \pm 1.786 \times 10^{-13}$
75	50	100	$1.570 \times 10^{-12} \pm 1.443 \times 10^{-13}$	$3.223 \times 10^{-12} \pm 1.287 \times 10^{-13}$
85	70	100	$2.408 \times 10^{-12} \pm 2.418 \times 10^{-13}$	$3.620 \times 10^{-12} \pm 1.375 \times 10^{-13}$

The water absorption coefficient of the PIR samples is determined according to ASTM Standard C1794: "Standard Test Methods for Determination of the Water Absorption Coefficient by Partial Immersion." Three test specimens, 100 mm × 100 mm with a thickness of 37 mm each, are used for the measurements. The lab conditions are 33.7 ± 0.2 °C temperature and 39.8 ± 3% relative humidity. Water was maintained at 20.6 ± 0.5 °C. The measured water absorption coefficients of the PIR are found to be 0.0004 kg/(m^2·s$^{1/2}$) and 0.0007 kg/(m^2·s$^{1/2}$) for fiberglass facer and no facer PIR boards, respectively.

The porosity of PIR samples are obtained by determining the full saturation weight of the samples following the water immersion test procedure. Based on these measurements of three replica specimens, the measured porosity values of no facer and fiberglass-faced of PIR boards are 31.37 ± 1.34% and 20.51 ± 0.39%, respectively.

Since the hygrothermal property of the PIR with fiberglass is dependent on both the hygrothermal properties of PIR core and the facer material, a hygrothermal property of fiberglass-faced PIR and the PIR core is measured. By subtracting the water vapor resistance of the fiberglass-faced PIR from PIR core, the permeability of the fiberglass facer can be easily calculated. Once the independent permeability values of the facer material and PIR core are found, the overall permeability values of the fiberglass-faced PIR is computed. Figure 1 shows the water vapor transmission rate of fiberglass-faced PIR and PIR core at different thicknesses.

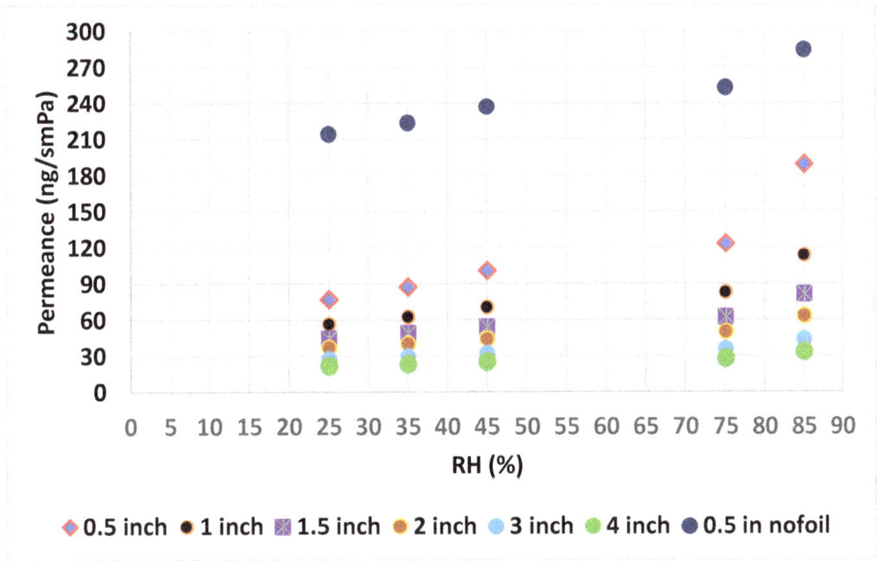

Figure 1. Water vapor permeance of fiberglass-faced Polyisocyanurate (PIR).

3. Hygrothermal Simulation

To study how the facing parts of rigid PIR exterior insulation affect the hygrothermal performance of a building envelope, two-year WUFI®simulations for a highly insulated rainscreen wall systems are conducted. This section discusses the simulation setup, initial and boundary conditions, simulation assumptions.

3.1. Simulation Set up

The indoor relative humidity and temperature conditions are set based on the ASHRAE standard 160 P Intermediate model. The input parameters varied to simulate the dynamic response of the wall system in different climates, with different types and thicknesses of exterior insulations, rain infiltration, and vapor barrier. The wall system that is considered for the study comprises the following layers of materials:

- regular Portland Stucco as an exterior cladding;
- 19 mm ($\frac{3}{4}$ in.) rainscreen air gap;
- different types and thicknesses of exterior insulations;
- spun-bonded polyolefin as a sheathing membrane (as the second plane of protection from precipitation and water intrusion);
- plywood sheathing board;
- fiberglass insulation in 2 × 6 wood frame studs;

- 6 mm (1/4 in.) polyethylene sheet as a vapor and air barrier;
- gypsum board as an interior finishing layer

Table 6 shows the variation in simulation parameters during this modeling work. Three types of facing options (aluminum foil, fiberglass facer, and no facer) and two insulation thicknesses (2 in. and 4 in.) are considered in the study to evaluate the effect of PIR thickness and different facer materials on the overall hygrothermal performance of a wall system. Even though using a PIR product without facers is uncommon, incorporating this parameter shows how an innovative vapor-open facer product can enhance the hygrothermal performance of an envelope. While the material properties of the extruded polystyrene (XPS) were obtained from the WUFI® database, the properties for the PIR insulation with different facer material properties are measured.

Table 6. Simulation parameters.

	Parameter
Climate	Vancouver, BC (Zone 4C) Winnipeg, MB (Zone 7)
Exterior insulation (type)	PIR with aluminum foil PIR with fiberglass facing PIR with no facing XPS
Exterior insulation (thickness)	2 in. 4 in.
Rain infiltration	0% 0.1% 0.5%

3.2. Boundary and Initial Conditions

The indoor conditions of relative humidity and temperature are set using the ASHRAE standard 160 P intermediate model. The external surface is exposed to the weather conditions of Vancouver, BC, to study the PIR performance with different facers and thickness under different moisture infiltration and combined effects of vapor barrier and exterior insulation. Weather data of two North American cities (Winnipeg, MB, Canada and Vancouver, BC, Canada) are used to study how a PIR exterior insulation of different facers performs from a hygrothermal perspective. The initial conditions of 20 °C and 80% RH are used for all wall component members and the simulation ran for two consecutive years. WUFI's weather data for cities of Winnipeg, MB, and Vancouver is used and a cold year data is selected.

3.3. Modeling Assumptions

Figure 2 shows the wall assembly considered in the study. Continuity of vapor/air barrier (polyethylene sheet) is assumed to be maintained in the modeling. Therefore, there is no airflow through the wall system. The PIR boards with fiberglass and aluminum facers hygrothermal properties are lumped together based on the measured values of the core and facer material. The wall system was assumed to be with no deficiency and the layers of materials to be in perfect contact, exhibiting no dimensional physical change with time. The material properties of the wall components other than the PIR boards which are used in the simulation are shown in Table 7.

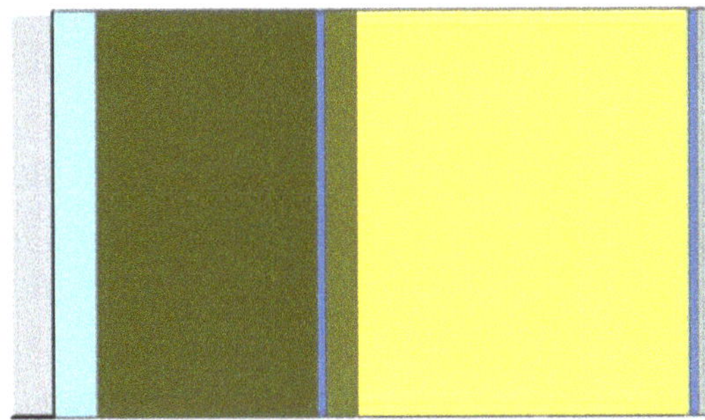

Components from left to right

1. Stucco
2. Air gap
3. Exterior insulation
4. WRB
5. Plywood
6. Fibreglass insulation
7. Vapor barrier
8. Gypsum board

Figure 2. The wall assembly.

Table 7. Material properties used for simulation.

Material	Density (kg/m^3)	Porosity (m^3/m^3)	Heat Capacity (J/kgK)	Thermal Conductivity (W/mK)	Diffusion Resistance Factor (-)
Stucco	1955.5	0.225	840	0.399	355.7
XPS	28.6	0.99	1470	0.025	170.56
Plywood	470	0.69	1880	0.084	1078.2
Spun bonded Polyolefin membrane	448	0.001	1500	2.4	328.4
Fiberglass	30	0.99	840	0.035	1.3

4. Hygrothermal Simulation Results and Discussions

In this section, based on the hygrothermal properties characterization of the PIR material, the long-term performance of a wood frame wall with PIR as exterior insulation under different thickness and facer materials for different climate zones is presented and discussed.

This section presents the results of a two-year hygrothermal assessment of four different types of exterior insulations, namely PIR with aluminum foil; PIR with fiberglass facer, PIR with no facer and XPS.

The effects of wind-driven rain, application of vapor barrier, different climates, and the overall performance of wall assemblies under different types and thicknesses of exterior insulations are presented.

4.1. *Effect of Wind-Driven Rain*

The Vancouver weather, a climate zone known for its heavy annual rainfall, is used to analyze how moisture infiltrated into an interior wall system due to wind-driven rain can affect the overall performance of a wood frame wall with exterior insulation.

Rain infiltration percentages of 0%, 0.1%, and 0.5% are assumed to reach the sheathing board (plywood). Figures 3a, 4a and 5a show the dynamic moisture content of plywood in wall systems with two-inch exterior insulation with different rain infiltration percentages. Similarly, Figures 3b, 4b and 5b represent moisture content values of plywood in wall systems with four-inch exterior insulation at 0%, 0.1%, and 0.5% infiltration percentages. In all cases, constant cavity ventilation of 100 ACH is assumed.

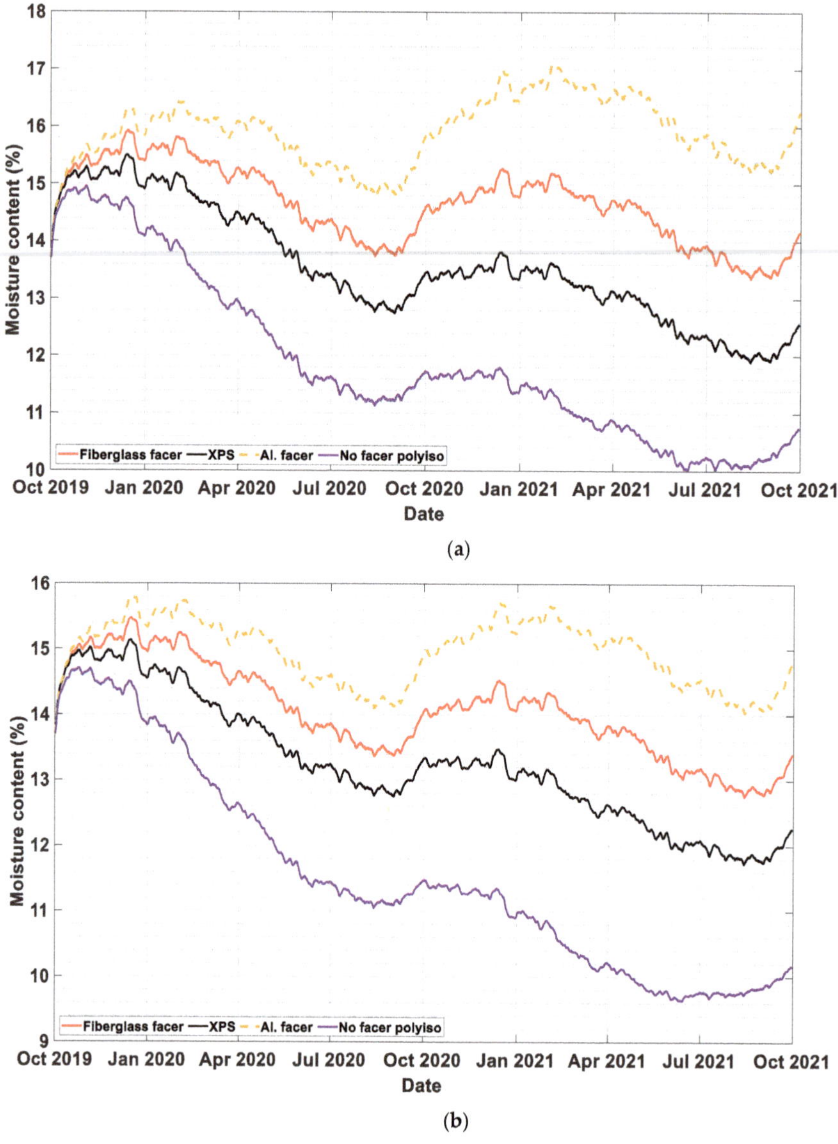

(a)

(b)

Figure 3. The moisture content of plywood with 0% rain infiltration in wall systems with (**a**) two-inch (**b**) four-inch exterior insulations.

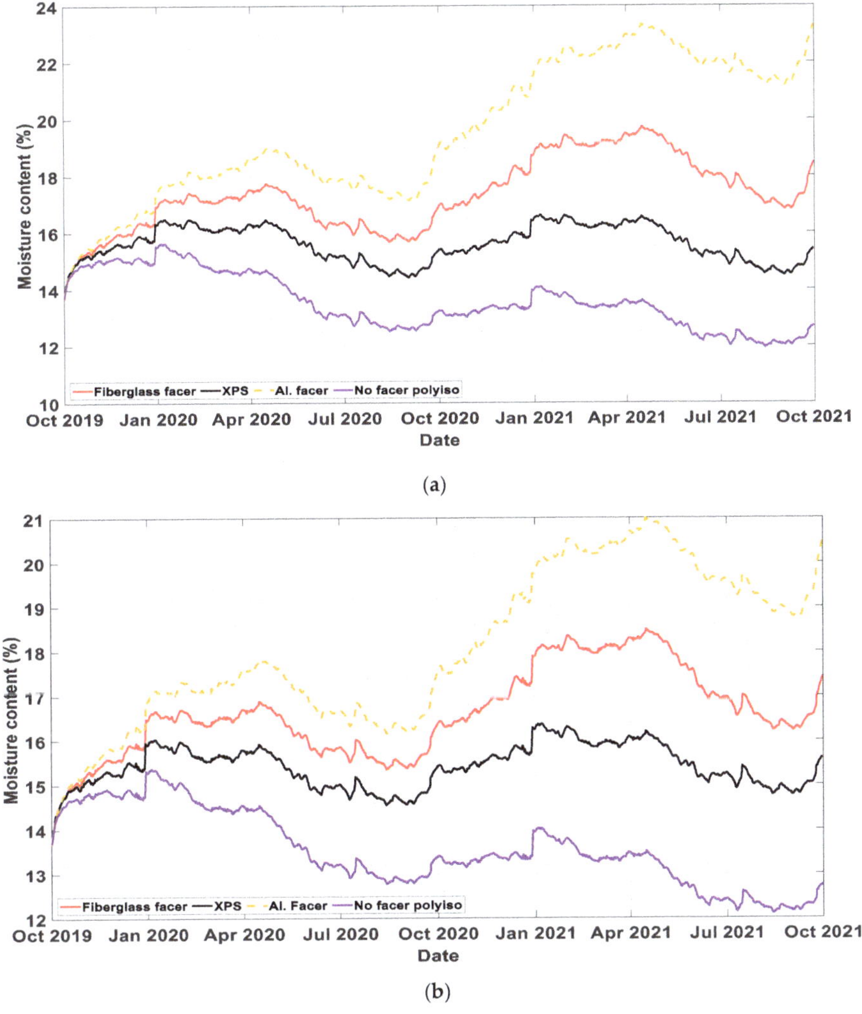

Figure 4. The moisture content of plywood with 0.1% rain infiltration in wall systems with (**a**) two-inch (**b**) four-inch exterior insulations.

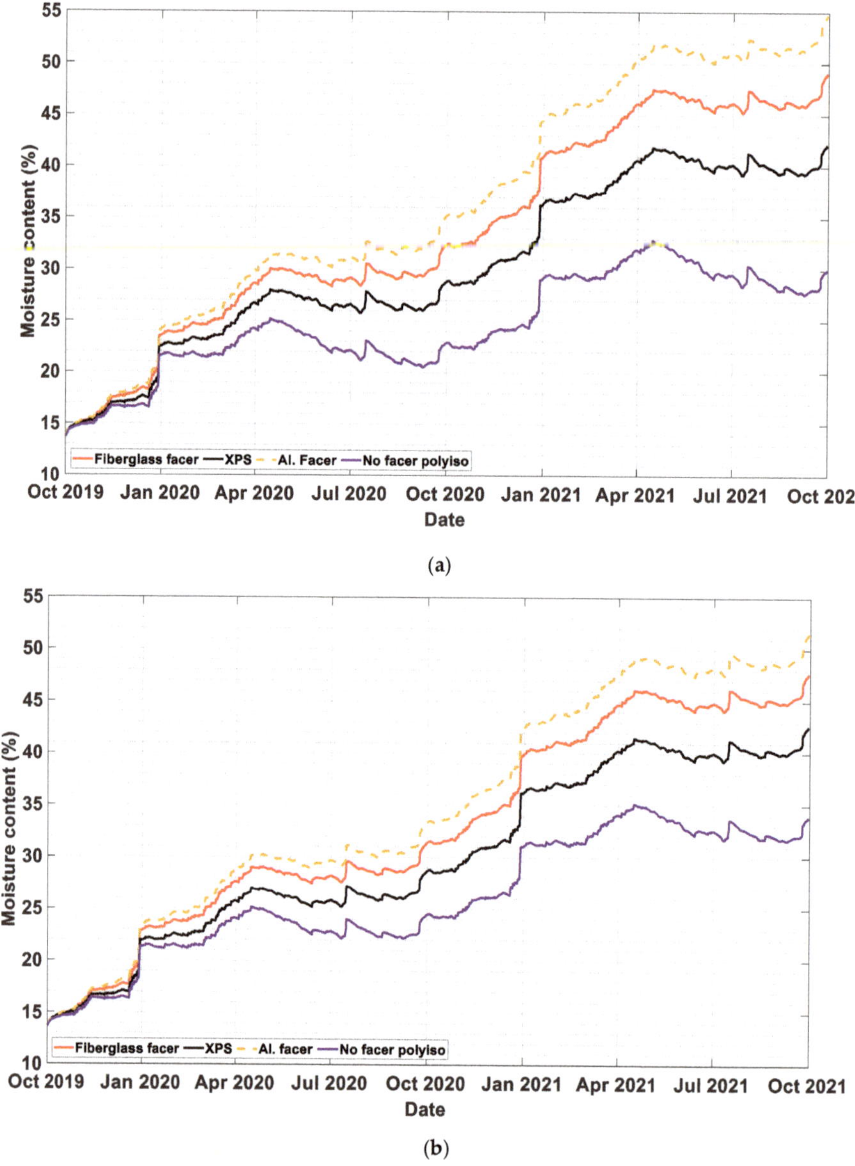

Figure 5. The moisture content of plywood with 0.5% rain infiltration in wall systems with (**a**) two-inch (**b**) four-inch exterior insulations.

In most cases, the performance ranking from the least to the most accumulation of moisture on the plywood during the simulation period was a PIR with no facer, XPS, PIR with fiberglass facer, and PIR with aluminum foil. In cases without rain infiltration, starting from the last quarter of the second year, the wall with PIR exterior insulations without facer outperformed the other walls with different exterior insulation by a bigger margin.

In cases where the rain infiltration was not the dominant moisture source, for all wall systems and exterior insulation types, envelope systems with four-inch insulation performed better than envelope systems with two-inch insulation. This is because the thicker exterior insulation keeps the plywood warmer, minimizing the potential for condensation.

The moisture content of the plywood remained at an acceptable level for all wall systems when there is no rain infiltration. In simulation cases where 0.1% of the wind-driven rain was added in the plywood, all wall systems, except the wall with PIR with aluminum foil insulation, remained below the critical moisture level of 18%, for both 2 in and 4 in continuous simulation as shown in Figure 4a,b, respectively. This result shows if the moisture infiltration is managed to an acceptable limit most foam insulations with the exception of vapor impermeable foam insulation can be applied.

To simulate an extreme case scenario, a rain infiltration of 0.5% is assumed. As Figure 5a,b show, most of the wall systems were subjected to very high moisture accumulation in the sheathing board. In this case the moisture control performance of the wall systems has failed in both 2-in and 4-in continuous insulations of the all simulated foam boards.

4.2. Combined Effects of Vapor Barrier and Exterior Insulation

Many building codes require an application of a vapor barrier to be used in building envelopes to enhance the moisture durability of wall systems. This study examines if the PIR exterior insulation can serve as a vapor barrier and what are the combined effects of vapor barrier and exterior insulation in wall systems.

Figure 6a,b show the effect of the vapor barrier on the hygrothermal performance of a wall system under different exterior insulations. Figure 6a shows when the wall includes a two-inch thick exterior insulation, the wall systems with vapor barrier performed better in all four exterior insulation cases. The walls with no vapor barrier and either exterior insulation of PIR with aluminum foil, PIR with fiberglass facing, or XPS registered a high moisture content. From the wall systems with the vapor barrier, the only wall exposed to a beyond-critical moisture content was the wall with PIR with aluminum foil insulation.

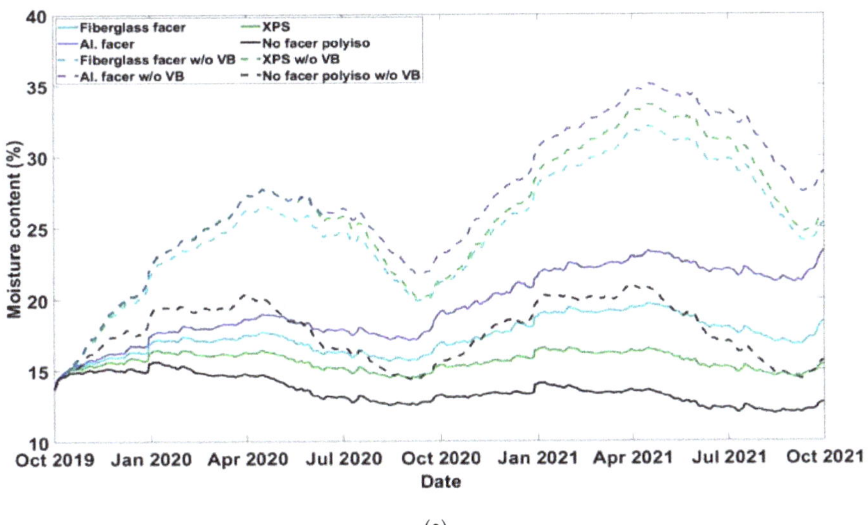

(a)

Figure 6. *Cont.*

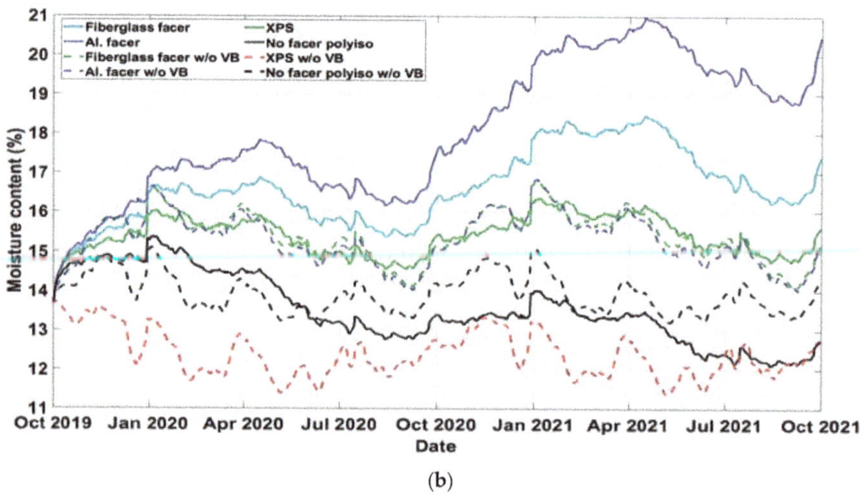

(b)

Figure 6. The moisture content of plywood in wall systems with and without vapor barrier and (**a**) two-inch and (**b**) four-inch exterior insulation thicknesses.

Figure 6b shows the insulation thickness increased from two inches to four inches for all four types of exterior insulations. In all types of exterior insulations, the wall systems without vapor barrier perform better than walls with a vapor barrier. This due to the thick exterior insulation facilitating a warm condition for the plywood, discouraging condensation. The lack of a vapor barrier provides a route for moisture from rain infiltration to dry to the inside. The plywood moisture content stays below 18% in all wall systems without a vapor barrier.

4.3. Hygrothermal Performance of Exterior Insulations under Different Climates

This study looked at the hygrothermal performance of the exterior insulation under the different climates of the cities of Winnipeg, MB and Vancouver, BC. These locations represent wet-coastal and cold-dry climates, respectively. This simulation assumed 0.1% rain infiltration reaching the plywood surface and constant air cavity ventilation of 100 ACH. Figure 7a,b show the moisture content of plywood for the different types of exterior insulations of two-inch and four-inch thicknesses, respectively. In all four exterior insulation cases, the plywood moisture content is lower in Winnipeg cases than those in Vancouver. This is due to Vancouver's higher annual rainfall. In both two-inch and four-inch thicknesses of PIR insulation with aluminum foil, the moisture content of the plywood exceeded 18%. As shown in Figure 7a, the maximum moisture content of the plywood in wall systems with two-inch fiberglass-faced PIR for Winnipeg and Vancouver was 17.78% and 19.71%, respectively. However, in the case of four-inch-thick fiberglass-faced PIR, the maximum moisture content of the plywood for Winnipeg and Vancouver cases was 16.70% and 18.47%, respectively, as shown in Figure 7b. The moisture content of the plywood in all wall systems with no-facer PIR insulation remained below 15% throughout the simulation period.

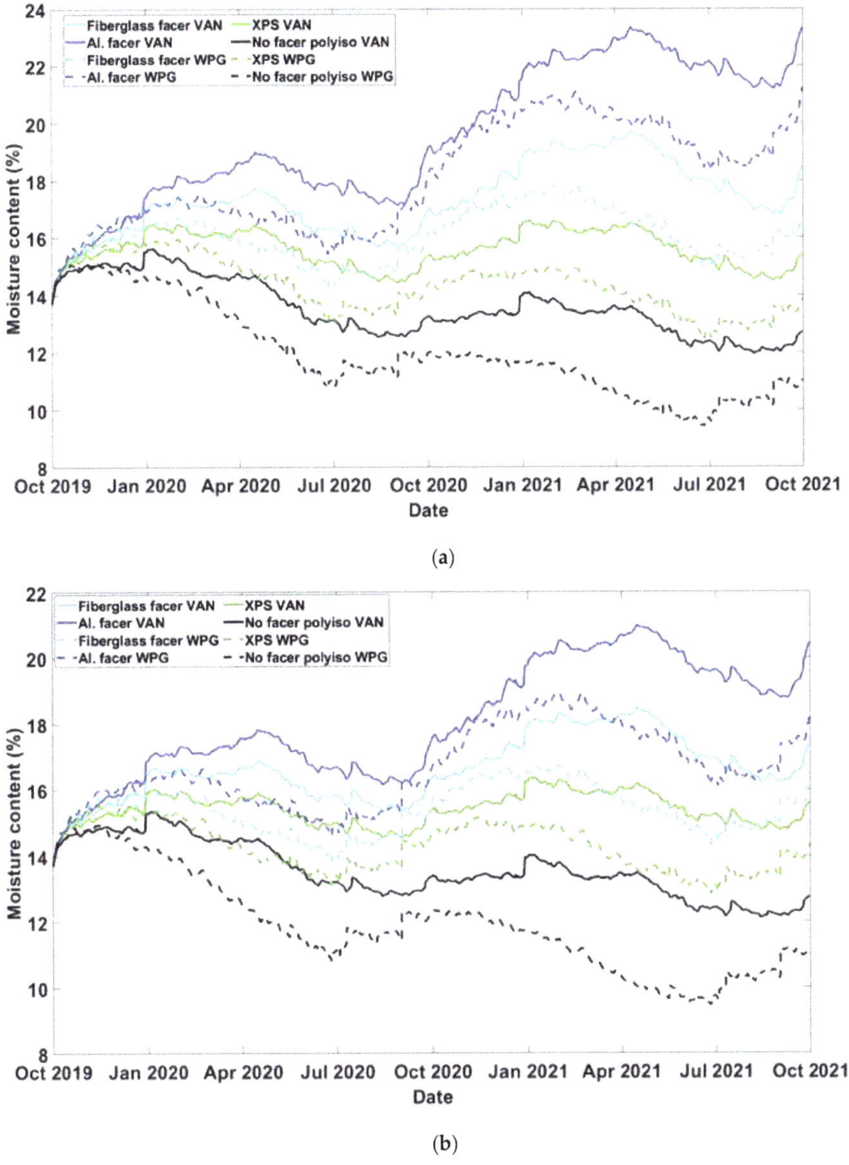

Figure 7. The moisture content of plywood in wall systems in Winnipeg, MB and Vancouver, BC with (**a**) two-inch and (**b**) four-inch exterior insulation thicknesses.

5. Conclusions

This study characterized the thermal and hygrothermal property of a PIR insulation with different facers. Based on the measured material property data, the long-term hygrothermal performance assessment of PIR insulation as rigid exterior insulation was examined. The experimental measurements show that most properties (such as density, water permeability and sorption isotherm) vary significantly as the thickness increases due to PIR being a composite of facer and PIR core. The material data presented here can be used in future modeling works to accurately simulate the hygrothermal property of PIR board with different thicknesses and facer materials.

In addition, this study examined specific application parameters such as thickness and facer types of the insulation core, simulations with varied rain infiltration rates, vapor barrier applications, and climates. Results show that thicker insulation provided a better moisture control strategy because it helped the sheathing board to stay warm. Facer materials used in the PIR insulation significantly affected the hygrothermal performance of wall systems. The hygrothermal performance of the PIR board PIR with unfaced insulation outperformed the XPS and the PIR Boards with aluminum and fiberglass facers. In light of its superior thermal performance in comparison to that of most foam insulation boards, PIR insulation with no facer or vapor-open facer material could contribute significantly to the current demands of high-performance, durable building construction practices.

Author Contributions: Conceptualization, E.I. and F.T.; Methodology, E.I. and F.T.; Software, E.I. Formal Analysis, E.I.; Investigation, E.I., W.Y.S. and F.T.; Resources, F.T.; Writing—Original Draft Preparation, E.I.; Writing—Review & Editing, E.I. and F.T. All authors have read the manuscript and agreed to publish it.

Funding: This manuscript has been partially authored by UT-Battelle, LLC, under Contract No. DE-AC05-00OR22725 with the US Department of Energy. The authors would like to acknowledge the funding provided by NSERC Engage grants and the school of Construction and Environment at British Columbia Institute of Technology.

Conflicts of Interest: The authors declare no conflict of interest. The funders had no role in the design of the study; in the collection, analyses, or interpretation of data; in the writing of the manuscript, or in the decision to publish the results.

References

1. Ibrahim, M.; Biwole, P.H.; Wurtz, E.; Achard, P. A study on the thermal performance of exterior walls covered with a recently patented silica-aerogel-based insulating coating. *Build. Environ.* **2014**, *81*, 112–122. [CrossRef]
2. Isaia, F.; Fantucci, S.; Capozzoli, A.; Perino, M. Vacuum insulation panels: Thermal bridging effects and energy performance in real building applications. *Energy Procedia* **2015**, *83*, 269–278. [CrossRef]
3. Kossecka, E.; Kosny, J. Effect of insulation and mass distribution in exterior walls on dynamic thermal performance of whole buildings. In Proceedings of the Thermal Performance of the Exterior Envelopes of Buildings VII, Clearwater, FL, USA, 6–10 December 1998; pp. 721–731.
4. Khoukhi, M. The combined effect of heat and moisture transfer dependent thermal conductivity of polystyrene insulation material: Impact on building energy performance. *Energy Build.* **2018**, *169*, 228–235. [CrossRef]
5. Kochkin, V.; Wiehagen, J. *Construction Guide to Next-Generation High-Performance Walls in Climate Zones 3-5-Part 1: 2x6 Walls (No. DOE/EE-1673-1)*; Home Innovation Research Labs: Upper Marlboro, MD, USA, 2017.
6. Li, Y.; Dang, X.; Xia, C.; Ma, Y.; Ogura, D.; Hokoi, S. The effect of air leakage through the air cavities of building walls on mold growth risks. *Energies* **2020**, *13*, 1177. [CrossRef]
7. Ferdyn-Grygierek, J.; Kaczmarczyk, J.; Blaszczok, M.; Lubina, P.; Koper, P.; Bulińska, A. Hygrothermal risk in museum buildings located in moderate climate. *Energies* **2020**, *13*, 344. [CrossRef]
8. Ge, H.; Straube, J.; Wang, L.; Fox, M.J. Field study of hygrothermal performance of highly insulated wood-frame walls under simulated air leakage. *Build. Environ.* **2019**, *160*, 106202. [CrossRef]
9. Ibrahim, M.; Wurtz, E.; Biwole, P.H.; Achard, P.; Sallee, H. Hygrothermal performance of exterior walls covered with aerogel-based insulating rendering. *Energy Build.* **2014**, *84*, 241–251. [CrossRef]
10. Fox, M.; Straube, J.; Ge, H.; Trainor, T. Field test of hygrothermal performance of highly insulated wall assemblies. In Proceedings of the 14th Canadian Conference on Building Science and Technology, Toronto, ON, Canada, 28–30 October 2014; Volume 2830, pp. 101–110.
11. Kočí, V.; Maděra, J.; Černý, R. Exterior thermal insulation systems for AAC building envelopes: Computational analysis aimed at increasing service life. *Energy Build.* **2012**, *47*, 84–90. [CrossRef]
12. Langmans, J.; Klein, R.; Roels, S. Hygrothermal risks of using exterior air barrier systems for highly insulated light weight walls: A laboratory investigation. *Build. Environ.* **2012**, *56*, 192–202. [CrossRef]
13. Pescari, S.; Tudor, D.; Tölgyi, S.; Măduţa, C. Study Concerning the Thermal Insulation Panels with Double-Side Anti-Condensation Foil on the Exterior and Polyurethane Foam or Polyisocyanurate on the Interior. *Key Eng. Mater.* **2015**, *660*, 244–248. [CrossRef]

14. Wang, L.; Ge, H. Stochastic modelling of hygrothermal performance of highly insulated wood framed walls. *Build. Environ.* **2018**, *146*, 12–28. [CrossRef]
15. *Handbook, ASHRAE: Fundamentals 2013*; ASHRAE: Atlanta, GA, USA, 2013.
16. Burch, D.M.; Desjarlais, A.O. *Water Vapor Measurements of Low Slope Roofing Materials*; Report Submitted to National Institute of Standards and Technology: Gaithersburg, MD, USA, 1995.
17. Kosmela, P.; Hejna, A.; Suchorzewski, J.; Piszczyk, Ł.; Haponiuk, J.T. Study on the structure-property dependences of rigid pur-pir foams obtained from marine biomass-based biopolyol. *Materials* **2020**, *13*, 1257. [CrossRef]
18. Borowicz, M.; Paciorek-Sadowska, J.; Lubczak, J.; Czupryński, B. Biodegradable, flame-retardant, and bio-based rigid polyurethane/polyisocyanurate foams for thermal insulation application. *Polymers* **2019**, *11*, 1816. [CrossRef]
19. Berardi, U.; Madzarevic, J. Microstructural analysis and blowing agent concentration in aged polyurethane and polyisocyanurate foams. *Appl. Therm. Eng.* **2020**, *164*, 114440. [CrossRef]
20. Mukhopadhyaya, P.; Bomberg, M.; Kumaran, M.; Drouin, M.; Lackey, J.; Van Reenen, D.; Normandin, N. Long-term thermal resistance of polyisocyanurate foam insulation with impermeable facers. In *Insulation Materials: Testing and Applications: 4th Volume*; ASTM International: West Conshohocken, PA, USA, 2009.
21. Marrucho, I.M.; Santos, F.M.; Oliveira, N.; Dohrn, R. Aging of Rigid Polyurethane Foams: Thermal Conductivity of N2 and Cyclopentane Gas Mixtures. *J. Cell. Plast.* **2005**, *41*, 207–224. [CrossRef]
22. Bogdan, M.; Hoerter, J.; Moore, F.O. Meeting the insulation requirements of the building envelope with polyurethane and polyisocyanurate foam. *J. Cell. Plast.* **2005**, *41*, 41–56. [CrossRef]
23. Stovall, T. *Closed Cell Foam Insulation: A Review of Long Term Thermal Performance Research*; Oak Ridge National Laboratory: Oak Ridge, TN, USA, 2012.
24. Makaveckas, T.; Bliūdžius, R.; Burlingis, A. The Influence of Different Facings of Polyisocyanurate Boards on Heat Transfer through the Wall Corners of Insulated Buildings. *Energies* **2020**, *13*, 1991. [CrossRef]
25. ASTM Standard C518-17. *Standard Test Method for Steady-State Thermal Transmission Properties by Means of the Heat Flow Meter Apparatus*; ASTM International: West Conshohocken, PA, USA, 2017.
26. ASTM C1794-19. *Standard Test Methods for Determination of the Water Absorption Coefficient by Partial Immersion*; ASTM International: West Conshohocken, PA, USA, 2019.
27. ASTM C1498-04a. *Standard Test Method for Hygroscopic Sorption Isotherms of Building Materials*; ASTM International: West Conshohocken, PA, USA, 2016.
28. ASTM E96/E96M-16. *Standard Test Methods for Water Vapor Transmission of Materials*; ASTM International: West Conshohocken, PA, USA, 2016.

Article

Optimization of Multilayered Walls for Building Envelopes Including PCM-Based Composites

Victor D. Fachinotti [1,*], Facundo Bre [1], Christoph Mankel [2], Eduardus A. B. Koenders [2] and Antonio Caggiano [2,3]

1 Centro de Investigación de Métodos Computacionales (CIMEC), Universidad Nacional del Litoral (UNL), Consejo Nacional de Investigaciones Científicas y Técnicas (CONICET), Santa Fe 3000, Argentina; facubre@cimec.santafe-conicet.gov.ar

2 Institut für Werkstoffe im Bauwesen, Technische Universität Darmstadt, 64287 Darmstadt, Germany; mankel@wib.tu-darmstadt.de (C.M.); koenders@wib.tu-darmstadt.de (E.A.B.K.); caggiano@wib.tu-darmstadt.de (A.C.)

3 Consejo Nacional de Investigaciones Científicas y Técnicas (CONICET), LMNI, INTECIN, Facultad de Ingeniería, Universidad de Buenos Aires, Ciudad Autónoma de Buenos Aires C1127AAR, Argentina

* Correspondence: vfachinotti@cimec.unl.edu.ar; Tel.: +54-342-451-1594 (ext. 7045)

Received: 18 May 2020; Accepted: 17 June 2020; Published: 20 June 2020

Abstract: This work proposes a numerical procedure to simulate and optimize the thermal response of a multilayered wallboard system for building envelopes, where each layer can be possibly made of Phase Change Materials (PCM)-based composites to take advantage of their Thermal-Energy Storage (TES) capacity. The simulation step consists in solving the transient heat conduction equation across the whole wallboard using the enthalpy-based finite element method. The weather is described in detail by the Typical Meteorological Year (TMY) of the building location. Taking the TMY as well as the wall azimuth as inputs, EnergyPlus^{TM} is used to define the convective boundary conditions at the external surface of the wall. For each layer, the material is chosen from a predefined vade mecum, including several PCM-based composites developed at the Institut für Werkstoffe im Bauwesen of TU Darmstadt together with standard insulating materials (i.e., EPS or Rockwool). Finally, the optimization step consists in using genetic algorithms to determine the stacking sequence of materials across the wallboard to minimize the undesired heat loads. The current simulation-based optimization procedure is applied to the design of envelopes for minimal undesired heat losses and gains in two locations with considerably different weather conditions, viz. Sauce Viejo in Argentina and Frankfurt in Germany. In general, for each location and all the considered orientations (north, east, south and west), optimal results consist of EPS walls containing a thin layer made of the PCM-based composite with highest TES capacity, placed near the middle of the wall and closer to the internal surface.

Keywords: thermal-energy storage (TES); phase change materials (PCMs); multilayered walls; building envelopes; non-linear optimization; genetic algorithms

---

1. Introduction

According to the International Energy Agency (IEA), the buildings and construction sector across the world was responsible for 36% of final energy use and 39% of carbon dioxide (CO_2) emissions in 2018 [1]. In Germany—and in Europe as well—buildings consume almost 40% of the total energy [2]. Far from Europe, in Argentina, a similar situation is faced: buildings are the largest energy consumers (33% of the total energy) [3].

Engineers, scientists, and decision makers all around the world are being appealed to promote best construction practices for using the least, cleanest, and/or most economical resources. The optimization of the energy efficiency in buildings plays a key role in this sense. It can be achieved by actuating on three main building components [4]: (i) building envelope (insulation, sealing, windows, and façades) [5,6]; (ii) technical building equipment (lighting, heating, and ventilation systems) [7,8]; and (iii) sustainable building systems (i.e., dealing with substantial aspects such as sustainable energies, automatization operations, and efficiency-centered planning) [9–11].

In this work, we are particularly interested in building envelopes, which are of key importance for controlling the heat loads in new buildings and/or retrofitted ones [12]. More particularly, we are interested in the use of phase change materials (PCMs) in building envelopes, making the so-called "energy dynamic building envelopes" [13]. These envelopes take advantage of the dynamic synergetic interplay between the thermal conductivity and the thermal-energy storage (TES) capacity of their components.

PCMs can be directly added to construction materials of building envelopes as microencapsulated PCMs [14,15] or raw PCMs into carriers (e.g., impregnated gypsum boards [16] or porous aggregates [17]). Building envelopes can also contain macro-encapsulated PCMs having several shapes (plates, spheres, tubes, pipes, etc.) [18].

Usually, an energy dynamic building envelope is a multilayer wall. Most researchers place PCM-based layers either in contact [19,20] or very close to the building interior [21,22], without making any quantitative comparative analysis of position and thickness. However, Shi et al [23], who did compare the effect of placing the PCM-based layer innermost, outermost or in the middle of the wall, recommend to place it in the middle based on thermal comfort criteria for Hong Kong.

Thus, the placement and thickness of PCM-based layers remain an open question. Among other open questions, we aim to answer the four following questions: (i) When is it convenient to use PCMs? (ii) Which PCMs should be chosen? (iii) With which standard (non-PCM) materials should they be combined? (iv) How many layers are needed and/or how thick should the wall be in total? Any possible question has multiple answers. This work aims to offer a specific answer to each one while considering the performance index (energy efficiency, cost, environmental impact, etc.), the solar azimuth, and the local weather conditions along a typical year.

To determine the performance of a multilayer wall that is part of the building envelope, the heat exchange between the outer and inner environments through this wall must be known. To this end, we solve the transient heat conduction equation across the wall. This equation is generally highly nonlinear in the presence of PCMs due to the peaks in their effective heat capacity during latent heat release/absorption. Besides nonlinearity, another great challenge arises when a whole year must be considered. In this case, the external boundary condition varies on an hourly basis according to the typical meteorological year (TMY) of the building location. For the robust solution of such nonlinear equation at all the time steps, we use the enthalpy-based finite element method with implicit time integration [24].

Then, to optimize the performance of a multilayered wall in the building envelope, let each layer be made of a material consciously chosen from a predefined vade mecum of appropriate candidates, which especially include cementitious composites containing different fractions of PCMs either microencapsulated [25] or embedded into recycled brick aggregates (RBAs) [17], developed at the Institut für Werkstoffe im Bauwesen (WiB) of TU Darmstadt.

This vade mecum may also contain standard construction materials such as concrete and insulation materials (EPS, Rockwool, etc.), thus it can offer a wide choice of material properties for optimization purposes. Having a rich enough vade mecum, the evaluation of all the possible designs can be excessively expensive. Since we aim to continuously update and enlarge this vade mecum all along the recently launched four-year-long NRG-STORAGE project [26], we are obliged to use efficient optimization algorithms. Given its effectiveness when dealing with many integer design variables and discontinuities induced by such variables together with its low computational cost (mainly due to its easy parallelization), among other properties that are greatly appreciated in building performance simulation, and following our expertise in previous works [27–29], we decided to use genetic algorithms (GA).

The paper is structured as follows. Section 2 introduces the optimization-based methodology for the design of multilayered walls for building envelopes to be optimal in the light of a given performance index (as example, a purely thermal criterion is defined here). Section 3 describes the thermal modeling of a multilayered wall in a building envelope, pointing out the influence of local weather conditions on boundary conditions and determining the temperature as a function of the design. Section 4 is devoted to describing the vade mecum of materials for building envelopes, with emphasis on the PCM-based composites it contains. Section 5 shows and discuss the results of applying this methodology to walls with different orientations located in two different climates. Finally, Section 6 addresses the concluding remarks and future work.

2. Optimization-Based Design of Wallboards for Building Envelopes

Let us consider the N-layer wallboard with total thickness L. As shown in Figure 1, $x \in [0, L]$ denotes the distance to the external surface and the layer $i = 1, 2, \ldots, L$ has thickness Δx_i and lies between x_{i-1} and $x_i = x_{i-1} + \Delta x_i$.

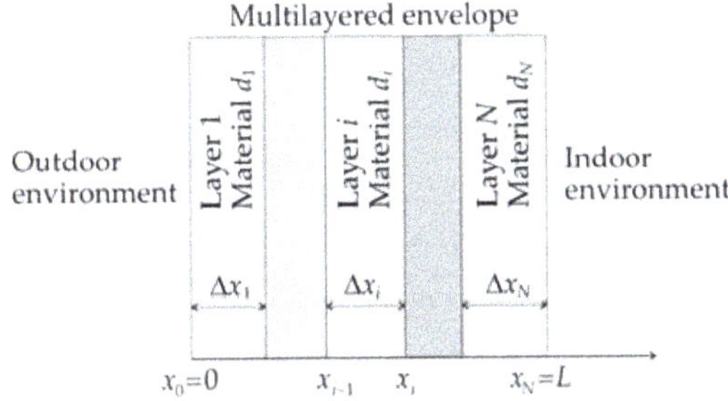

Figure 1. Schema of a building envelope using a multilayered wall.

To optimize the performance of the wall, each layer is allowed to be made of a different material. Let the integer d_i identify the material at layer $i = 1, 2, \ldots, N$. In the context of optimization, d_i plays the role of a *categorical design variable* that can take a finite number of *levels*. Each level is an integer that identifies a candidate material. In general, each d_i can take its own levels; in other words, the layer i can have its particular set of material choices; for instance, layer 1 (the external one) may only allow hydrophobic and non-flammable materials.

Each possible design of the N-layered wall is identified by the set $\boldsymbol{d} = [d_1, d_2, \ldots, d_N]$. Then, we propose to determine the optimal design, say $\boldsymbol{d}^{\text{opt}}$, by solving the following optimization problem (Equation (1)):

$$\boldsymbol{d}^{\text{opt}} = \text{argmin} C(\boldsymbol{d}), \tag{1}$$

where C is the cost function representing the performance of the wall.

Several choices for C can be found in the literature: the energy consumption for comfort in air-conditioned rooms and the degree-hours of discomfort in naturally-ventilated rooms, which can be considered either as multiple objectives [28,29] or combined in a weighted sum as a unique objective [27], the life cycle cost [30], the environmental impact [30,31], and a sum of the initial investment and the energy bill minus the resale value [32]. In these works, the cost function involves the whole building, but they can perfectly serve to characterize the performance of the wallboards enveloping these buildings.

In general, the problem in Equation (1) is subject to inequality constraints that serve to prescribe specific thresholds for the thermal transmittance (U-value) [32], the structural compliance, the weight of the whole wall, etc. The objective C and these may be interchangeable. For instance, one could either minimize the energy demand without exceeding a given budget or minimize the budget without exceeding a given energy demand.

For example, let us consider a purely thermal performance criterion adopting as cost function the total undesired heat loads all along the 8760 h of a typical meteorological year (TMY) (Equation (2)):

$$C(\boldsymbol{d}) = \underbrace{\sum_{h=1}^{8760} h_{\text{int}} \langle T_{\text{surfint}}^{(h)}(\boldsymbol{d}) - T_{\text{tgt}}^{\max} \rangle}_{\text{undesired gains } C_{\text{gain}}} + \underbrace{\sum_{h=1}^{8760} h_{\text{int}} \langle T_{\text{tgt}}^{\min} - T_{\text{surfint}}^{(h)}(\boldsymbol{d}) \rangle}_{\text{undesired losses } C_{\text{loss}}} \tag{2}$$

where h_{int} is the heat convection coefficient between the wall and the indoor environment, $T_{\text{surfint}}^{(h)}$ is the temperature at the internal wall surface at hour h, $T_{\text{tgt}}^{\max}$ and $T_{\text{tgt}}^{\max}$ are the maximum and minimum target (ideal) indoor temperatures (maybe time-dependent), and $\langle u \rangle = (u + |u|)/2$ is the ramp function such that there is no contribution to the undesired heat gains C_{gain} if $T_{\text{int}}^{(h)} < T_{\text{tgt}}^{\max}$ or to the undesired heat losses if C_{loss} if $T_{\text{surfint}}^{(h)} > T_{\text{tgt}}^{\min}$. In the above equation, h_{int}, $T_{\text{tgt}}^{\max}$, and $T_{\text{tgt}}^{\min}$ are assumed to be given, while $T_{\text{surfint}}^{(h)}$ is determined by the design $\boldsymbol{d}$ in a way to be defined in the next section.

3. Temperature Evolution Across the Wall

The temperature T at a distance $x \in [0, L]$ from the external surface of the wall at the time instant $t > 0$ is governed by the heat conduction Equation (3)

$$\rho c_{\text{eff}} \dot{T} - \frac{\partial}{\partial x}\left(k \frac{\partial T}{\partial x}\right) = 0 \tag{3}$$

subject to the initial conditions (Equation (4))

$$T(x, 0) = T_0 \tag{4}$$

and the boundary conditions (Equations (5) and (6))

$$k \frac{\partial T}{\partial x} = q_{\text{ext}} \quad \text{at } x = 0 \text{ (external surface)} \tag{5}$$

$$-k \frac{\partial T}{\partial x} = h_{\text{int}}(T - T_{\text{room}}) \quad \text{at } x = L \text{ (internal surface)} \tag{6}$$

In the above equations $\dot{T} = \partial T / \partial t$ is the temperature rate, ρc_{eff} is the effective heat capacity, k is the thermal conductivity, q_{ext} is the heat flux from the outdoor environment through the external wall surface (which evolves in time following the local weather conditions as described in Section 3.1), h_{int} is the heat convection coefficient at the internal wall surface (the same as in Equation (2)) and T_{room} is the indoor room temperature.

Given the multilayered nature of the wall, the physical properties k and ρc_{eff} are layer-wise defined: at a distance $x \in (x_{i-1}, x_i)$ from the external surface, they are those of the material in layer i, that is the material d_i in the vade mecum (see Figure 1).

In general, material properties depend on temperature (Equations (7)). Further, the effective heat capacity in PCMs is rate-dependent during phase changes [33]. Thus,

$$\left.\begin{aligned} k &= k(d_i, T) \\ \rho c_{\text{eff}} &= \rho c_{\text{eff}}\big(d_i, T, \dot{T}\big) \end{aligned}\right\} \quad \text{at } x \in (x_{i-1}, x_i) \text{ (layer } i\text{)} \tag{7}$$

The temperature dependence of material properties makes the heat Equation (3) nonlinear. This nonlinearity becomes severe in presence of PCMs due to the peaks in their heat capacity during phase changes. Then, it is crucial to develop a robust solver of the heat conduction Equation (3). Here, recourse is made to the enthalpy-based finite element formulation proposed by Morgan et al. [24]. For the sake of conciseness, the reader interested in the finite element implementation is referred to the just cited work.

Finally, since the thermal properties k and ρc_{eff} at each layer i are dependent on d_i, it becomes apparent that the temperature T at any point x across the wall at any time instant t depends on the whole design $\boldsymbol{d} = [d_1, d_2, \ldots, d_N]$, namely $T = T(x, t, \boldsymbol{d})$. Particularly, the temperature at the internal surface at the hour h is $T^{(h)}_{\text{surfint}}(\boldsymbol{d}) = T(L, h \text{ hours}, \boldsymbol{d})$, making explicit the influence of the design $\boldsymbol{d}$ on the performance of the wall as represented by the cost function $C(\boldsymbol{d})$ given by Equation (2).

3.1. External Boundary Conditions

The heat flux from the outdoor environment is defined as follows (Equations (8)):

$$q_{\text{ext}}(t) = \alpha q_{\text{solar}}(t) + h_{\text{ext}}(t)(T_{\text{out}}(t) - T_{\text{surfext}}(t)) \tag{8}$$

where q_{solar} is the absorbed short-wave (direct and diffuse) solar radiation, α is the solar absorbance (assumed equal to 0.6 h_{ext} is the external heat transfer coefficient T_{out} is the outdoor temperature, and $T_{\text{surfext}}(t) = T(0, t, \boldsymbol{d})$ is the temperature at the external surface.

The heat flux q_{ext} depends on t not only via $T_{\text{surfext}}(t)$ but also through T_{out}, h_{ext} and q_{solar}, which change following the instantaneous local weather conditions as defined by the Typical Meteorological Year (TMY).

Regarding T_{out}, this coincides with either the dry bulb temperature T_{db} for $h_{\text{ext}} < 1000$ W/(m^{2}K) or the wet-bulb temperature T_{wb} for $h_{\text{ext}} = 1000$ W/(m^{2}K), with T_{db} and T_{wb} directly taken from the TMY, where they vary on an hourly basis. On the other hand, q_{solar} and the convection coefficient h_{ext} are computed using EnergyPlus™ [34]. To take into account not only the local weather conditions but also the influence of the surface facing angle on these variables, EnergyPlus is applied to a square 4-m-wide, 4-m-deep, and 3-m-high thermal zone with walls facing north, east, south, and west. Regarding h_{ext}, it is computed using the so-called *AdaptiveConvectionAlgorithm* in EnergyPlus, which takes into account the wind direction and magnitude and sets h_{ext} to an arbitrarily high value (1000 W/(m^{2}K)) at the wall exposed to the wind when it is raining, forcing to assume $T_{\text{out}} = T_{\text{db}}$ at that instant.

4. Vade Mecum of Materials for Building Envelopes

To improve the performance of a multilayer wall, let each layer be built of a material thoughtfully chosen from the henceforth called vade mecum of materials for building envelopes. This is a database that should contain a wide choice of materials in terms of thermal properties (conductivity, specific heat, and thermal energy storage capacity, etc.) as well as non-flammability, water and air tightness, weight, cost, and embedded energy, among others, such that it offers a large enough design space for optimization purposes. Mathematically speaking, each material in the vade mecum is a level of the categorical design variable d_i. Note that a vade mecum containing M materials gives rise to N^M different designs for an N-layered wall.

As a first step, the current vade mecum is built on the basis of purely thermal criteria. Particularly, we are interested in using two cement-based PCM-composites, henceforth referred to as MPCM-p and RBA-p, with p related to the PCM content.

The MPCM-p is a concrete with $w/c = 0.45$, 70 vol% of normal aggregates (granitic crushed stones) and 30 vol% of a PCM paste containing p =0, 10, or 20 vol% of microencapsulated PCM. The PCM is Micronal® DS 5038 X type, which is a powder of microencapsulated paraffin wax developed by BASF, with a melting point of around 26 °C and a heat storage capacity of 145 kJ/kg [35]. The effective heat capacity of MPCM-p mixtures as distinct temperature-dependent functions for either heating or cooling are shown on the left of Figure 2 (see [25] for more details on these mixtures).

The RBA-p is a concrete with $w/c = 0.50$, 30 vol% of surrounding cement paste, and 70 vol% of recycled brick aggregates (RBA), which are filled with $p = 0$, 65, and 80 vol% of PCM. In this case, the PCM is the non-encapsulated paraffin wax RT 25 HC® developed by RUBITHERM®, with a melting point of around 25 °C and a heat storage capacity of 210 kJ/kg [36]. The effective heat capacity of MPCM-p mixtures are also temperature-dependent functions that differ for heating and cooling, and are depicted on the right of Figure 2. More details on RBA-p composites are given in [17].

To enlarge the choice of thermal properties, the vade mecum also contains two widely used insulating materials: rockwool and expanded polystyrene (EPS). The current vade mecum is summarized in Table 1. Let us note that this is a first version, to be continuously enriched with additional properties (cost, embedded energy, etc.) and enlarged all along the recently launched project NRG-STORAGE, mainly to account for other performance indexes (cost, environmental aspects, etc.). Further, the wall must exhibit fire safety and water tightness, among other essential requirements for real life applications. Considering fire safety for instance, PCM-based concretes offer a non-flammable choice; on the contrary, EPS and wools exhibit a high flammability, which is a well-known critical issue yet to be solved. Thus, although indispensable, these additional requirements may be detrimental to the performance of the wall; therefore, in the optimization problem in Equation (1), they are not represented by the cost function C but by inequality constraints.

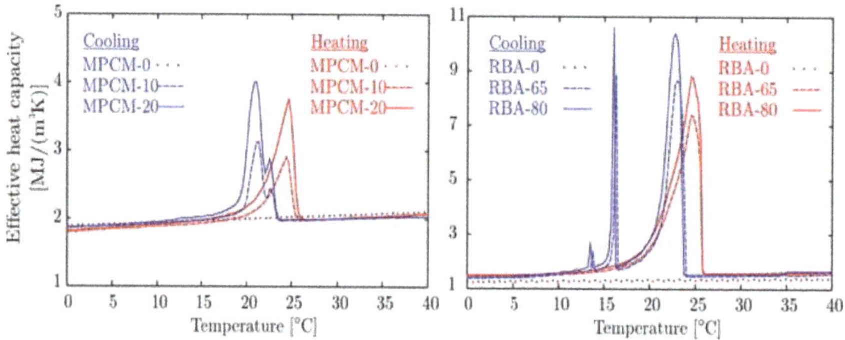

Figure 2. Effective heat capacities of MPCM-p concretes (on the **left**) and RBA-p concretes (on the **right**).

Table 1. Vade mecum of materials for building envelopes.

Level	Name	Effective Thermal Conductivity [W/(mK)]	Effective Heat Capacity [J/(m^{3}K)]
1	MPCM-0	2.336	Temperature dependent *
2	MPCM-10	2.311	Temperature dependent *
3	MPCM-20	2.311	Temperature dependent *
4	RBA-0	0.910	Temperature dependent *
5	RBA-65	0.769	Temperature dependent *
6	RBA-80	0.769	Temperature dependent *
7	Rockwool	0.042	33,750
8	EPS	0.030	60,000

* see Figure 2.

5. Numerical Results

This section reports the results of the optimization of multilayered systems for external walls, considering four different orientations (N, S, W, and E) and two different locations (Sauce Viejo in Argentina and Frankfurt in Germany). Sauce Viejo has a humid subtropical climate, which is considered as Cfa in the Köppen–Geiger classification; the average temperatures along typical summer and winter weeks are 24.4 and 13.8 °C, respectively. Frankfurt has a warm temperate climate, Cfb in the Köppen–Geiger classification, with average temperatures of 18.0 and 1.7 °C, respectively, for typical summer and winter weeks. For later discussions concerning orientation, let us keep in mind that Sauce Viejo and Frankfurt are in the Southern and Northern hemispheres, respectively.

For an accurate evaluation of the performance of a building envelope all along a year, the weather at each location is described by its typical meteorological year (TMY). This TMY is a database with relevant weather variables (including dry-bulb and dew-point temperatures, wind speed and direction, relative humidity, total sky cover, ceiling height, atmospheric pressure, global horizontal solar radiation, diffuse and direct normal solar radiation, precipitation, etc.) given on an hourly basis along twelve typical meteorological months, which were chosen from different years following the Sandia method [37]. The TMYs for Sauce Viejo and Frankfurt can be downloaded for free from Climate.OneBuilding [38]; that at Sauce Viejo was recently generated by Bre and Fachinotti [39].

As pointed out in Section 2, the cost function $C(\boldsymbol{d})$ (to be minimized) represents the undesired heat loads along a TMY defined by Equation (2). We further assume that: (i) the room temperature T_{room} appearing in the boundary condition at the internal surface (Equation (6)) is ideally maintained at 24 °C; (ii) the maximal and minimal temperatures involved in the definition of $C(\boldsymbol{d})$ (Equation (2)) are set to the same value, i.e., $T_{\text{tgt}}^{\min} \equiv T_{\text{tgt}}^{\min} \equiv T_{\text{room}} = 24$ °C; and (iii) the convection coefficient of the internal surface in Equations (2) and (6) is set to the typical value $h_{\text{int}} = 8.24\ \text{W}/\left(\text{m}^2\text{K}\right)$, as adopted by Biswas et al. [21,22] for non-reflective vertical interior walls.

5.1. Reference Solutions on Homogenous Walls

For comparison purposes, let us start by considering a homogeneous 20-cm-thick external wall made of one of the candidate materials found in the vade mecum (Table 2).

Table 2. Vade mecum of multilayered systems for building envelopes.

Location	Orientation	Layer *																			
		1	2	3	4	5	6	7	8	9	10	11	12	13	14	15	16	17	18	19	20
Sauce Viejo	N	8	8	8	8	8	8	8	8	8	6	6	6	6	8	8	8	8	8	8	8
	S	8	8	8	8	8	8	8	8	8	6	6	8	8	8	8	8	8	8	8	8
	E	8	8	8	8	8	8	8	8	8	8	8	8	6	6	8	8	8	8	8	8
	W	8	8	8	8	8	8	8	8	8	8	6	6	8	8	8	8	8	8	8	8
Frankfurt	N	8	8	8	8	8	8	8	8	8	8	6	8	8	8	8	8	8	8	8	8
	S	8	8	8	8	8	8	8	8	8	8	6	8	8	8	8	8	8	8	8	8
	E	8	8	8	8	8	8	8	8	8	8	6	8	8	8	8	8	8	8	8	8
	W	8	8	8	8	8	8	8	8	8	8	6	8	8	8	8	8	8	8	8	8

* Layers 1 and 20 are the outermost and the innermost, respectively. Materials 6 and 8 are RBA-80 and EPS, respectively.

The transient heat conduction Equation (3) was solved using the enthalpy-based finite element method [24] with Euler-backward (implicit) time stepping. A previous analysis was carried out to determine the best deal between and computational cost, from which we decided to use four linear finite elements per layer and constant time steps of half-an-hour (i.e., 17,520 time steps along the TMY). To take due account of the steep variation of the effective heat capacity in PCM-based composites, six Gauss points were used in the corresponding finite elements, while two Gauss points (as usual) were used in the remaining finite elements.

Then, 64 problems were solved, one per each location, material, and orientation. From now on, for the sake of simplicity, let us refer simply as N to the case of a wall facing N, and so on for E-, S, and W-facing cases.

At Sauce Viejo, as shown in Figure 3, the undesired heat gains C_{gain} and losses C_{loss} are balanced. Using the PCM–concrete composites, C_{gain} is maximal for the N and minimal for S; for W and E, it is almost as prejudicial as for N. Regarding C_{loss}, its maximum and minimum occur at S and N, respectively. Once again, E and W are closer to the worst case (S). In general, considering the total undesired loads $C = C_{\text{gain}} + C_{\text{loss}}$, the worst case is W, followed by E, N and S, in that order. Using insulating materials (either rockwool or EPS), the total C considerably reduced. Best performances are achieved using EPS (the least conductive material of the vade mecum). For EPS, C attains its minimum value for S.

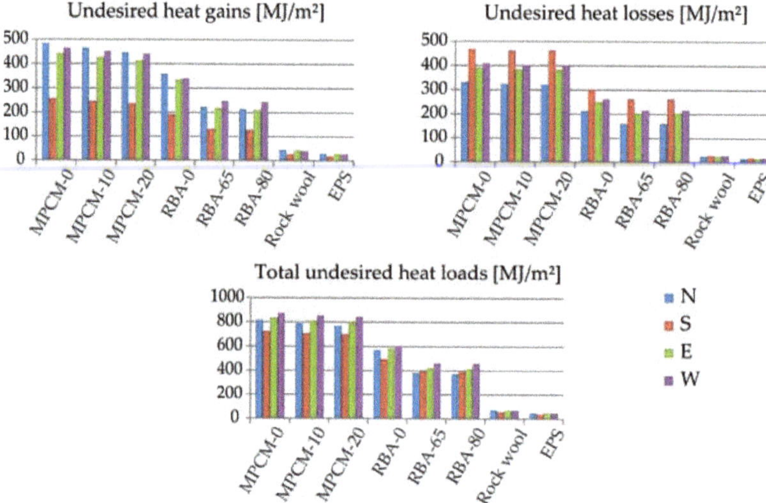

Figure 3. Undesired heat gains, heat losses and total heat loads in 20-cm-thickness external walls made of different materials at Sauce Viejo (Argentina) for different orientations.

At Frankfurt, as shown in Figure 4, C_{loss} is considerably higher than C_{gain}. This made the conclusions for C_{loss} also valid for the total heat loads C. By using the PCM–concrete composites, the performance of the wall is not as sensitive to the orientation as it is for Sauce Viejo. In general, N is the worst case, but E, W and S are not much better. Once again, the performance is greatly improved by using insulating materials.

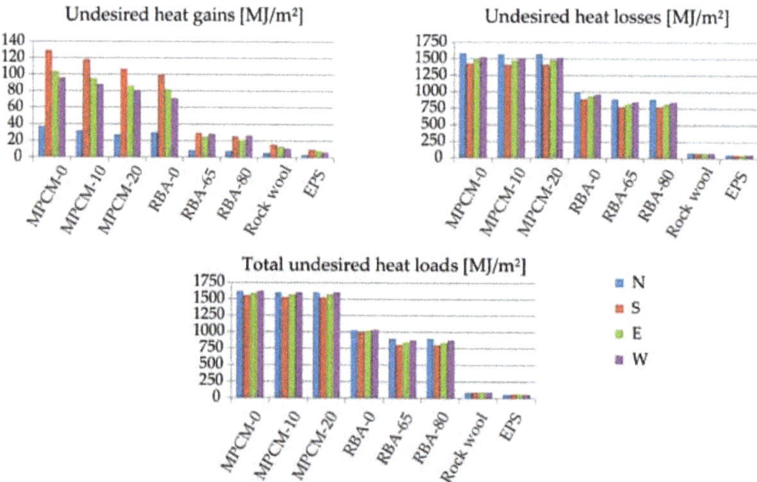

Figure 4. Undesired heat gains, heat losses and total heat loads in 20-cm-thickness external walls made of different materials at Frankfurt (Germany) for different orientations.

5.2. Optimal Multilayered Walls

Given the 20-cm-thick external wall of the preceding section, let it be made of 20 equally thick layers, and each layer is allowed to be made of one the eight materials in the vade mecum (Table 1). In this case, a design is actually a stacking sequence defined by $\boldsymbol{d} = [d_1, d_2, \ldots, d_{20}]$, where d_i identifies the material in layer i.

Then, the optimal stacking sequence is determined by solving the nonlinear, integer programming problem given by Equation (1) by using genetic algorithms (GA). Note that a homogeneous wall made of the material m in the vade mecum represents a possible design $\boldsymbol{d}$ of the current multilayered wall, where all the layers are made of the same material m, i.e., $d_i = m = constant$ for $i = 1, 2, \ldots, N$. The best homogeneous wall from the previous section (that made of EPS for all the cases) has been included in the initial population for GA.

The optimal solutions for both locations and the four orientations are shown in Table 2. Not surprisingly, only two materials, among the eight possible ones in the vade mecum, are present in all the optimal solutions: that with the lowest conductivity (level 8 = EPS) and that with the highest effective thermal energy storage capacity (level 6 = RBA-80). The great majority of the layers are made of EPS (16 or 18 for Sauce Viejo and 19 for Frankfurt).

At Sauce Viejo (see Figure 5), despite the little use of RBA-80, the energy performance of the optimal multilayered wall is considerable better than that of the EPS (insulating only) wall: C is reduced between 23.4% and 45.6%, bounds corresponding to S and N, respectively.

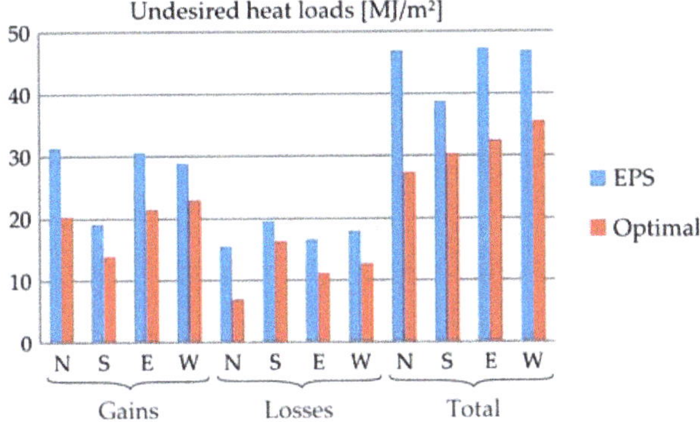

Figure 5. Undesired heat loads in a 20-cm-thick external wall at Sauce Viejo (Argentina) for different orientations: comparison between the EPS insulating wall and the optimal multilayered wall.

At Frankfurt, as shown in Figure 6, only Layer 11 is made of RBA-80 for all the considered orientations. For this optimal wall, C_{gain} is greatly reduced (from 70.6% for W to 78.9% for N). However, the weight of C_{gain} into C is considerably lower than that of C_{loss}, for which there is a modest improvement: from 0.4% for N to 8.3% for S. At the end, C is reduced between 4.8% for N to 18.8% for S.

In general, for Sauce Viejo (in the Southern hemisphere) as well as for Frankfurt (in the Northern one), the most beneficial effect of RBA-80 is observed at the most sun-exposed wall. Furthermore, the use of RBA-80 leads to considerably higher improvements in Sauce Viejo, where the temperatures remain a longer time in the phase-change range of RBA-80 (from 20 °C to 26 °C, approximately).

Finally, let us remark that Table 2 is the initial version of a new vade mecum of multilayered wall systems for building envelopes. For the time being, it offers a quick answer to the question of minimizing the undesired thermal loads at locations climatically close to those analyzed here. Once again, it is a goal of the recently launched NRG-STORAGE project to continuously enlarge this vade mecum to offer quick solutions for more climates considering different performance indexes.

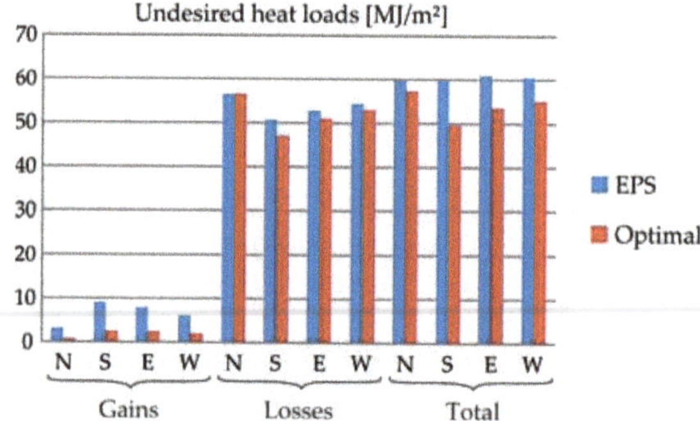

Figure 6. Undesired heat loads in a 20-cm-thick external wall at Frankfurt (Germany) for different orientations: comparison between the EPS insulating wall and the optimal multilayered wall.

6. Conclusions

This article introduces an optimization-based methodology to improve the performance of multilayer walls to be used in building envelopes. This is done in the following steps:

i. A vade mecum of materials is built to be used in building envelopes, including particularly materials with thermal energy storage capacity and insulating properties, which should offer a wide choice of material responses.
ii. Given the location and the wall azimuth, EnergyPlus is used to translate the local typical meteorological year (TMY) into time-dependent boundary conditions for the heat conduction equation.
iii. Given a multilayered wall "design", i.e., a specific stacking sequence of materials chosen from the vade mecum of Step i, the heat conduction equation, subject to the boundary conditions from Step ii, is solved along a whole Typical Meteorological Year (TMY) using a robust finite element method to determine the temperature across a given multilayered wall.
iv. The temperature evolution at the internal wall surface, resulting from Step iii, serves to determine the thermal performance index of the current design, here defined by a cost function representing the undesired thermal loads along a TMY.
v. Genetic algorithms are used to make the designs evolve until achieving optimal performance.
vi. Steps ii–v must be repeated first for the remaining orientations at the same location and then for different locations.

Here, the methodology was applied to optimize thermal performance of envelopes in Sauce Viejo (Argentina) and Frankfurt (Germany), having humid subtropical and warm temperate climates, respectively. Further, the walls were assumed to face north, east, south, and west. Considering a 20-cm-thick external wall, the optimal solution in any case is mostly made of EPS (the best insulating material in the current vade mecum), including a 1–4-cm-thick layer of a PCM-based composite (the material with the highest thermal energy storage capacity in the current vade mecum). In general, This PCM-based layer is placed next to the middle of the wall, closer to the internal surface. In this way, this methodology defines not only the proper placement but also the proper thickness of the PCM-based layer considering weather and orientation.

Furthermore, despite the little use of PCMs, the undesired heat loads were reduced in comparison to a 20-cm-thick EPS wall: up to 18.8% and 45.6% for Frankfurt and Sauce Viejo, respectively. The better performance at Sauce Viejo is explained by the fact that local temperatures remain for longer periods closer to the phase-change temperatures of the PCM-based composites available in the current vade mecum.

Further steps which follow from this research will address the enrichment of the vade mecum, including particularly PCM-based composites with various phase-change temperature ranges and materials.

Once the vade mecum of materials proves to be wide enough, optimal solutions can be obtained by applying the current methodology within a wide range of climates and different performance indexes, which will be taken as inputs for a new vade mecum of building envelopes.

Author Contributions: Conceptualization, V.D.F., F.B., C.M., and A.C.; data curation, all authors; formal analysis, V.D.F. and F.B.; writing—original draft preparation, V.D.F. and A.C.; writing, review and editing, all authors; supervision, V.D.F. and A.C.; project administration, V.D.F. and A.C.; and funding acquisition, A.C. and E.A.B.K. All authors have read and agreed to the published version of the manuscript.

Funding: This work represents a study of the research activities realized within the framework of the NRG-STORAGE project (number 870114, 2020-2024, https://nrg-storage.eu/), financed by the European Union H2020 Framework under the LC-EEB-01-2019 call, H2020-NMBP-ST-IND-2018-2020/H2020-NMBP-EEB-2019, IA type. The support to networking activities provided by the PoroPCM Project (part of the EIG CONCERT-Japan funding, http://concert-japan.eu/) is also gratefully acknowledged.

Conflicts of Interest: The authors declare no conflict of interest.

References

1. Brian, D.; John, D.; Ksenia, P.; Peter, G. *Towards Zero-Emission Efficient and Resilient Buildings. Global Status Report*; Global Alliance for Buildings and Construction: Paris, France, 2016.
2. European Commission. Energy Use in Buildings. Available online: https://ec.europa.eu/energy/eu-buildings-factsheets-topics-tree/energy-use-buildings_en (accessed on 9 April 2020).
3. Argentine Secretariat for Energy. *National Energy Balance—Year 2018—Revision 2*; Argentine Secretariat for Energy: Buenos Aires, Argentina, 2018; Available online: https://www.argentina.gob.ar/produccion/energia/hidrocarburos/balances-energeticos (accessed on 19 June 2020). (In Spanish)
4. German Federal Ministry for Economic Affairs and Energy. Available online: https://www.german-energy-solutions.de/GES/Navigation/EN/Energy-Solutions/EnergyEfficiencyInBuildings/energy-efficiency-in-buildings.html (accessed on 9 April 2020).
5. Sharma, V.; Rai, A.C. Performance assessment of residential building envelopes enhanced with phase change materials. *Energy Build.* **2020**, *208*, 109664. [CrossRef]
6. Li, Y.; Chen, L. A study on database of modular façade retrofitting building envelope. *Energy Build.* **2020**, *214*, 109826. [CrossRef]
7. Eleftheriadis, G.; Hamdy, M. The Impact of Insulation and HVAC Degradation on Overall Building Energy Performance: A Case Study. *Buildings* **2018**, *8*, 23. [CrossRef]
8. Grohman, W.M. System for Controlling HVAC and Lighting Functionality. U.S. Patent 10,248,088, 2 April 2019.
9. Moreno, M.V.; Zamora, M.A.; Skarmeta, A. User-centric smart buildings for energy sustainable smart cities. *Trans. Emerg. Telecommun. Technol.* **2013**, *25*, 41–55. [CrossRef]
10. Mlakar, U.; Stropnik, R.; Koželj, R.; Medved, S.; Stritih, U. Experimental and numerical analysis of seasonal solar-energy storage in buildings. *Int. J. Energy Res.* **2019**, *43*, 6409–6418. [CrossRef]
11. Buonomano, A.; Forzano, C.; Kalogirou, S.; Palombo, A. Building-façade integrated solar thermal collectors: Energy-economic performance and indoor comfort simulation model of a water based prototype for heating, cooling, and DHW production. *Renew. Energy* **2019**, *137*, 20–36. [CrossRef]
12. Sadineni, S.B.; Madala, S.; Boehm, R.F. Passive building energy savings: A review of building envelope components. *Renew. Sustain. Energy Rev.* **2011**, *15*, 3617–3631. [CrossRef]
13. Pisello, A.L.; Cotana, F.; Nicolini, A.; Buratti, C. Effect of dynamic characteristics of building envelope on thermal-energy performance in winter conditions: In field experiment. *Energy Build.* **2014**, *80*, 218–230. [CrossRef]
14. Thiele, A.M.; Sant, G.; Pilon, L. Diurnal thermal analysis of microencapsulated PCM–concrete composite walls. *Energy Convers. Manag.* **2015**, *93*, 215–227. [CrossRef]
15. Ramos, J.S.; Navarro, L.; Pisello, A.L.; Olivieri, L.; Bartolomé, C.; Sánchez, J.; Álvarez, S.; Tenorio, J.A. Behaviour of a concrete wall containing micro-encapsulated PCM after a decade of its construction. *Sol. Energy* **2019**. [CrossRef]

16. Castell, A.; Farid, M. Experimental validation of a methodology to assess PCM effectiveness in cooling building envelopes passively. *Energy Build.* **2014**, *81*, 59–71. [CrossRef]
17. Mankel, C.; Caggiano, A.; Koenders, E. Thermal energy storage characterization of cementitious composites made with recycled brick aggregates containing PCM. *Energy Build.* **2019**, *202*, 109395. [CrossRef]
18. Liu, Z.; Yu, Z.; Yang, T.; Qin, D.; Li, S.; Zhang, G.; Haghighat, F.; Joybari, M.M. A review on macro-encapsulated phase change material for building envelope applications. *Build. Environ.* **2018**, *144*, 281–294. [CrossRef]
19. Evola, G.; Marletta, L.; Sicurella, F. A methodology for investigating the effectiveness of PCM wallboards for summer thermal comfort in buildings. *Build. Environ.* **2013**, *59*, 517–527. [CrossRef]
20. Gowreesunker, B.; Tassou, S.A. Effectiveness of CFD simulation for the performance prediction of phase change building boards in the thermal environment control of indoor spaces. *Build. Environ.* **2013**, *59*, 612–625. [CrossRef]
21. Biswas, K.; Lu, J.; Soroushian, P.; Shrestha, S.S. Combined experimental and numerical evaluation of a prototype nano-PCM enhanced wallboard. *Appl. Energy* **2014**, *131*, 517–529. [CrossRef]
22. Biswas, K.; Shukla, Y.; DesJarlais, A.; Rawal, R. Thermal characterization of full-scale PCM products and numerical simulations, including hysteresis, to evaluate energy impacts in an envelope application. *Appl. Therm. Eng.* **2018**, *138*, 501–512. [CrossRef]
23. Shi, X.; Memon, S.A.; Tang, W.; Cui, H.; Xing, F. Experimental assessment of position of macro encapsulated phase change material in concrete walls on indoor temperatures and humidity levels. *Energy Build.* **2014**, *71*, 80–87. [CrossRef]
24. Morgan, K.; Lewis, R.W.; Zienkiewicz, O.C. An improved algrorithm for heat conduction problems with phase change. *Int. J. Numer. Methods Eng.* **1978**, *12*, 1191–1195. [CrossRef]
25. Mankel, C.; Caggiano, A.; Ukrainczyk, N.; Koenders, E. Thermal energy storage characterization of cement-based systems containing microencapsulated-PCMs. *Constr. Build. Mater.* **2019**, *199*, 307–320. [CrossRef]
26. European Commission. Integrated Porous Cementitious Nanocomposites in Non-Residential Building Envelopes for Green Active/Passive Energy STORAGE (https://nrg-storage.eu/). Available online: https://cordis.europa.eu/project/id/870114 (accessed on 14 May 2020).
27. Bre, F.; Silva, A.S.; Ghisi, E.; Fachinotti, V. Residential building design optimisation using sensitivity analysis and genetic algorithm. *Energy Build.* **2016**, *133*, 853–866. [CrossRef]
28. Bre, F.; Fachinotti, V. A computational multi-objective optimization method to improve energy efficiency and thermal comfort in dwellings. *Energy Build.* **2017**, *154*, 283–294. [CrossRef]
29. Bre, F.; Roman, N.; Fachinotti, V.D. An efficient metamodel-based method to carry out multi-objective building performance optimizations. *Energy Build.* **2020**, *206*, 109576. [CrossRef]
30. Echenagucia, T.M.; Capozzoli, A.; Cascone, Y.; Sassone, M. The early design stage of a building envelope: Multi-objective search through heating, cooling and lighting energy performance analysis. *Appl. Energy* **2015**, *154*, 577–591. [CrossRef]
31. Fesanghary, M.; Asadi, S.; Geem, Z.W. Design of low-emission and energy-efficient residential buildings using a multi-objective optimization algorithm. *Build. Environ.* **2012**, *49*, 245–250. [CrossRef]
32. Cascone, Y.; Capozzoli, A.; Perino, M. Optimisation analysis of PCM-enhanced opaque building envelope components for the energy retrofitting of office buildings in Mediterranean climates. *Appl. Energy* **2018**, *211*, 929–953. [CrossRef]
33. Castellón, C.; Günther, E.; Mehling, H.; Hiebler, S.; Cabeza, L.F. Determination of the enthalpy of PCM as a function of temperature using a heat-flux DSC—A study of different measurement procedures and their accuracy. *Int. J. Energy Res.* **2018**, *32*, 1258–1265. [CrossRef]
34. U.S. Department of Energy. *EnergyPlus™ Version 9.3.0 Documentation—Engineering Reference*; U.S. Department of Energy: Washington, DC, USA, 2020.
35. BASF. *Datenblatt Micronal PCM DS 5038 X*; BASF: Ludwigshafen, Germany, 2013.
36. Rubitherm Technologies GmbH. *Datasheet of RT25HC*; Rubitherm Technologies GmbH: Berlin, Germany, 2018.
37. Hall, I.J.; Prairie, R.R.; Anderson, H.E.; Boes, E.C. *Generation of a Typical Meteorological Year (No. SAND-78-1096C; CONF-780639-1)*; Sandia Labs: Albuquerque, NM, USA, 1978.

38. Crawley, D.B.; Lawrie, L.K. Climate.OneBuilding.Org. Available online: http://climate.onebuilding.org/sources/ (accessed on 10 December 2019).
39. Bre, F.; Fachinotti, V. Generation of typical meteorological years for the Argentine Littoral Region. *Energy Build.* **2016**, *129*, 432–444. [CrossRef]

Article

A Study on the Thermal Properties of High-Strength Concrete Containing CBA Fine Aggregates

In-Hwan Yang * and Jihun Park

Department of Civil Engineering, Kunsan National University, Jeonbuk 54150, Korea; jhpark3@kunsan.ac.kr
* Correspondence: ihyang@kunsan.ac.kr; Tel.: +82-63-469-4752; Fax: +82-63-469-4791

Received: 17 February 2020; Accepted: 23 March 2020; Published: 25 March 2020

Abstract: The thermal conductivity of concrete is a key factor for efficient energy consumption in concrete buildings because thermal conductivity plays a significant role in heat transfer through concrete walls. This study investigated the effects of replacing fine aggregates with coal bottom ash (CBA) and the influence of curing age on the thermal properties of high-strength concrete with a compressive strength exceeding 60 MPa. The different CBA aggregate contents included 25%, 50%, 75%, and 100%, and different curing ages included 28 and 56 days. For concrete containing CBA fine aggregate, the thermal and mechanical properties, including the unit weight, thermal conductivity, compressive strength, and ultrasonic velocity, were measured. The experimental results reveal that the unit weight and thermal conductivity of the CBA concrete were highly dependent on the CBA content. The unit weight, thermal conductivity, and compressive strength of the concrete decreased as the CBA content increased. Relationships between the thermal conductivity and the unit weight, thermal conductivity and compressive strength of the CBA concrete were proposed in the form of exponential functions. The equations proposed in this study provided predictions that were in good agreement with the test results. In addition, the test results show that there was an approximately linear relationship between the thermal conductivity and ultrasonic velocity of the CBA concrete.

Keywords: thermal conductivity; coal bottom ash; unit weight; compressive strength; ultrasonic velocity

---

1. Introduction

Currently, due to the increasing frequency of extremely hot weather conditions, efficient energy consumption is required in the construction field. In particular, there is an increasing demand for energy-efficient buildings, in which the internal temperature can be optimized [1–3]. One of the key factors for optimizing energy efficiency is thermal conductivity. When a building is constructed from materials with high thermal conductivity, a great amount of energy is consumed for cooling and heating [4,5]. To ensure the internal temperature of structures, materials with low thermal conductivity are recommended for constructing concrete structures. Accordingly, concrete with low thermal conductivity is preferable for efficient energy consumption in residential and commercial concrete buildings.

Regarding concrete with low thermal conductivity, some experimental studies have been performed [6–11]. Aghdam et al. [6] performed an experimental study to estimate the effects of carbon nanotubes on the thermal conductivity of steel fiber-reinforced concrete. The test results show that the addition and the increasing length of the carbon nanotubes significantly improved the thermal conductivity of steel fiber-reinforced concrete. Wang et al. [7] studied the thermal conductivity of concrete with expanded perlite, which is a porous material, and concluded that the mechanical strength and thermal conductivity of the concrete decreased after the expanded perlite was added to the concrete. Nguyen et al. [8] studied the influence of moisture content and temperature on the thermal properties of lightweight concrete, for which expanded clay, expanded shale, and pumice were used for the fabrication of lightweight aggregates. This study reported that the thermal conductivity of the concrete specimens, including expanded clay, expanded shale, and pumice, showed a great dependence on the moisture content. In addition, Brooks et al. [9] also investigated the effect of different lightweight fillers, including expanded polystyrene beads, dry-expanded thermoplastic microspheres, hollow glass microspheres, and lightweight hollow spheres made of fly ash, on the thermal properties of lightweight cementitious composites. The authors indicated that the thermal properties of lightweight concrete were greatly affected by the type and volume fraction of lightweight filler. Lower-density hollow glass microspheres, expanded polystyrene beads, and dry-expanded thermoplastic microspheres are more suitable for nonstructural thermal insulating components.

The demand for electricity is also an increasingly prevalent issue, and a thermal power plant is one of the methods to supply electricity. However, coal-fired thermal power plants create an enormous amount of bottom ash and fly ash [12–17]. Coal bottom ash (CBA) is an industrial waste produced at the bottom of a coal furnace in coal-fired thermal power plants. CBA is a kind of porous material with some advantages, such as low thermal conductivity and low specific density, which can be used in the concrete industry [18,19]. Accordingly, the material properties of CBA concrete have been examined in previous studies [20–22]. Mangi et al. [23,24] investigated the effect of CBA on the concrete strength properties under sulfate and chloride environments. Balapour et al. [25] performed an experimental program to investigate the potential use of CBA for the internal curing of concrete systems. They indicated that CBA exhibited a low oven dry-specific gravity, which makes it capable of storing the amount of water needed for concrete internal curing. In addition, Khongpermgoson et al. [26] reported that the compressive strength of concretes mixed with ground CBA and other binders increased with increasing curing age.

However, most previous studies investigated the mechanical properties of CBA concrete with a normal compressive strength of less than 40 MPa, which included CBA as an aggregate replacement. Moreover, few studies have assessed the thermal properties of high-strength CBA concrete. Therefore, to develop concrete with low thermal conductivity for energy efficiency, an experimental study must be performed to investigate the effects of the partial or total replacement of natural aggregates with CBA on the thermal properties of concrete.

In this experimental study, the thermal and mechanical properties of high-strength CBA concrete with a target compressive strength exceeding 60 MPa were investigated. The CBA concrete included 0%, 25%, 50%, 75%, and 100% replacement of natural fine aggregates with CBA aggregates. The unit weight, thermal conductivity, compressive strength, and ultrasonic velocity of the produced concrete were measured. In addition, relationships between the thermal conductivity and the unit weight, compressive strength, and ultrasonic velocity for the CBA concrete were proposed.

2. Experimental Program

2.1. Materials

Both the fine and coarse aggregates used in this study were crushed materials. Due to the depletion of natural resources, crushed fine aggregates have been used favorably in Korea. The crushed fine aggregate is shown in Figure 1a. The particle size distribution of the crushed fine aggregate is shown in Figure 2. The minimum and maximum sizes of the coarse aggregates used in this study were 5 and 20 mm, respectively. The density, water absorption, and fineness modulus of the crushed fine and coarse aggregates used are shown in Table 1. The densities of the fine and coarse aggregates were 2.60 and 2.61 g/cm^3, respectively, and the water absorption of the fine and coarse aggregates were 0.69 and 1.44, respectively.

Figure 1. Crushed fine and CBA aggregates. (**a**) Crushed fine aggregate; (**b**) CBA aggregate.

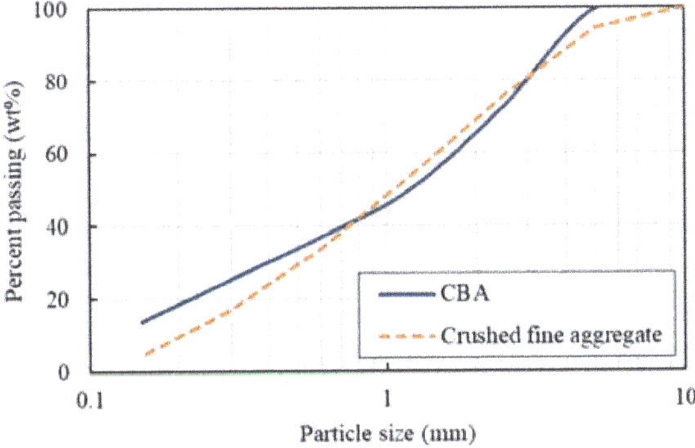

Figure 2. Grading curve of CBA and crushed fine aggregates.

Table 1. Physical properties of fine, coarse and coal bottom ash (CBA) aggregates.

Material \ Property	Fineness Modulus	Water Absorption (%)	Density (g/cm^3)
Crushed fine aggregate	3.17	0.69	2.60
Coarse aggregate	6.77	1.44	2.61
CBA	3.83	6.87	1.84

The CBA was collected from a thermal power plant company (Korea South-East Power Co., Ltd, Yeongheung Power Division, Yeongheung, Korea). The chemical components of the CBA used were determined through energy-dispersive spectroscopy (EDS, Hitachi High-Technologies Corporation, Tokyo, Japan), and the analysis results are shown in Table 2. The three major components in the CBA were SiO_2, Al_2O_3, and Fe_2O_3, which had contents of 60.03%, 20.25%, and 9.80%, respectively, thereby comprising greater than 90% of the CBA.

Table 2. Chemical components of CBA and OPC.

Component	CBA (%)	OPC (%)
SiO_2	60.03	31.90
Al_2O_3	20.25	8.97
Fe_2O_3	9.80	0.87
CaO	5.58	46.95
Na_2O	1.95	0.38
MgO	1.44	3.25
K_2O	0.95	0.96
SO_3	-	5.25

The CBA aggregate was screened to remove particles greater than 5.0 mm and to retain particles greater than 0.15 mm. The CBA used in this study is shown in Figure 1b, and the particle size distribution of the CBA is also presented in Figure 2. A scanning electron microscopy (SEM) image of the CBA is given in Figure 3, and the image shows the presence of voids in the CBA particles. As shown in Table 1, the density of the CBA was smaller than that of the crushed fine aggregate, which were 1.84 and 2.60 g/cm^3, respectively. On the other hand, the CBA water absorption was much higher than that of the crushed fine aggregate, which were 6.87% and 0.69%, respectively.

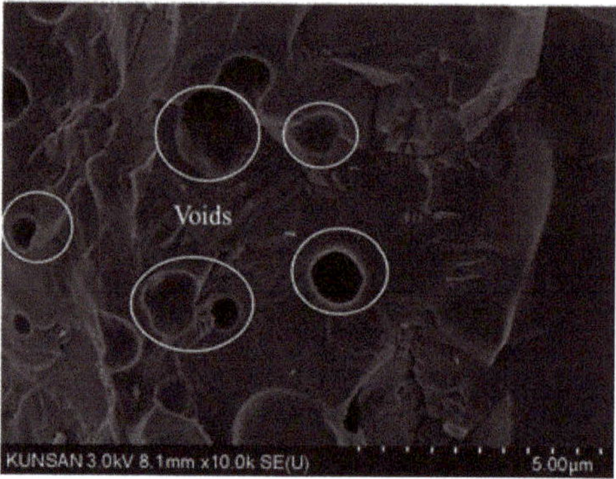

Figure 3. SEM image of CBA aggregate.

Ordinary Portland cement (OPC) used in this study was type I in accordance with KS L 5201 [27]. The specific gravity of the OPC used was 3.15, and the chemical components of the OPC are shown in Table 2. To enhance the workability and reduce the water-cement ratio of the CBA concrete, a superplasticizer with a dosage of 3.6 kg/m^3, which corresponded to 0.6% of the weight of the OPC, was used.

2.2. *Mixing Proportions*

The mixing proportions of the control concrete and CBA concrete are provided in Table 3. A concrete mix was designed with a target compressive strength of 60 MPa at a curing age of 28 days. In the experimental study, five different series of concrete mixtures were prepared with various percentages of CBA as crushed fine aggregate replacement. The crushed fine aggregate in the concrete was replaced with CBA at five different volume fractions of 0%, 25%, 50%, 75%, and 100%, and the corresponding CBA concrete mixtures were named CBA00, CBA25, CBA50, CBA75, and CBA100, respectively. The amounts of cement and coarse aggregate were constant for each concrete mixture at 595.0 and 878.5 kg/m^3, respectively. A water-cement ratio of 0.3 was applied in all of the concrete mixtures.

Table 3. Mixing proportions of the CBA concrete.

Mixtures	CBA Content (%)	W/C	Water	Unit Content (kg/m^3)				
				Cement (OPC) [a]	Coarse Aggregate	Crushed Fine Aggregate	CBA	Superplasticizer (0.6% × Cement)
CBA00	0	0.3	178.5	595.0	878.5	663.0	0.0	3.6
CBA25	25	0.3	178.5	595.0	878.5	497.2	117.7	3.6
CBA50	50	0.3	178.5	595.0	878.5	331.5	235.3	3.6
CBA75	75	0.3	178.5	595.0	878.5	165.7	353.0	3.6
CBA100	100	0.3	178.5	595.0	878.5	0.0	470.7	3.6

[a] OPC: ordinary Portland cement.

2.3. Specimen Preparation and Test Procedures

All of the concrete specimens were fabricated in a laboratory mixer. Cylindrical specimens with dimensions of 100 mm × 200 mm were fabricated to determine the unit weight, thermal conductivity, and compressive strength of the different mixtures. After casting the concrete, the concrete specimens were covered with plastic wrap and moist-cured for one day. Thereafter, the specimens were demolded at an age of 24 ± 1 h and then cured under submersed conditions at 23 ± 2 °C in a water tank until the ages of 28 and 56 days after the casting of concrete.

The unit weight, compressive strength, thermal conductivity, and ultrasonic velocity of the CBA concrete were measured at curing ages of 28 and 56 days. Both end surfaces of the cylinders for compressive strength tests were ground before implementing each experiment.

The unit weight (bulk density) of hardened CBA concrete was measured by using the cylindrical specimens after curing for 28 and 56 days, respectively. The unit weight was determined by dividing the mass of the cylindrical specimen by the volume of the specimen.

The compressive strength of the cylindrical specimens was tested with a universal testing machine (UTM). Loading was applied under displacement control using a UTM with a capacity of 2000 kN. The mean values of three specimens were recorded to obtain the material properties of the concrete.

There are several testing methods and their related devices for the measurement of the thermal conductivity of concrete. First, the thermal conductivity test can be carried out in accordance with ASTM D 5334-05 [28]. The ASTM method is based on the concept that the temperature rise in the heat source depends on the thermal conductivity of the medium into which it is inserted. The probe consists of a heating wire and a temperature measuring unit, and it should be inserted into a hole drilled in the concrete specimen. Similarly, Kim et al. [29] also used the two linear parallel probe (TLPP) method to determine the thermal conductivity of concrete. For the TLPP method, two probes are inserted into two parallel holes drilled in the specimen, where one probe is used as a heating source and the other is used as a temperature sensor.

The transient plane source (TPS) method has been explained in detail by Gustafsson [30] and Log and Gustafsson [31], and its consideration was summarized by He [32]. For the TPS method, the probe is sandwiched between the cast sides of two specimens or the cut faces of two elements of a concrete specimen, whereas probe rods are inserted into holes in the concrete specimen when using the ASTM method and TLPP method. The TPS method has been widely used to measure the thermal conductivity of solid materials such as concrete [33,34].

In this study, the measurement of the thermal conductivity was based on the TPS method. The thermal conductivity of the CBA specimens was measured using a TPS1500 testing device supplied from Hot Disk Ltd. (Gothenburg, Sweden) as shown in Figure 4a. In the TPS method, to ensure that the sensor was exposed to fine and coarse aggregates and cement paste, the cylindrical specimens were cut into two halves at the middle section of the cylinder, as shown in Figure 4a, and then the sensor was sandwiched between the two half cylinders, as shown in Figure 4b. The cut surfaces had planeness to ensure contact between the concrete specimen and the sensor.

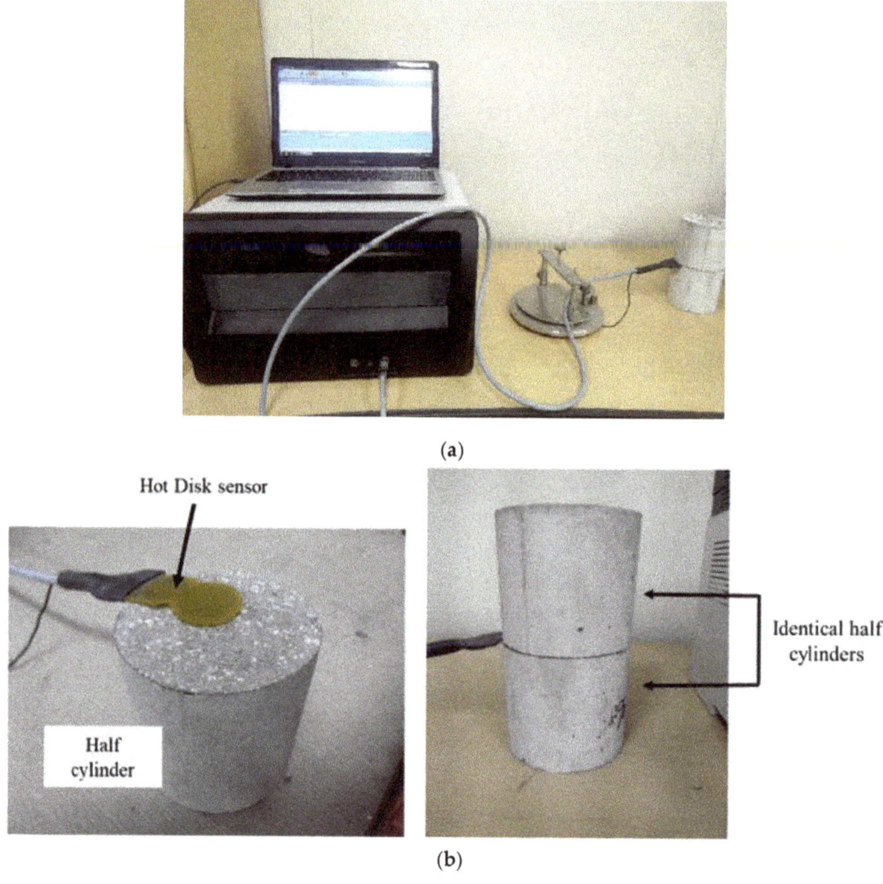

Figure 4. Test setup for the thermal conductivity measurements. (**a**) Transient plane source (TPS) measurement system; (**b**) concrete specimens used for the thermal conductivity experiments.

The sensor contained a nickel double spiral that applied a heating pulse to the specimen. The concrete specimen was controlled to satisfy thermal equilibrium before the measurements. After a thermal equilibrium time of at least 90 min under laboratory temperature conditions, measurements were made with an applied heating power of 0.3 W. Three measurements were taken for each specimen to ensure accurate test results.

3. Test Results and Discussion

3.1. Properties of Fresh Concrete

To investigate the workability of the fresh concrete, a slump test was performed. The slump test results are shown in Figure 5. The slump of the CBA concrete mixtures decreased as the CBA fine aggregate content increased. The slumps of CBA concrete mixtures CBA25, CBA50, CBA75, and CBA100 were 75, 68, 57, and 47 mm, respectively, whereas that of the control concrete mixture was 79 mm. The decrease in the workability of concrete is mainly due to the irregular shapes and the increase in the surface area of the aggregates used in concrete. The use of CBA as fine aggregates affected the concrete texture, which had more irregular and porous particles than the control concrete.

Therefore, the friction between particles in CBA concrete increased the obstruction of the workability of the fresh concrete and then led to a decrease in the slump of the CBA concrete.

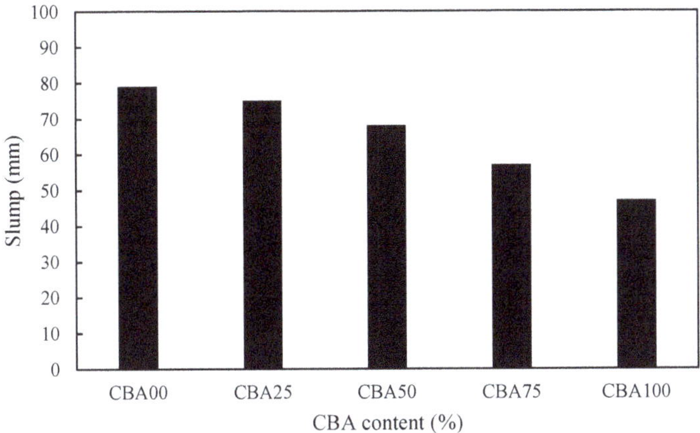

Figure 5. Slump test results.

3.2. Unit Weight

The unit weights of the concrete specimens with different CBA replacement ratios are shown in Figure 6. The figure indicates that the unit weight of the CBA concrete decreased as the CBA replacement ratio increased in the concrete, which was the expected response. At a curing age of 28 days, the unit weight of the control mixture CBA00 was 2370.2 kg/m^3, whereas the unit weight of the mixture containing 100% CBA fine aggregate decreased to as low as 2190.2 kg/m^3. Specifically, the unit weights of the CBA concrete mixtures CBA25, CBA50, CBA75, and CBA100 were 2.1%, 3.2%, 5.3%, and 7.6% less than that of the control mixture, respectively. This decrease in the unit weight of the CBA concrete mixtures occurred because the unit weight of CBA fine aggregate was lower than that of crushed fine aggregate, as shown in Table 1. The porosity of each CBA concrete specimen was estimated by using the mercury intrusion porosity (MIP) method. After performing the compressive strength test, the crushed concrete specimen was broken into small samples to be placed in the MIP dilatometer. For the small samples, coarse aggregates were eliminated from the samples. Then, the MIP test was carried out by using the concrete piece samples. The porosity in the sample consisted of the contribution from cement paste and that from CBA fine aggregates. Accordingly, it could be considered that the porosity from the MIP test was affected by the CBA fine aggregate contents under the conditions of a constant water-cement ratio and the use of the same type of cement. The MIP test results in the present study show that the porosity of mixture CBA100 was higher than that of the reference mixture (CBA00). Specifically, the porosities for mixtures CBA00, CBA25, CBA50, CBA75, and CBA100 at a curing age of 28 days were 8.5%, 9.6%, 10.4%, 12.3%, and 15.6%, respectively. Thus, the porosity increased as the CBA aggregate content increased.

At a curing age of 56 days, the unit weight of the control mixture was 2386.5 kg/m^3, whereas the unit weight of mixture CBA100 decreased to as low as 2225.7 kg/m^3. Specifically, the unit weights of the CBA concrete mixtures CBA25, CBA50, CBA75, and CBA100 were 2.0%, 3.7%, 5.8%, and 6.7% less than that of the control mixture, respectively.

Figure 6 also compares the measured unit weights of the concrete specimens at different curing ages. At a curing age of 56 days, the unit weights of the concrete with 0%, 25%, 50%, 75%, and 100% CBA replacement of crushed fine aggregate were 0.7%, 0.9%, 0.2%, 0.1%, and 1.6% higher than the corresponding values at a curing age of 28 days, respectively; hence, these increases were insubstantial.

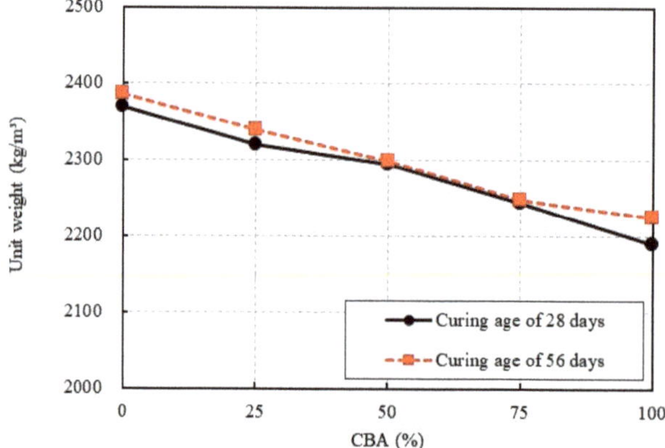

Figure 6. Unit weight test results.

3.3. *Thermal Conductivity*

The thermal conductivities of the concrete specimens with different CBA contents are presented in Figure 7. The thermal conductivities of CBA concrete decreased as the CBA content increased at a curing age of 28 days. The thermal conductivities of the CBA concrete mixtures CBA25, CBA50, CBA75, and CBA100 were 6.4%, 11.7%, 14.2%, and 22.5% less than that of the control concrete mixture CBA00 (1.87 W/mK), respectively. At a curing age of 56 days, the thermal conductivity of 1.45 W/mK in the CBA concrete mixture CBA100 was 31.2% less than the value of 2.04 W/mK in the control concrete mixture. The thermal conductivity of the CBA concrete with a 100% CBA content was significantly less than that of the control concrete mixture. It is known that thermal conductivity highly depends on the pore structure of the concrete, and subsequently the density of the concrete. Hence, the pore structure was one of the key elements affecting thermal conductivity [10,35]. As already discussed in the previous section, the porosity of the CBA concrete specimen for each mixture increased as the CBA aggregate content increased. As the CBA content increased, the total porosity increased, so the thermal conductivities of the concrete decreased. For this reason, the observed decline in thermal conductivity could be explained by the increase in the CBA aggregate content in the concrete.

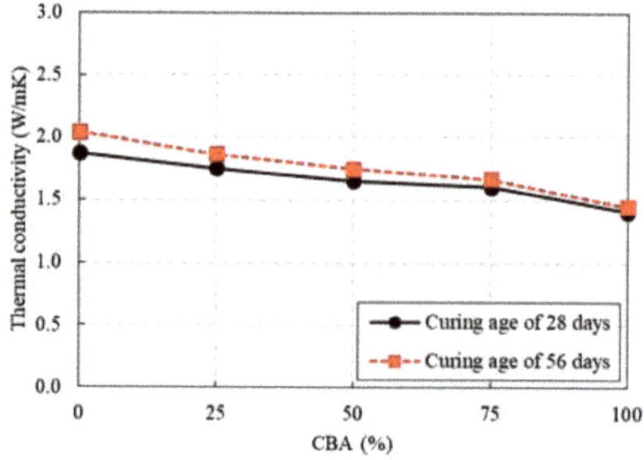

Figure 7. Thermal conductivity test results.

A decrease in the thermal conductivity of the CBA concrete would increase the thermal insulation provided by the concrete and reduce the heating and cooling costs for buildings constructed from these materials. Therefore, the test result of the thermal conductivities for CBA concrete in this study implies that CBA could be utilized to fabricate high-strength concrete with low thermal conductivity for efficient energy consumption.

The effect of curing age on the thermal conductivity of the CBA concrete specimens is also shown in Figure 7. When the concrete curing age increased from 28 to 56 days, the thermal conductivities of the concrete with CBA contents of 0%, 25%, 50%, 75% and 100% CBA increased by 9.2%, 6.5%, 5.7%, 4.0%, and 3.1%, respectively. With the increase in curing age, the pores in the concrete matrix were filled by hydration products and calcium silicate hydrate (CSH) gel [36]. Heat is transferred faster in solid materials than in porous materials. Therefore, the thermal conductivity for the well packed concrete specimens at a curing age of 56 days will be higher than that at a curing age of 28 days.

The relationship between the thermal conductivity and the unit weight of the CBA concrete with two different curing ages is shown in Figure 8a. As the unit weights of the CBA concrete increased, the thermal conductivity of the CBA concrete increased. Moreover, the thermal conductivity of the CBA concrete is nearly linearly proportional to the unit weight. This phenomenon occurred because the substitution of CBA as fine aggregate increased the porosity in the concrete, thereby reducing the thermal conductivity and unit weight of the CBA concrete. The smallest unit weight was nearly consistent with the smallest thermal conductivity of CBA concrete.

Asadi et al. [11] proposed Equation (1) to predict the thermal conductivity of CBA concrete by using the unit weight. Their proposed equation was an exponential function that was derived based on the test data of thermal conductivities of lightweight concrete available in the literature. The test data used in their equation did not contain only CBA but also pumice, expanded polystyrene, and expanded perlite.

$$k = 0.0625e^{0.0015\rho} \tag{1}$$

where k is the thermal conductivity (W/mK) and ρ is the unit weight of the CBA concrete (kg/m^3).

ACI committee 213 R-03 [37] proposed Equation (2) to estimate the thermal conductivity of lightweight concrete.

$$k = 0.0864e^{0.00125\rho} \tag{2}$$

Wongkeo [35] and Zhang et al. [38] carried out an experimental program on CBA concrete, and their studies indicated that the thermal conductivities had a close relation with the unit weight. This study proposes Equation (3), which is based on the test results from the present study and the results from Wongkeo [35] and Zhang et al. [38]. A comparison of the three equations is also shown in Figure 8b.

$$k = 0.0725e^{0.0013\rho} \tag{3}$$

The equation from Asadi et al. [11] overestimated the thermal conductivities because it was derived based on concrete including different kinds of lightweight aggregates. In contrast, the equation proposed in this study provides predictions that are in close agreement with the test results of the CBA concrete.

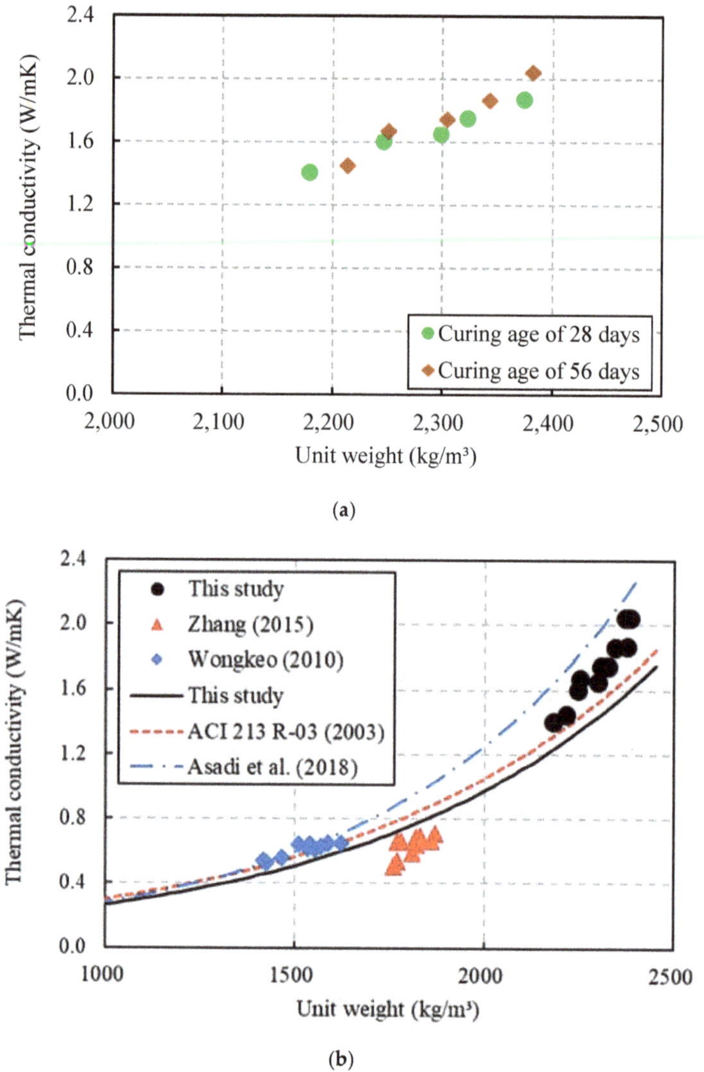

Figure 8. Relationship between the thermal conductivity and the unit weight. (**a**) Test results in this study; (**b**) comparison of the predictions in this study and previous studies

3.4. Compressive Strength

The test results of the compressive strength of the concrete specimens with different CBA contents are shown in Figure 9. At a curing age of 28 days, the compressive strength of the CBA mixtures decreased as the CBA replacement increased. The compressive strength values of CBA concrete mixtures CBA50, CBA75, and CBA100 were 3.0%, 4.6%, and 8.8% less than that of the control concrete mixture (CBA00), respectively. However, the compressive strength of the CBA concrete mixture CBA25 was only 1.2% higher than that of the control concrete mixture. At a curing age of 56 days, the compressive strength values of the CBA concrete mixtures CBA25, CBA50, CBA75 and CBA100 were 2.0%, 3.0%, 4.8%, and 6.2% less than that of the control concrete mixture (CBA00), respectively.

This compressive strength loss could be explained by the increase in the porosity of the concrete. These pores might have an adverse influence on the compressive strength of the CBA concrete specimens [39].

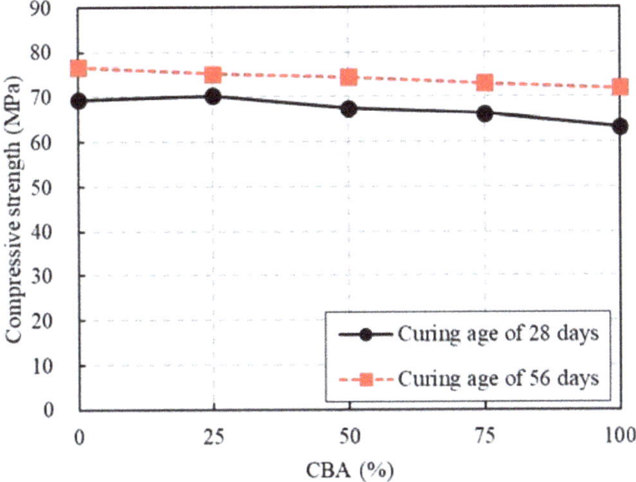

Figure 9. Compressive strength test results.

Figure 9 also illustrates the effects of curing ages of 28 and 56 days on compressive strength. At a curing age of 56 days, the concrete compressive strength values with CBA contents of 0%, 25%, 50%, 75%, and 100% were 110.4%, 106.9%, 110.4%, 110.2%, and 113.6% of the corresponding values at a curing age of 28 days, respectively. The substantial increase in the compressive strength of the CBA concrete mixtures after curing for 56 days might result from the pozzolanic reaction of the CBA. According to the study of Abdulmatin et al. [36], due to the pozzolanic activity, secondary CSH and calcium aluminate hydrate (CAH) form; therefore, the porosity of the concrete matrix is filled with these materials. In addition, $Ca(OH)_2$ is transformed into CSH. These phenomena are why the concrete compressive strength increased with the increase in curing age.

The relationship between the compressive strength and the unit weight of the CBA concrete under two different curing ages is presented in Figure 10a. The compressive strength of the CBA concrete has a nearly direct relationship with the unit weight of the CBA concrete. The compressive strength of the CBA concrete increased as the unit weight of the CBA concrete increased. This phenomenon occurred because both the unit weight and the compressive strength of the CBA concrete were affected by the replacement of a stronger material (crushed sand) with a weaker material (CBA) and the increase in pore volume in the CBA concrete, as described in the previous section.

The relationship between the compressive strength and the unit weight of the CBA concrete, based on the test results in this study and in previous studies [35,38], is shown in Figure 10b. The proposed exponential equation for predicting the relationship between the compressive strength and the unit weight of the CBA concrete is expressed as follows:

$$f_c = 1.217e^{0.0018\rho} \tag{4}$$

where f_c is the compressive strength (MPa) and ρ is the unit weight (kg/m^3).

This equation underestimated the compressive strength values when the unit weight ranged from approximately 1700 to 1900 kg/m^3, whereas it overestimated the compressive strength values when the unit weight ranged from approximately 2300 to 2400 kg/m^3.

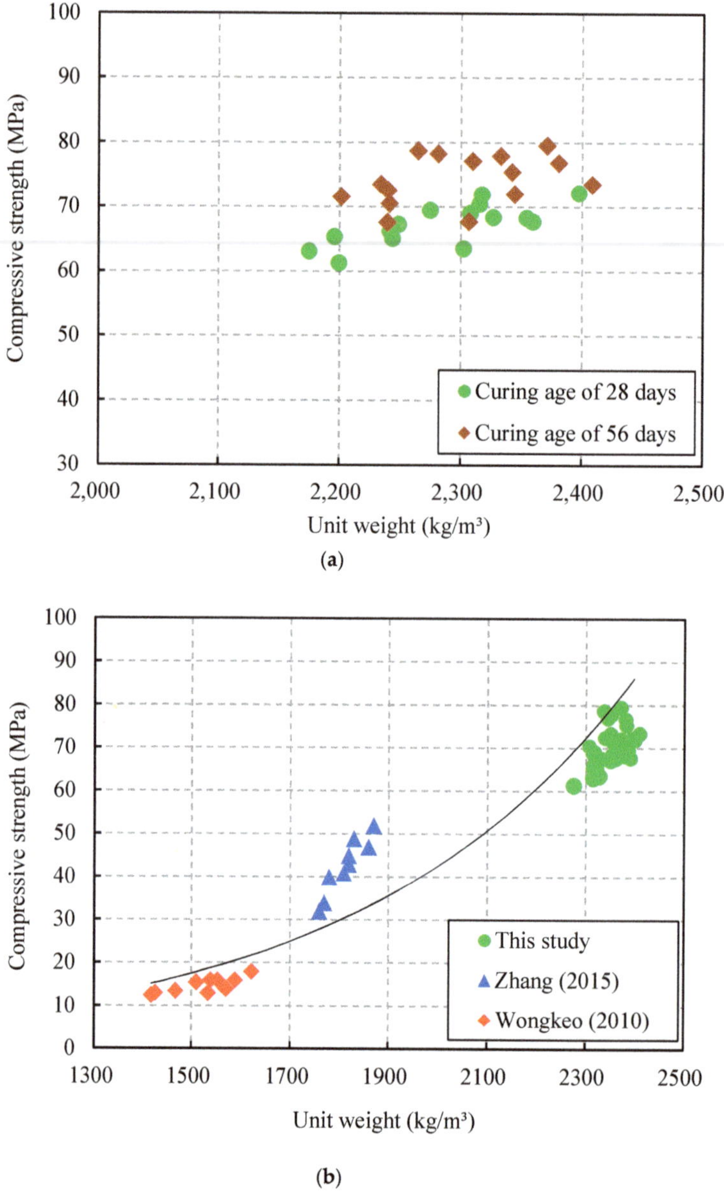

Figure 10. Relationship between the compressive strength and the unit weight. (**a**) Test results in this study; (**b**) relationship based on the results in this study and previous studies.

In addition, the relationship between the thermal conductivity and the compressive strength is shown in Figure 11a. The figure shows that this relationship tendency was similar to that between the thermal conductivity and the unit weight of the CBA concrete. The compressive strength of the CBA concrete was affected by the unit weight. Therefore, the thermal conductivity had a close relationship with the compressive strength of the CBA concrete. In this study, the thermal conductivity of the CBA concrete varied from 1.41 to 2.04 W/mK when the compressive strength ranged from 63.3 to 76.7 MPa. Albayrak et al. [40] also reported that the compressive strength values and thermal conductivities of lightweight concrete decreased with decreasing density.

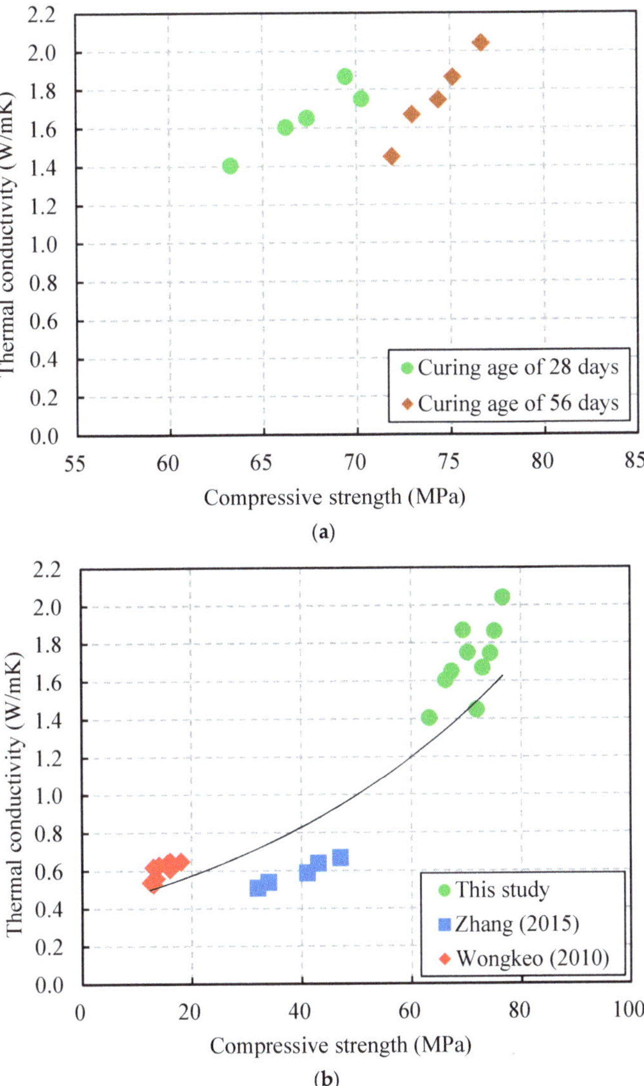

Figure 11. Relationship between the thermal conductivity and the compressive strength. (**a**) Test results in this study; (**b**) relationship based on the results in this study and previous studies.

Moreover, the relationship between the thermal conductivity and the compressive strength, which is based on the test results in this study and in previous studies [35,38], is shown in Figure 11b. The exponential equation form used to predict the relationship between the thermal conductivity and the unit weight of CBA concrete was also applied to predict the relationship between the thermal conductivity and compressive strength as follows:

$$k = 0.3976e^{0.0184f_c} \tag{5}$$

where k is the thermal conductivity (W/mK) and f_c is the compressive strength (MPa).

Overall, the thermal conductivity test results in the figure had some deviations with various CBA concrete compressive strength values. Therefore, the equation overestimated the thermal conductivities when the measurements of compressive strength ranged from approximately 30 to 50 MPa, whereas it underestimated the thermal conductivities when the measurements of compressive strength ranged from approximately 60 to 80 MPa.

3.5. Ultrasonic Velocity

An ultrasonic pulse velocity test was carried out to assess the material characterization of the CBA concrete. The measured quantity of this experiment was the travel time of the ultrasonic pulse between the transducers that were held on each surface of a concrete specimen, as shown in Figure 12; the pulse velocity was calculated by dividing the distance between the transducers by the travel time.

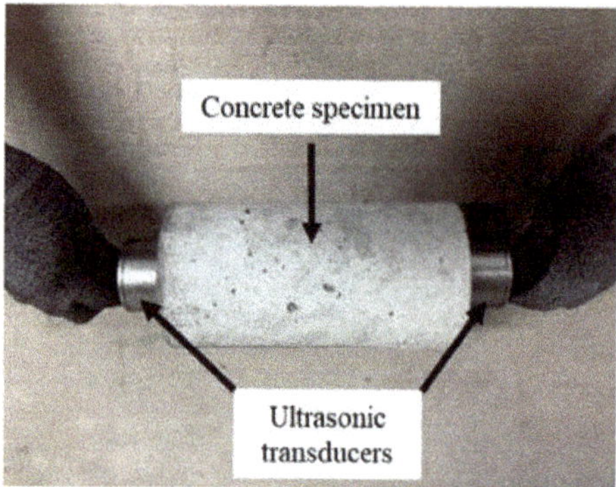

Figure 12. Measurement of ultrasonic velocity.

According to the ASTM C597-09 [41], the frequency of the ultrasonic pulse test procedure should be greater than 50 kHz to achieve accurate transit-time measurements and greater sensitivity for the short measured path. In the previous studies by Ashrafian et al. [42] and Nik et al. [43], the pulse frequency of 54 kHz was applied for measuring the ultrasonic velocity of the concrete cubes of 100 mm. In this study, therefore, an ultrasonic instrument with a frequency of the transducers of 54 kHz was used to measure the ultrasonic velocity.

The relationship between the thermal conductivity and the ultrasonic velocity is shown in Figure 13. In this study, the thermal conductivity of the concrete specimen ranged from 1.41 to 2.04 W/mK when the ultrasonic velocity ranged from 4256 to 4415 m/s. The test results in this study show that there was an approximately linear relationship between the thermal conductivity and the ultrasonic velocity of CBA concrete. Solid materials transfer sound faster than porous materials. Higher ultrasonic velocity indicates that the concrete has greater continuity, whereas lower ultrasonic velocity indicates that the concrete contains more voids and defects (e.g., cracks). The thermal conductivity and ultrasonic velocity are mainly dependent on the density of the concrete [44]. The addition of the CBA aggregate reduced the density of the CBA concrete, which resulted in a reduction in both the thermal conductivity and ultrasonic velocity.

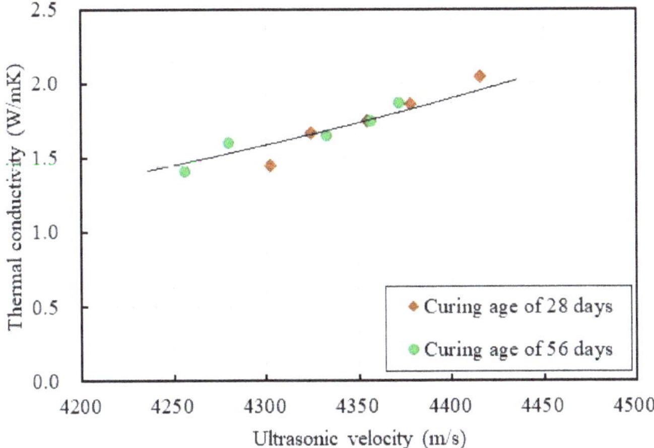

Figure 13. Relationship between the thermal conductivity and the ultrasonic velocity.

4. Conclusions

In this research, an experimental study was performed to investigate the thermal conductivity of CBA concrete with a compressive strength exceeding 60 MPa. The research findings are summarized as follows:

- The unit weight of CBA concrete decreased as the replacement of CBA as fine aggregate increased. This decrease in the unit weight of the CBA concrete mixtures occurred because the CBA had a lower unit weight and a higher porosity than the crushed fine aggregate.
- The thermal conductivity of the CBA concrete was highly dependent on the CBA content. In addition, overall, the thermal conductivity of the CBA concrete increased as the curing age increased. When the curing age increased from 28 to 56 days, the thermal conductivity of the concrete increased by 3.1~6.5%.
- The relationship between the thermal conductivity and the unit weight of the CBA concrete was modeled with an exponential function. The results indicate that the equation proposed in this study provides predictions that are in good agreement with the test results.
- The compressive strength of the CBA concrete decreased as the CBA content in the concrete increased. In addition, an equation relating the thermal conductivity of the CBA concrete to the compressive strength was proposed. The equation overestimates the thermal conductivity of moderate-strength concrete (30~50 MPa), whereas it underestimates the thermal conductivity of high-strength concrete (60~80 MPa) because the test results used as a basis for the equation have some deviations.
- The ultrasonic velocity of the CBA concrete decreased as the amount of CBA fine aggregate in the concrete increased. Moreover, the test results show that there was an approximately linear relationship between the thermal conductivity and ultrasonic velocity of the CBA concrete.

Data Availability: The data used to support the findings in this study are available from the corresponding author upon request.

Author Contributions: Conceptualization, I.-H.Y.; Data Curation, J.P.; Investigation, I.-H.Y. and J.P.; Formal analysis, J.P.; Methodology, I.-H.Y. and J.P.; Resources, I.-H.Y.; Writing—original draft preparation, J.P.; Writing—review and editing, I.-H.Y.; Funding acquisition, I.-H.Y. All authors have read and agree to the published version of the manuscript.

Funding: This research was supported by a grant (19CTAP-C151889-01) from the Technology Advancement Research Program (TARP), which is funded by the Ministry of Land, Infrastructure and Transport of the Korean government.

Conflicts of Interest: The authors declare that they have no conflicts of interest regarding the publication of this manuscript.

References

1. Danish, M.S.S.; Senjyu, T.; Ibrahimi, A.M.; Ahmadi, M.; Howlader, A.M. A managed framework for energy-efficient building. *J. Build. Eng.* **2019**, *21*, 120–128. [CrossRef]
2. Najjar, M.; Figueiredo, K.; Hammad, A.W.A.; Haddad, A. Integrated optimization with building information modeling and life cycle assessment for generating energy efficient buildings. *Appl. Energy* **2019**, *250*, 1366–1382. [CrossRef]
3. Sayadi, S.; Tsatsaronis, G.; Morosuk, T.; Baranski, M.; Sangi, R.; Muller, D. Exergy-based control strategies for the efficient operation of building energy systems. *J. Clean. Prod.* **2019**, *241*, 118277. [CrossRef]
4. Elrahman, M.A.; Chung, S.Y.; Sikora, P.; Rucinska, T.; Stephan, D. Influence of nanosilica on mechanical properties, sorptivity, and microstructure of lightweight concrete. *Materials* **2019**, *12*, 3078. [CrossRef]
5. Khoukhi, M.; Abdelbaqi, S.; Hassan, A. Yearly energy performance assessment of employing expanded polystyrene with variable temperature and moisture–thermal conductivity relationship. *Materials* **2019**, *12*, 3000. [CrossRef]
6. Aghdam, M.K.H.; Mahmoodi, M.J.; Safi, M. Effect of adding carbon nanotubes on the thermal conductivity of steel fiber-reinforced concrete. *Compos. Part B* **2019**, *174*, 106972. [CrossRef]
7. Wang, L.; Liu, P.; Jing, Q.; Liu, Y.; Wang, W.; Zhang, Y.; Li, Z. Strength properties and thermal conductivity of concrete with the addition of expanded perlite filled with aerogel. *Constr. Build. Mater.* **2018**, *188*, 747–757. [CrossRef]
8. Nguyen, L.H.; Beaucour, A.L.; Ortola, S.; Noumowe, A. Experimental study on the thermal properties of lightweight aggregate concretes at different moisture contents and ambient temperatures. *Constr. Build. Mater.* **2017**, *151*, 720–731. [CrossRef]
9. Brooks, A.L.; Zhou, H.; Hanna, D. Comparative study of the mechanical and thermal properties of lightweight cementitious composites. *Constr. Build. Mater.* **2018**, *159*, 316–328. [CrossRef]
10. Tasdemir, C.; Sengul, O.; Tasdemir, M.A. A comparative study on the thermal conductivities and mechanical properties of lightweight concretes. *Energy Build.* **2017**, *151*, 469–475. [CrossRef]
11. Asadi, I.; Shafigh, P.; Hassan, Z.F.B.A.; Mahyuddin, N.B. Thermal conductivity of concrete—A review. *J. Build. Eng.* **2018**, *20*, 81–93. [CrossRef]
12. Dwivedi, A.; Jain, M.K. Fly ash—Waste management and overview: A review. *Recent Res. Sci. Technol.* **2014**, *6*, 30–35.
13. Fu, B.; Liu, G.; Mian, M.M.; Sun, M.; Wu, D. Characteristics and speciation of heavy metals in fly ash and FGD gypsum from Chinese coal-fired power plants. *Fuel* **2019**, *251*, 593–602. [CrossRef]
14. Cicek, T.; Cincin, Y. Use of fly ash in production of light-weight building bricks. *Constr. Build. Mater.* **2015**, *94*, 521–527. [CrossRef]
15. Rathnayake, M.; Julnipitawong, P.; Tangtermsirikul, S.; Toochinda, P. Utilization of coal fly ash and bottom ash as solid sorbents for sulfur dioxide reduction from coal fired power plant: Life cycle assessment and applications. *J. Clean. Prod.* **2018**, *202*, 934–945. [CrossRef]
16. Munawer, M.E. Human health and environmental impacts of coal combustion and post-combustion wastes. *J. Sustain. Min.* **2018**, *17*, 87–96. [CrossRef]
17. Ramsey, A.B.; Szynkiewicz, A. Coupled chemical-isotope assessment of potential metal releases to the water column from river sediments impacted by coal ash spill. *Appl. Geochem.* **2019**, *107*, 34–44. [CrossRef]
18. Singh, N.; Bhardwaj, A. Reviewing the role of coal bottom ash as an alternative of cement. *Constr. Build. Mater.* **2020**, *233*, 117267. [CrossRef]
19. Singh, N.; Mithulraj, M.; Arya, S. Utilization of coal bottom ash in recycled concrete aggregates based self-compacting concrete blended with metakaolin. *Resour. Conserv. Recycl.* **2019**, *144*, 240–251. [CrossRef]
20. Muthusamy, K.; Rasid, M.H.; Jokhio, G.A.; Budiea, A.M.A.; Hussin, M.W.; Mirza, J. Coal bottom ash as replacement in concrete: A review. *Constr. Build. Mater.* **2020**, *236*, 117507. [CrossRef]

21. More, S.R.; Bhatt, D.V.; Menghani, J.V. Failure analysis of coal bottom ash slurry pipeline in thermal power plant. *Eng. Fail. Anal.* **2018**, *90*, 489–496. [CrossRef]
22. Zhou, H.; Bhattarai, R.; Li, Y.; Li, S.; Fan, Y. Utilization of coal fly and bottom ash pellet for phosphorus adsorption: Sustainable management and evaluation. *Resour. Conserv. Recycl.* **2019**, *149*, 372–380. [CrossRef]
23. Mangi, S.A.; Ibrahim, M.H.W.; Jamaluddin, N.; Arshad, M.F.; Jaya, R.P. Short-term effects of sulphate and chloride on the concrete containing coal bottom ash as supplementary cementitious material. *Eng. Sci. Technol. Int. J.* **2019**, *22*, 515–522. [CrossRef]
24. Mangi, S.A.; Ibrahim, M.H.W.; Jamaluddin, N.; Arshad, M.F.; Shahidan, S. Performances of concrete containing coal bottom ash with different fineness as a supplementary cementitious material exposed to seawater. *Eng. Sci. Technol. Int. J.* **2019**, *22*, 929–938. [CrossRef]
25. Balapour, M.; Zhao, W.; Garboczi, E.J.; Oo, N.Y.; Spatari, S.; Hsuan, Y.G.; Billen, P.; Farnam, Y. Potential use of lightweight aggregate (LWA) produced from bottom coal ash for internal curing of concrete systems. *Cem. Concr. Compos.* **2020**, *105*, 103428. [CrossRef]
26. Khongpermgoson, P.; Abdulmatin, A.; Tangchirapat, W.; Jaturapitakkul, C. Evaluation of compressive strength and resistance of chloride ingress of concrete using a novel binder from ground coal bottom ash and ground calcium carbide residue. *Constr. Build. Mater.* **2019**, *214*, 631–640. [CrossRef]
27. Korea Industrial Standards. *Portland Cement*; KS L 5201; Korea Industrial Standards: Seoul, Korea, 2016.
28. American Society for Testing and Materials (ASTM). *Standard Test Method for Determination of Thermal Conductivity of Soil and Soft Rock by Thermal Needle Probe Procedure*; ASTM D5334-05; ASTM: West Conshohocken, PA, USA, 2005.
29. Kim, K.-H.; Jeon, S.-E.; Kim, J.-K.; Yang, S. An experimental study on thermal conductivity of concrete. *Cem. Concr. Res.* **2003**, *33*, 363–371. [CrossRef]
30. Gustafsson, S.E. Transient plane source techniques for thermal conductivity and thermal diffusivity measurements of solid materials. *Rev. Sci. Instrum.* **1991**, *62*, 797–804. [CrossRef]
31. Log, T.; Gustafsson, S.E. Transient plane source (TPS) technique for measuring thermal transport properties of building materials. *Fire Mater.* **1995**, *19*, 43–49. [CrossRef]
32. He, Y. Rapid thermal conductivity measurement with a hot disk sensor: Part 1. Theoretical considerations. *Thermochim. Acta* **2005**, *436*, 122–129. [CrossRef]
33. Bentz, D.P.; Peltz, M.A.; Durán-Herrera, A.; Valdez, P.; Juárez, C.A. Thermal properties of high-volume fly ash mortars and concretes. *J. Build. Phys.* **2010**, *34*, 263–275. [CrossRef]
34. Yuan, H.-W.; Lu, C.-H.; Xu, Z.-Z.; Ni, Y.-R.; Lan, X.-H. Mechanical and thermal properties of cement composite graphite for solar thermal storage materials. *Sol. Energy* **2012**, *86*, 3227–3233. [CrossRef]
35. Wongkeo, W.; Chaipanich, A. Compressive strength, microstructure and thermal analysis of autoclaved and air cured structural lightweight concrete made with coal bottom ash and silica fume. *Mater. Sci. Eng. A* **2010**, *527*, 3676–3684. [CrossRef]
36. Abdulmatin, A.; Tangchiratpa, W.; Jaturapitakkul, C. An investigation of bottom ash as a pozzolanic material. *Constr. Build. Mater.* **2018**, *186*, 155–162. [CrossRef]
37. ACI Committee 213. ACI 213.R-03. Guide for structural lightweight-aggregate concrete. In *ACI Manual of Concrete Practice, Part 1*; American Concrete Institute: Farmington Hills, MI, USA, 2003.
38. Zhang, B.; Poon, C.S. Use of furnace bottom ash for producing lightweight aggregate concrete with thermal insulation properties. *J. Clean. Prod.* **2015**, *99*, 94–100. [CrossRef]
39. Singh, M.; Siddique, R. Strength properties and micro-structural properties of concrete containing coal bottom ash as partial replacement of fine aggregate. *Constr. Build. Mater.* **2014**, *50*, 246–256. [CrossRef]
40. Albayrak, M.; Yorukoglu, A.; Karahan, S.; Atlıhan, S.; Aruntas, H.Y.; Girgin, I. Influence of zeolite additive on properties of autoclaved aerated concrete. *Build. Environ.* **2007**, *42*, 3161–3165. [CrossRef]
41. American Society for Testing and Materials (ASTM). *Standard Test Method for Pulse Velocity through Concrete*; ASTM C597-02; ASTM: West Conshohocken, PA, USA, 2002.
42. Ashrafian, A.; Taheri Amiri, M.J.; Rezaie-Balf, M.; Ozbakkaloglu, T.; Lotfi-Omran, O. Prediction of compressive strength and ultrasonic pulse velocity of fiber reinforced concrete incorporating Nano silica using heuristic regression methods. *Constr. Build. Mater.* **2018**, *190*, 479–494. [CrossRef]

43. Sadeghi Nik, A.; Lotfi Omran, O. Estimation of compressive strength of self-compacted concrete with fibers consisting Nano-SiO2 using ultrasonic pulse velocity. *Constr. Build. Mater.* **2013**, *44*, 654–662. [CrossRef]
44. Neville, A.M. *Properties of Concrete*, 4th ed.; Addison Wesley Longman Ltd.: Essex, UK, 1995.

Article

Modelling the Thermal Energy Storage of Cementitious Mortars Made with PCM-Recycled Brick Aggregates

Christoph Mankel [1], Antonio Caggiano [1,2,*], Andreas König [3], Diego Said Schicchi [1], Mona Nazari Sam [1] and Eddie Koenders [1]

1 Institut für Werkstoffe im Bauwesen, Technische Universität Darmstadt, Franziska-Braun-Straße 3, 64287 Darmstadt, Germany; mankel@wib.tu-darmstadt.de (C.M.); dmsaid@gmail.com (D.S.S.); sam@wib.tu-darmstadt.de (M.N.S.); koenders@wib.tu-darmstadt.de (E.K.)

2 CONICET, LMNI, INTECIN, Facultad de Ingeniería, Universidad de Buenos Aires, C1053 Buenos Aires, Argentina

3 Department of Prosthodontics and Dental Materials Science, Leipzig University, 04109 Leipzig, Germany; akoenig@uni-leipzig.de

* Correspondence: caggiano@wib.tu-darmstadt; Tel.: +49-6151-16-22210

Received: 8 January 2020; Accepted: 17 February 2020; Published: 27 February 2020

Abstract: This paper reports a numerical approach for modelling the thermal behavior and heat accumulation/liberation of sustainable cementitious composites made with Recycled Brick Aggregates (RBAs) employed as carriers for Phase-Change Materials (PCMs). In the framework of the further development of the fixed grid modelling method, classically employed for solving the well-known Stefan problem, an enthalpy-based approach and an apparent calorific capacity method have been proposed and validated. More specifically, the results of an experimental program, following an advanced incorporation and immobilization technique, developed at the *Institut für Werkstoffe im Bauwesen* for investigating the thermal responses of various combinations of PCM-RBAs, have been considered as the benchmark to calibrate/validate the numerical results. Promising numerical results have been obtained, and temperature simulations showed good agreement with the experimental data of the analyzed mixtures.

Keywords: thermal-energy storage; enthalpy method; apparent calorific capacity method; recycled brick aggregates; meso-scale; PCMs; paraffin waxes; Stefan Problem

1. Introduction

The energy demand for heating and cooling the global building stock represents a massive part of the total energy consumption around the world (≈ 40%) [1]. In the EU, it accounts for about half of all energy consumption [2]. To attenuate this number, thermal efficiency of construction and building elements, like walls, roofs, and floors, has become the most important measure to enhance energy savings of the new and existing building stock. Recently, with the introduction of the 2019/2021 EU Buildings Directive, it has been additionally emphasized that new buildings of the EU-Member States have to be designed as "Nearly-Zero-Energy Buildings" from the beginning of 2021 onward [3]. For public non-residential buildings, this obligation should become active in 2019 and will be permanently anchored in future building legislation. Moreover, the German Energy Saving Regulation (EnEV) 2014 has proclaimed 25% stricter requirements for energy savings in their 2016 issue, which affects the set of most important regulations of new and existing German buildings. These regulations define the maximum annual primary energy demand and the maximum permissible loss of transmission heat through the building envelope, on the basis of a reference building. In order to comply with the European targets on energy savings, the German federal government will merge the rules that currently apply in parallel—(i) the Energy Saving Act (EnEG) [4], (ii) the Energy Saving Regulation (EnEV) [5], and (iii) the Renewable Energy Heat Act (EEWärmeG) [6]—into a new Building Energy Act (GEG) [7].

Possible innovative solutions include the potential of materials to embody large amounts of thermal energy, which would stabilize the inner thermal comfort of either residential or non-residential buildings. Particularly, this can be achieved through an effective use of Phase Change Materials (PCMs), which minimizes the additional need of primary energy for heating/cooling [8,9]. Numerous articles on PCMs have been reported in literature and in several fields of applications. The experimental-based research is predominantly addressing the thermo-hydro-chemo-mechanical properties of cementitious materials with PCMs [10–12]. In those works, the investigated PCMs are characterized by a melting temperature that varies between 19 °C and 26 °C, which corresponds to a standard temperature range for comfortable living [13]. According to their chemical compositions, PCMs can be categorized as organic, inorganic and eutectics [14], while according to their phase transition mode as liquid–gas, solid–gas, solid–liquid, and solid–solid. Solid–liquid is the most preferred mode for energy and building applications [15,16].

A large number of review studies are already available in literature, which are very helpful for understanding the State of the Art (SoA) of the current research, and also to evaluate the potential directions for future investigations. In a very recent review paper, performed at TU-Darmstadt [17], the authors reviewed the various numerical tools that are used for modelling phase change phenomena in problems of materials science and engineering. The available modelling strategies for PCM-based cementitious composites, at different scale levels (micro-, meso-, macro-, and multiscale [18]), have been deeply explored, evaluated and discussed.

Numerical tools and theoretical approaches, available in scientific literature for modelling Thermal Energy Storage (TES) and heat accumulation/liberation of PCM-based applications, deal with solving the so-called Stefan problem [19] or an extended version of it. The Stefan problem assumes the existence of two different domains, one representing the solid phase (Ω_S) and another representing the liquid phase (Ω_L). These are separated by a sharp front, identified as a moving interface (Γ), which is the location where the phase transformation takes place. Such sharp surface front is characterized by a temperature field that equals the melting temperature T_m. Several authors, under simplified assumptions (like 1D conduction-only heat problem in both the solid and liquid phase), have proposed analytical solutions for this problem [20]. While, more complex models can be taken into account for further evaluation of the classical Stefan problem: e.g., by introducing a so-called "mushy zone" representing non-isothermal conditions and dealing with a finite thickness of this moving interface zone, where the liquid and solid phases may coexist [21]. The solution method of the Stefan problem and/or its extended formulation, can generally be subdivided into three different categories [22]: (i) the fixed grid method, where the grid of spatial nodes, used for discretizing the problem, remains fixed during time, thus the phase change is traced through auxiliary constitutive formulations and state functions; (ii) the deformed grid method, where the nodes forming the grid may move to explicitly follow the moving sharp front that occurs during the melting and/or solidification; (iii) and hybrid methods, which are a combination of (i) and (ii). Most classical examples are those based on a further development of the fixed grid method through following the enthalpy-based approach (EA) [23], which mainly boils down into two alternative solutions: (i) the apparent calorific capacity method (ACCM) [24] and (ii) the heat source method (HSM) [25].

In this context, the present study investigated the TES of mortars made of Recycled Brick Aggregates containing PCMs. The main scope was to assess a numerical procedure for simulating the effects of paraffin waxes on the thermal energy responses of mortars produced with different types and amounts of PCM-RBAs. The results of a wide series of thermal tests [26] (DSC measurements, thermal conductivity and special spherical tests, labelled DKK—*Dynamische Kugel Kalorimetrie*), carried out on PCM, aggregates, plain mortars, and PCM-RBA mortars are summarized. The aforementioned experimental results are employed in an inverse identification procedure for unveiling the key parameters that drive the TES of the mixtures.

Following this short introductive literature review, the paper is structured as follows. Section 2 addresses the enthalpy-based model and the ACCM resolution of it, employed for simulating the phase transformation phenomena of the PCM-RBA mortars. Experimental results, outlined in terms of TES results, are shown in Section 3. Section 4 outlines the spherical-based solution and numerical implementation of the model procedure described in Section 2. In Section 5, numerical results and comparisons against the experimental data are reported and discussed to demonstrate the soundness and capability of the numerical procedure. Concluding remarks and future research outlook are addressed in Section 6.

2. The Enthalpy-Based Model and Apparent Calorific Capacity

This section reports the enthalpy-based model that will be employed for predicting the phase transformation phenomena of PCM-RBA mortars.

2.1. Basic Principles: Thermodynamics and Thermal Energy Storages

The enthalpy H of a thermodynamic system can be defined as

$$H = U + pV \tag{1}$$

with U the internal energy, p the pressure and V the volume of the system.

Then, by introducing the first law of thermodynamics, for closed systems and an infinitesimal process, the following equation can be stated as

$$dU = \delta Q - \delta W \tag{2}$$

i.e., the variation of the internal system energy (dU) is equal to an infinitesimal amount of heat added (δQ) minus an infinitesimal amount of work performed (δW).

Under the hypothesis that only p and V spend work (quasi-static process), $\delta W = p\, dV$.

It follows that

$$dU = \delta Q - p\, dV. \tag{3}$$

Now, by evaluating dH from Equation (1) and using Equation (3), the following equation will be achieved:

$$dH = dU + d(pV) = dU + dpV + pdV = \delta Q - pdV + dpV + pdV = \delta Q + dpV. \tag{4}$$

For a constant pressure, which represents most cases of PCM-concrete systems of construction and building applications, it can be assumed that a small variation of enthalpy is equal to a small amount of heat added:

$$dH = \delta Q. \tag{5}$$

2.2. Enthalpy Description and Apparent Calorific Capacity Method (ACCM)

In accordance with the fixed grid modelling method and with its discretization through the ACCM approach (see reference [17]), the classical equation for describing a heat conduction problem can written as follows:

$$\frac{\partial Q}{\partial t} = \nabla.(\lambda \nabla T) + \dot{q}_v \qquad \forall\, \mathbf{x} \in \Omega \tag{6}$$

where Q is the heat of the system, t the time, $\lambda = \lambda(T, \mathbf{x})$ the thermal conductivity of the material, which depends on the temperature T and position vector $\mathbf{x}$ (of the considered body Ω), $\dot{q}_v$ is the possible source term, while $\nabla.$ and ∇ are the divergence and gradient tensorial operators.

Substituting Equation (5) into Equation (6) leads to the following equation:

$$\frac{\partial H}{\partial t} = \nabla.(\lambda \nabla T) + \dot{q}_v \qquad \forall\, \mathbf{x} \in \Omega \tag{7}$$

which is the mostly adopted equation for solving phase changes in simulations for construction and building applications, and it is known in literature as the enthalpy-based method.

Applying the chain rule to $\frac{\partial H}{\partial t}$ of Equation (7),

$$\frac{\partial H}{\partial t} = \frac{\partial H}{\partial T}\frac{\partial T}{\partial t} \tag{8}$$

and by introducing the concept of the Apparent Calorific Capacity Method (ACCM), the following temperature-dependent (apparent of effective) heat capacity expression can be written [17]:

$$\frac{\partial H}{\partial T} = \rho C_{eff}(T) \tag{9}$$

Thus, Equation (7) modifies into the following heat transfer equation.

$$\rho C_{eff}(T)\frac{dT}{dt} = \nabla.(\lambda \nabla T) + \dot{q}_v \qquad \forall\, \mathbf{x} \in \Omega \tag{10}$$

Finally, the following initial condition (IC) and boundary conditions (BCs) complete the description of the phase change problem:

$$\textbf{IC:} \quad T(\mathbf{x}, t = 0) = T_0(\mathbf{x}) \qquad \forall\, \mathbf{x} \in \Omega \tag{11}$$

$$\textbf{BCs:} \quad \begin{array}{ll} T(\mathbf{x}, t) = T_D & \forall\, \mathbf{x} \in \Gamma_T \\ (\lambda \nabla T) \cdot \mathbf{n} = q & \forall\, \mathbf{x} \in \Gamma_q \\ (\lambda \nabla T) \cdot \mathbf{n} = h(T_{env} - T) & \forall\, \mathbf{x} \in \Gamma_c \\ (\lambda \nabla T) \cdot \mathbf{n} = \kappa(T_{rad} - T) & \forall\, \mathbf{x} \in \Gamma_r \end{array} \tag{12}$$

where Γ_T, Γ_q, Γ_c and Γ_r are the essential and natural (flux, convection and radiation) boundary conditions of the domain Ω. T_D is the specified temperature imposed at the essential boundary, q the heat flux, h the heat convection and κ the radiation coefficients, respectively, T_{env} the environmental temperature, and, finally, T_{rad} the external radiation source temperature.

3. Overview of the Experimental Test Data

This section reports the results of an experimental program performed for characterizing the thermal-energy response of PCM-RBA mortar mixtures and their components. The experimental data are used for validation of the numerical simulations discussed in Section 5.

Six mixtures were considered, having a w/c ratio of 0.5 and various amounts of PCM-RBA volume fractions (Table 1). All mixtures were prepared according to EN 196-1 [27]. Recycled bricks (labelled "SB") and high porosity Poroton®fired-clay blocks (labelled "PB") have been considered [28]. These building materials, processed in the form of medium/coarse aggregates and provided by a local company (SHW GmbH, Messel, Germany), were used as carriers (containers) for storing a predefined volume of PCM.

RT 25 HC [29] paraffin waxes were used as PCM. They are characterized by a high crystallinity and possess an excellent heat store capacity during phase changes, from solid to liquid and vice versa. The thermo-physical properties of the paraffin wax, considered in this research, are a melting temperature of 25 °C, a storage capacity of approximately 230 kJ/kg, a latent heat capacity of almost 200 kJ/kg, thermal conductivity (in both phases) of 0.20 W/m × K, and densities of 880 (liquid)/770 (solid) kg/m^3. The PCM-RBAs were thus produced following an advanced encapsulation technique, proposed and patented by the *Institut für Werkstoffe im Bauwesen* of TU-Darmstadt (Darmstadt, Germany) [30].

Table 1. Overview of the six PCM-RBA mortars.

Labels	REF-SB	SB-65	SB-80	REF-PB	PB-65	PB-80
Cement [kg/m^3]	701.5	694.2	692.9	700.2	691.7	691.7
Water [kg/m^3]	350.8	347.1	346.5	350.1	345.9	345.9
PCM-RBA [kg/m^3]	655.2	719.3	734.0	664.0	728.9	743.9
Air cont. [V.-%]	2.3	2.9	3.0	2.4	3.1	3.1
w/c ratio [-]	0.50					

The mixture names (labels) in Table 1 aim at providing the key information on the amount of PCM, filling the RBA's open porosity (expressed in volume fraction of the open capillary pore space), and the type of RBAs considered in the mixture. For example, the label "REF-SB" refers to the reference mixture (without PCM) using SB type of bricks; or "SB-65" indicates a mixture using SB-bricks and a filling degree of PCM of 65 V.-% of the total SB-RBA open capillary porosity. The complete description (materials, methods, results and discussion) of the experimental campaign is available in Mankel et al. [28].

3.1. DSC Measurements: Aggregates, Paste, and PCMs

Differential Scanning Calorimetry (DSC) tests were performed for each component used in the investigated PCM-RBA mortar mixtures. Their heat storage capacity has been expressed in terms of bulk density times the specific heat capacity, i.e., $\rho \times C_p$. Three samples per each component were analyzed and the mean value for each of them was presented in this section. Particularly, the DSC thermograms under either heating or cooling, for the cement paste with a w/c of 0.50 and for both SB- and PB-RBAs, have been shown in Figure 1a. They were done within the temperature range of 10 °C to 40 °C and using a heating/cooling rate of 10 K × min^{-1}. From these results, it can be observed that the sensible behavior of both RBAs is almost similar, whereby a slightly higher sensible heat storage capacity can be detected for the cement paste.

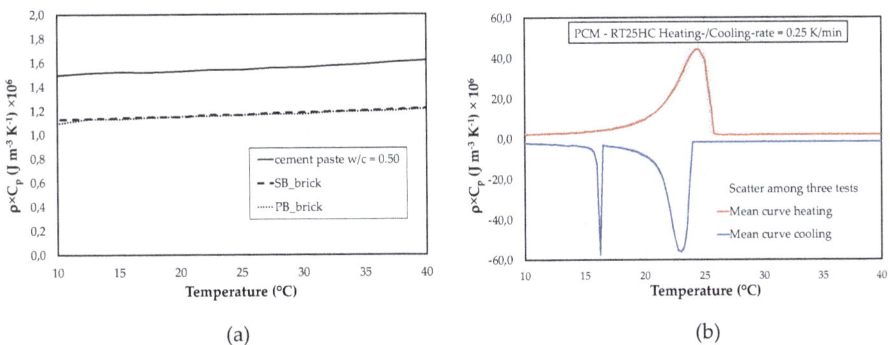

Figure 1. $\rho \times C_p$ results of DSC measurements of (**a**) w/c = 0.50 cement pastes, SB and PB obtained with a heating/cooling rate of 10 K × min^{-1} and (**b**) Paraffin RT25HC with a heating/cooling rate of 0.25 K × min^{-1}.

In Figure 1b, the DSC results of the used paraffin wax (Rubitherm®RT25HC, Rubitherm Technologies GmbH, Berlin, Germany) is shown. The test procedure was conducted in accordance with the IEA DSC 4229 PCM Standard [31] to determine the final heating/cooling rate for the considered dynamic DSC measurements. The adopted heating/cooling rate was 0.25 K × min^{-1} on the final measured results. This value represented a compromise between accuracy of the results and acceptable mitigation of heating rate measurements, fulfilling the IEA DSC 4229 requirements [31].

DSC curves of the considered PCM, in Figure 1b, shows a sensible heat storage character of the material in those temperature ranges, far from the phase change responses (e.g., in the solid and liquid stages), and shows a pronounced latent peak in the region close to the temperature where the phase change occurs (T_m = 24.5 °C for heating and T_m = 22.95 °C for cooling).

3.2. Thermal Conductivity of the PCM-RBA Mortars

The thermal conductivity of PCM-RBA mortar mixtures was determined using the Hot-Disk transient plane source method [28]. For this aim, three samples of 150 mm × 150 mm × 80 mm cuboids were tested. The measurements were done with a 9.9 mm diameter sensor, in three different specimen sides. Steady-state conditions with a temperature of 20 °C were considered. In Figure 2, it can be observed that all mixtures have quite comparable thermal conductivities, which range between 0.696 (min. value) and 0.846 (max. value) W/mK. As a general trend, it can be observed that the mixtures with PB RBAs deal with slightly lower conductivities than the SB RBA ones. This can be attributed to the higher porosity of the SB RBAs, which affect the overall conductivity of the composites. It could also be observed that for higher PCM contents, the thermal conductivity is lower.

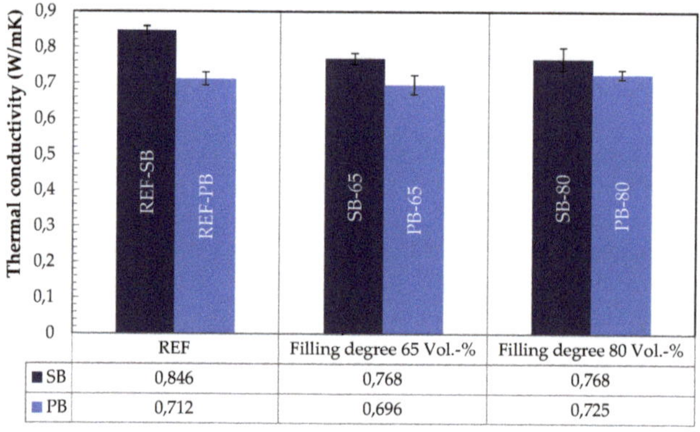

	REF	Filling degree 65 Vol.-%	Filling degree 80 Vol.-%
SB	0,846	0,768	0,768
PB	0,712	0,696	0,725

Figure 2. Thermal conductivity of the SB- and PB-RBA mortars, with and without PCMs.

Moreover, it can be also observed that the thermal conductivities are only slightly affected by the different PCM volume fractions. A little increase of the thermal conductivity was measured for those mixtures with a higher pore filling ratio (i.e. PB-80 in comparison with PB-65). This could be the result of the allocation of paraffin wax into the capillary pore space of the composites, which leads to a slight increase in heat conductivity of the PCM-RBAs since the considered PCMs are more conductive than air. However, it can finally be concluded that immobilizing PCMs into the RBA porous structure does basically not (e.g. SB-80 vs. SB-65), or only slightly (see, PB-80 vs. PB-65), modify the overall thermal conductivity of RBA mortars.

3.3. Thermal-Energy Storage DKK Tests in Spherical Samples

Spherical-shaped specimens were used to monitor the time-dependent temperature evolution of the PCM-RBA mortars. The adopted and patented non-conventional testing technique (namely *Dynamische Kugel Kalorimetrie*), DKK [26]) was followed by the authors for the TES identification of the composite materials under investigation. For each considered mixture, three spherical samples were produced, and two thermocouples were positioned in the center of each sphere and at its outer surface, respectively. Heating tests were done by using an isothermal conditioned oven with a fixed temperature of 49 °C. Cooling tests were done with a climatic chamber fixing the temperature at ca. 9 °C.

The graphs of Figure 3 show the average results (from three independent spherical specimens) of the measured temperature evolution, in the center of the spherical samples, versus time. Heating and cooling results are plotted for both PCM SB-brick and PB-brick mixtures. It can be seen that, for the mortar mixtures casted with either SB or PB bricks, a delay of the temperature development takes place when PCM-RBAs, with PCM filling degrees of 65 V.-% and 80 V.-%, are analyzed. The presence of PCM and their melting/solidification behavior actually shifts the temperature curve into the right/down direction for heating response (Figure 3a,c) and up/left for cooling (Figure 3b,d). Particularly, a quasi-horizontal plateau of the temperature evolution data can be appreciated during both temperature rise (heating) or temperature decrease (cooling), i.e. in the range between 21 and 26 °C the PCM and vice versa. By taking into consideration the effect of PCM volume fractions, it appeared that almost no thermal differences between the considered RBAs and PCM additions, i.e., 65 and 80%, exist. Comparison of the temperature evolutions SB-65 with SB-80 or PB-65 with PB-80 show an almost equal response. For each mixture and/or sample, the complete phase change of the paraffin was always fully occurring, meaning that all PCMs were in their final state at the end of each heating or cooling test. This is also shown in Figure 3 where the center of the spheres at the end of each cooling test has a temperature less than 10 °C and under heating close to 50 °C, while the melting points range between 21–26 °C.

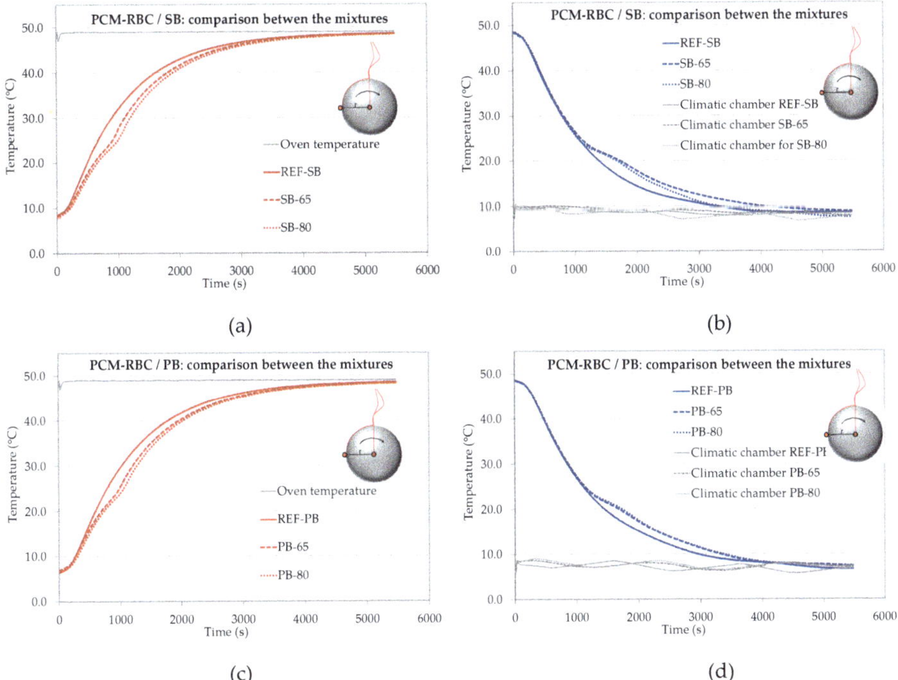

Figure 3. Temperature evolutions of the DKK tests: (**a**) heating and (**b**) cooling of PCM-RBA mortars with SB bricks, (**c**) heating and (**d**) cooling of PCM-RBA mortars with PB bricks. These results represent the average of three measurements in the center of the spherical samples.

For a thorough discussion on the present experimental data, with emphasis on the DSC analyses, conductivity measurements and DKK tests, reference is made to Mankel et al. 2019 [28].

4. Numerical Implementation and Spherical-based Solution

4.1. D Spherical-Based Solution for the ACCM

Equation (10) is now transferred into spherical coordinates for predicting the thermal energy storage behavior and temperature evolution in the tested specimens presented in Section 3.3.

From this, the following relation can be derived:

$$\rho C_{eff}(T)\frac{dT}{dt} = \frac{1}{r^2}\frac{\partial}{\partial r}\left(\lambda r^2\frac{\partial T}{\partial r}\right) + \frac{1}{r^2\sin^2\theta}\frac{\partial}{\partial\phi}\left(\lambda\frac{\partial T}{\partial\phi}\right) + \frac{1}{r^2\sin\theta}\frac{\partial}{\partial\theta}\left(\lambda\sin\theta\frac{\partial T}{\partial\theta}\right) + \dot{q}_v \tag{13}$$

where (r,θ,ϕ) are the radial distance and polar and azimuthal angles. Making use of the spherical symmetry, the previous relationship can be significantly simplified as follows:

$$\rho C_{eff}(T)\frac{dT}{dt} = \frac{1}{r^2}\frac{\partial}{\partial r}\left(\lambda r^2\frac{\partial T}{\partial r}\right) + \dot{q}_v. \tag{14}$$

4.2. Schematization and Discretization

The ACCM model, under the assumptions of spherical geometry and symmetry, was solved by means of the finite difference method. The heat-diffusion through the PCM-RBA mortar systems was calculated by solving the differential equation previously described in Equation (14), having a 1D spherical-based hypothesis and by adopting heat convection (Robin) boundary conditions for describing the environmental surface conditions of either a furnace or a climate chamber.

Thus, the boundary condition for the sample core at node "1" (Figure 4) was adiabatic, meaning that the heat flux, q_1, is null.

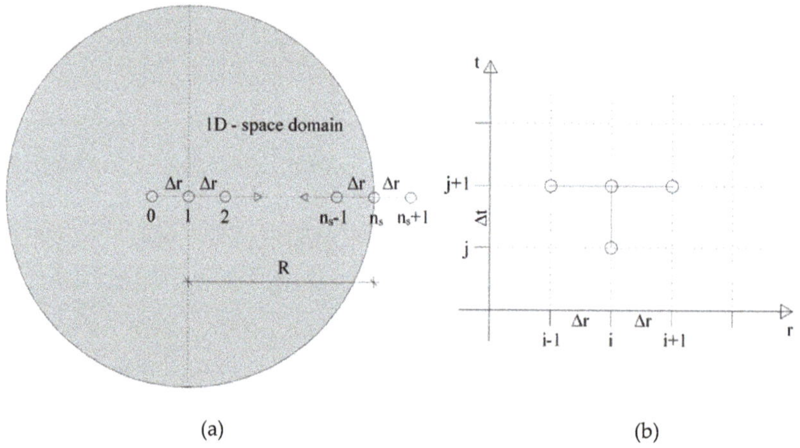

Figure 4. (**a**) Finite difference (FD) space discretization scheme of a 1D-heat transfer in spherical specimens; (**b**) FD molecule for implicit time using the Backward Euler Method and the Space Central Method.

$$\text{Sample core}: \qquad q_1 = -\lambda\left(\frac{\partial T}{\partial r}\right)\bigg|_{r=0} = 0 \tag{15}$$

while the Robin boundary condition is applied at the outer surface (node "n_s")

$$\text{Ambient condition}: \qquad q_{n_s} = -\lambda\left(\frac{\partial T}{\partial r}\right)\Bigg|_{r=R} = h\left(T_f - T_r\right) \tag{16}$$

where T_f is the ambient temperature (fixed by the oven or climatic chamber), T_r is the surface temperature at $r = R$, while h is the heat transfer coefficient.

In this context, the Finite Difference (FD) space domain was discretized into n_s spaces (namely Finite Differences), leading to $n_s + 1$ nodes and n_t time discretization steps. Furthermore, a fully implicit Euler Method for the transient problem was applied.

The differential equation shown in Equation (14), within a space domain with a length of R between the node 1 to n_s (see Figure 4a), can be solved as shown in the following scheme:

$$\rho C_{eff}(T)\frac{dT}{dt} = \frac{2\,\lambda}{r}\frac{\partial T}{\partial r} + \lambda\frac{\partial^2 T}{\partial r^2} \text{ with } r = (i-1)\Delta r \tag{17}$$

By using the implicit backward Euler Method for time (j) and central for space (i) (see Figure 4b), Equation (17) can be discretized to

$$\rho C_{eff}\left(T_i^{j+1}\right)\frac{T_i^{j+1} - T_i^j}{\Delta t} = \lambda\frac{T_{i+1}^{j+1} - T_{i-1}^{j+1}}{(i-1)\Delta r^2} + \lambda\frac{T_{i-1}^{j+1} + 2T_i^{j+1} + T_{i+1}^{j+1}}{\Delta r^2}. \tag{18}$$

The boundary condition, for the sample core at $r = 0$ ($i = 1$), has been further developed. The right-hand side of Equation (17) can be simplified as follows [32]

$$\lim_{r \to 0}\left(\frac{2\,\lambda}{r}\frac{\partial T}{\partial r} + \lambda\frac{\partial^2 T}{\partial r^2}\right) = 3\lambda\frac{\partial^2 T}{\partial r^2} \tag{19}$$

Then, by adopting the central difference approximation of the adiabatic boundary condition expressed of Equation (15) and by using one ghost node (namely node "0"), the following expression can be stated:

$$-\lambda\left(\frac{\partial T}{\partial r}\right)\Bigg|_{r=0} = -\lambda\frac{T_2^{j+1} - T_0^{j+1}}{2\Delta r} = 0 \tag{20}$$

It follows that $T_2^{j+1} = T_0^{j+1}$ and combining it into Equations (18) and (19), it can be easily achieved the following expression of the adiabatic boundary condition:

$$\rho C_{eff}\left(T_1^{j+1}\right)\frac{T_1^{j+1} - T_1^j}{\Delta t} = 6\lambda\frac{T_2^{j+1} + T_1^{j+1}}{\Delta r^2} \tag{21}$$

For the implementation of the Robin boundary condition at the outer surface of the specimens, e.g. node $r = R$ ($i = n_s$), Equation (15) can be discretized in the following way:

$$-\lambda\frac{T_{n_s-1}^{j+1} - T_{n_s+1}^{j+1}}{2\Delta r} = h\left(T_f^{j+1} - T_r^{j+1}\right) \tag{22}$$

In Equation (22), the temperature at the ghost node $r = R + 1$ ($i = n_s + 1$) is known, $T_{n_s+1}^{j+1} = T_f$, thus the following expression can be rewritten:

$$T_{n_s+1}^{j+1} = \frac{2\,\Delta r\,h}{\lambda}\left(T_{n_s}^{j+1} - T_f^{j+1}\right) + T_{n_s-1}^{j+1} \tag{23}$$

Thus, Equation (16) can be easily written as

$$\rho C_{eff}\left(T_{n_s}^{j+1}\right)\frac{T_{n_s}^{j+1}-T_{n_s}^{j}}{\Delta t} = \lambda\frac{\left(\frac{2\,\Delta r\,h}{\lambda}\left(T_{n_s}^{j+1}-T_{f}^{j+1}\right)+T_{n_s-1}^{j+1}\right)-T_{n_s-1}^{j+1}}{(i-1)\Delta r^2}+ \lambda\frac{T_{n_s-1}^{j+1}+2T_{n_s}^{j+1}\left(\frac{2\,\Delta r\,h}{\lambda}\left(T_{n_s}^{j+1}-T_{f}^{j+1}\right)+T_{n_s-1}^{j+1}\right)}{\Delta r^2} \tag{24}$$

and after some mathematical elaborations, the following expression can be achieved for the Robin natural boundary condition:

$$\rho C_{eff}\left(T_{n_s}^{j+1}\right)\frac{T_{n_s}^{j+1}-T_{n_s}^{j}}{\Delta t} = 2\left[h\left(\frac{T_{n_s}^{j+1}-T_{f}^{j+1}}{(i-1)\Delta r}+\frac{T_{n_s}^{j+1}-T_{f}^{j+1}}{\Delta r}\right)+\lambda\left(\frac{T_{n_s-1}^{j+1}+T_{n_s}^{j+1}}{\Delta r^2}\right)\right] \tag{25}$$

5. Numerical Results and Comparisons

This section reports the description of the numerical results and shows their comparisons against the experimental data, which has been briefly outlined in Section 3. The numerical simulations are based on the assumption that the PCM-RBA mortars can be considered as a continuum and homogenous media. In this context, homogenized (meso-scale based) parameters were considered in the selection of the input data for the numerical prediction.

5.1. Homogenized Macroscopic C_{eff} Model for the PCM-RBA Mixtures

A homogenization technique was employed for evaluating the effective specific heat capacity C_{eff} of the PCM-RBA mortars. It is based on the mixture theory by using the volume percentages of each individual component such as RBAs, cement paste and PCMs. More precisely, the model smears out the specific heat capacity of the RBA $C_{RBA}(T)$, cement paste $C_{paste}(T)$ and the apparent specific heat capacity of the PCM $C_{app,PCM}(T)$ through adopting the volume fraction of each component as the smeared out (weighting) factor χ.

The specific heat capacities of each component were experimentally determined with DSC measurements (see Section 3.1) and are shortly summarized in Table 2.

Table 2. Overview of the six PCM-RBA mortars.

Cement Paste	RBA	PCM	
C_{paste} Solid phase	C_{RBA} Solid phase	$C_{app,PCM}$	Liquid phase Phase change Solid phase
Figure 1a	Figure 1a	Figure 1b	

RBAs and cement pastes have only the sensible heat storage part (Figure 1a), while the PCMs have an apparent specific heat capacity $C_{app,PCM}(T)$ that incorporates the additional latent behavior within the temperature range of the phase change (during melting and solidification, as shown in Figure 1b).

The evaluation of the specific heat capacities was determined in two consecutive steps. First, at aggregate level, where the PCM-RBAs were considered as lumped components of RBAs ($C_{RBA}(T)$) plus PCM ($C_{app,PCM}(T)$) and weighting their volume fractions χ to achieve the smeared $C_{eff,PCM\text{-}RBA}(T)$:

$$\text{PCM-RBAs}: \quad C_{eff,PCM-RBA}(T) = \chi_{RBA}\times C_{RBA}(T) + \chi_{PCM}\times C_{app,PCM}(T) \tag{26}$$

where χ_{RBA} and $\chi_{PCM}$are the volume fraction of the recycled bricks and the filled PCMs, respectively.

Then, at a composite level (i.e., PCM-RBA mortar) a homogenized overall system of the effective specific heat capacity, C_{eff}, was determined by weighting the heat capacities $C_{eff,PCM\text{-}RBA}(\theta)$, evaluated through Equation (26), and $C_{paste}(\theta)$ of the individual material components by their volume fractions ψ:

$$\text{PCM-RBA mixtures}: \; C_{eff}(T) \; = \psi_{paste} \times C_{paste}(T) \; + \psi_{PCM-RBA} \times C_{eff,PCM-RBA}(T) \tag{27}$$

The exact volume fraction ratio between PCM-RBAs and cement paste, of the investigated mixtures, were investigated by performing µ3D-XCT-scans [33] of the spherical specimens. Nine 2D-slices were extracted from each 3D body of the scanned spherical specimens (Figure 5) and mean volume fractions of PCM-RBAs and cement paste were accurately determined by image analyses. An average area ratio was evaluated for each slice by applying a recoloring of the surface area through white and red pictures (Figure 5). The results of these analyses are shown in Figure 6 where the volume fractions of paste and PCM-RBA were determined in both PCM.RBA mortars SB and PB.

The effective specific heat capacity of the composite systems, $C_{eff}(T)$, were thus modelled using Equations (26) and (27) by taking into account the determined volume fractions between cement paste (including air voids) and PCM-RBAs (shown in Figure 6) and considering the specific heat capacities of each component as summarized in Table 2.

Figures 7 and 8 shows the afore described $C_{eff}(T)$ further multiplied by the bulk density ρ of the PCM-RBA mortar systems (ρ = 1829.6 kg/m^3 for REF-SB, 1710.37 kg/m^3 for SB-65, 1671.6 kg/m^3 for SB-80, 1853.5 kg/m^3 for REF-PB, 1767.25 kg/m^3 for PB-65 and 1739.8 kg/m^3 for PB-80).

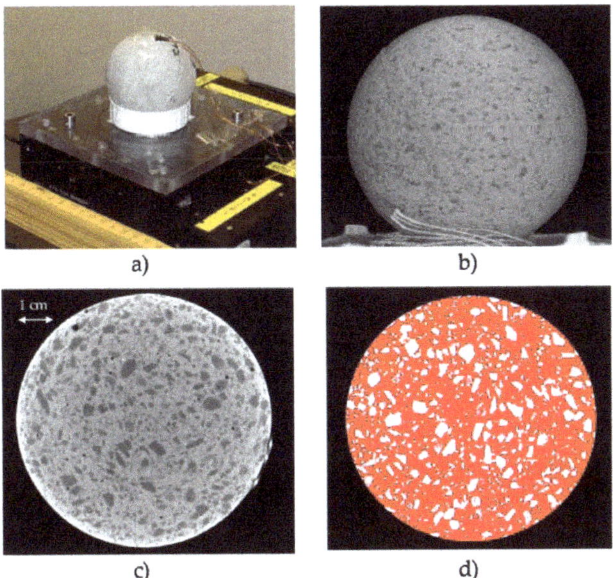

Figure 5. µ3D-XCT-scans setup (**a**,**b**) and slices cross including image-segmentation (**c**,**d**) of the spherical specimens.

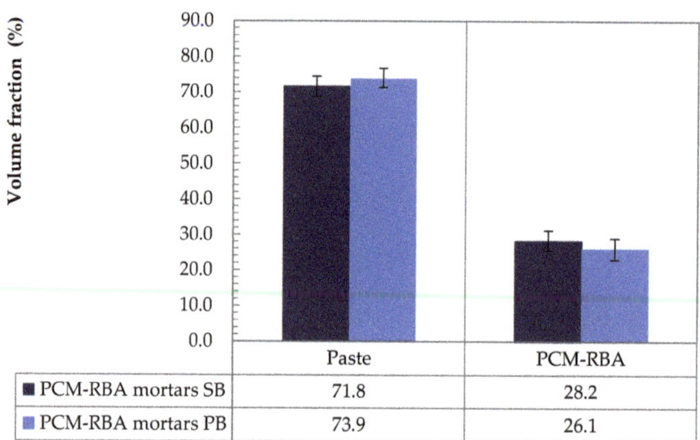

Figure 6. Volume fractions of paste and PCM-RBA determined by image-analysis.

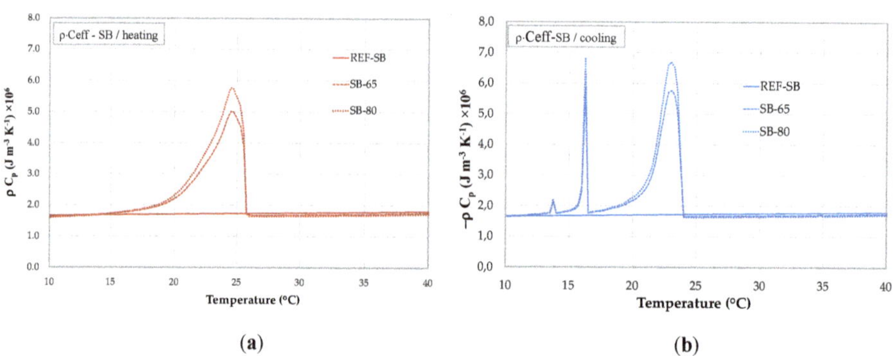

(**a**) (**b**)

Figure 7. ρ × Ceff of PCM-RBA mortar systems with SB bricks: (**a**) heating and (**b**) cooling.

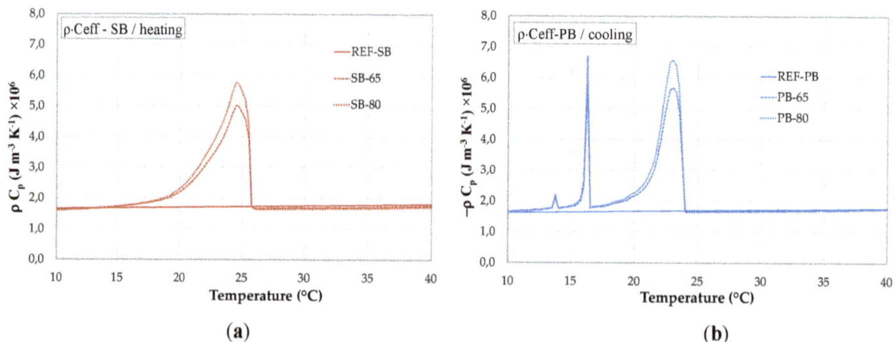

(**a**) (**b**)

Figure 8. ρ × Ceff of PCM-RBA mortar systems with PB bricks: (**a**) heating and (**b**) cooling.

5.2. Numerical Prediction and Comparison

By implementing the ACCM procedure described in Sections 2 and 4, temperature evolutions were simulated and compared with the experimental data reported in Section 3. More in detail, the spherical samples made of RBA mortars (with and without PCMs) were simulated with the proposed heat flow model. The input values were obtained from the conducted experimental measurements, as well as from the $C_{eff}(T)$ curves described in Section 5.1. All input parameters employed in the aforementioned simulations are summarized in Tables 3 and 4.

Table 3. Overview of the numerical parameters assumed for the SB mixtures.

Numerical Parameters	REF-SB	SB-65	SB-80
$\rho\ C_{eff}$ (J cm^{-3} K^{-1})		Figure 7	
λ (W/m K) (Figure 2)	0.846	0.768	0.768
h (W m^{-2} K^{-1})		25.0	

Table 4. Overview of the numerical parameters assumed for the PB mixtures.

Numerical Parameters	REF-PB	PB-65	PB-80
$\rho\ C_{eff}$ (J cm^{-3} K^{-1})		Figure 8	
λ (W/m K) (Figure 2)	0.712	0.696	0.725
h (W m^{-2} K^{-1})		25.0	

The thermal conductivities were assumed as temperature-independent and measured through Hot-Disk tests, as outlined in Section 3.2. Then, the effective specific heat capacity $C_{eff}(T)$ was modelled by using the homogenized model as described before in this section. The number of FD space discretization was chosen 100 while the number of time steps selected was 1000, in all simulations. Moreover, the calibrated Robin heat transfer coefficient (h), representing the heat transfer conditioning coefficient was the same for each mixture and reported in Tables 3 and 4.

The numerical simulations, based on the input parameters above mentioned, have been compared against the experimental data of Section 3.3. In Figure 9, the six graphs show the temperature evolutions for the control mixtures (Figure 9a,b), and for that one having PCM: i.e., SB-65 (Figure 9c), SB-80 (Figure 9c), PB-65 (Figure 9d), and PB-80 (Figure 9f). The experimental scatter of the result data for each mixture have been also plotted in grey.

It can be observed that the modeling approach was able to simulate the experimental temperature evolutions very accurately. In particular, the simulations of the reference mixtures represented in Figure 9a,b show a very good agreement with the experimental results. Then, the simulations of the mixtures TES enhanced with PCM also show good comparisons and trends as the experimental values. A slightly overestimated latent effect can be observed, which results in a slightly amplified shoulder in the temperature evolution (see Figure 9c–f). A reason for this effect can be attributed to micro- and/or meso-structural effect, which can influence the effective thermal conductivity and the melting/solidification activations of the integrated PCMs. This could lead to a slight deviation of the latent effects.

In this context, it may be important to remark that almost all input parameters were chosen from the characterizations of the thermal tests as well as from the homogenized C_{eff} model, which is actually represented by the experimentally determined heat capacities of the individual material components weighted by their volume fractions (see Section 5.1). Thus, with this set of input parameters, a sound numerical prediction, for all six PCM-RBA mortar systems, could be achieved, without the need of applying re-calibration and/or optimizations. The results also show that, with a unique heat transfer coefficient, described as the h (Robin) parameter, the simulations for the reference PCM-RBA mortar systems, REF-SB and REF-PB, are almost in perfect agreement with the corresponding experiments. This supports the assumption that both the spherical symmetry and the hypothesis that the composite PCM-RBA mortar systems could be considered by a homogeneous medium, were both effectively correct.

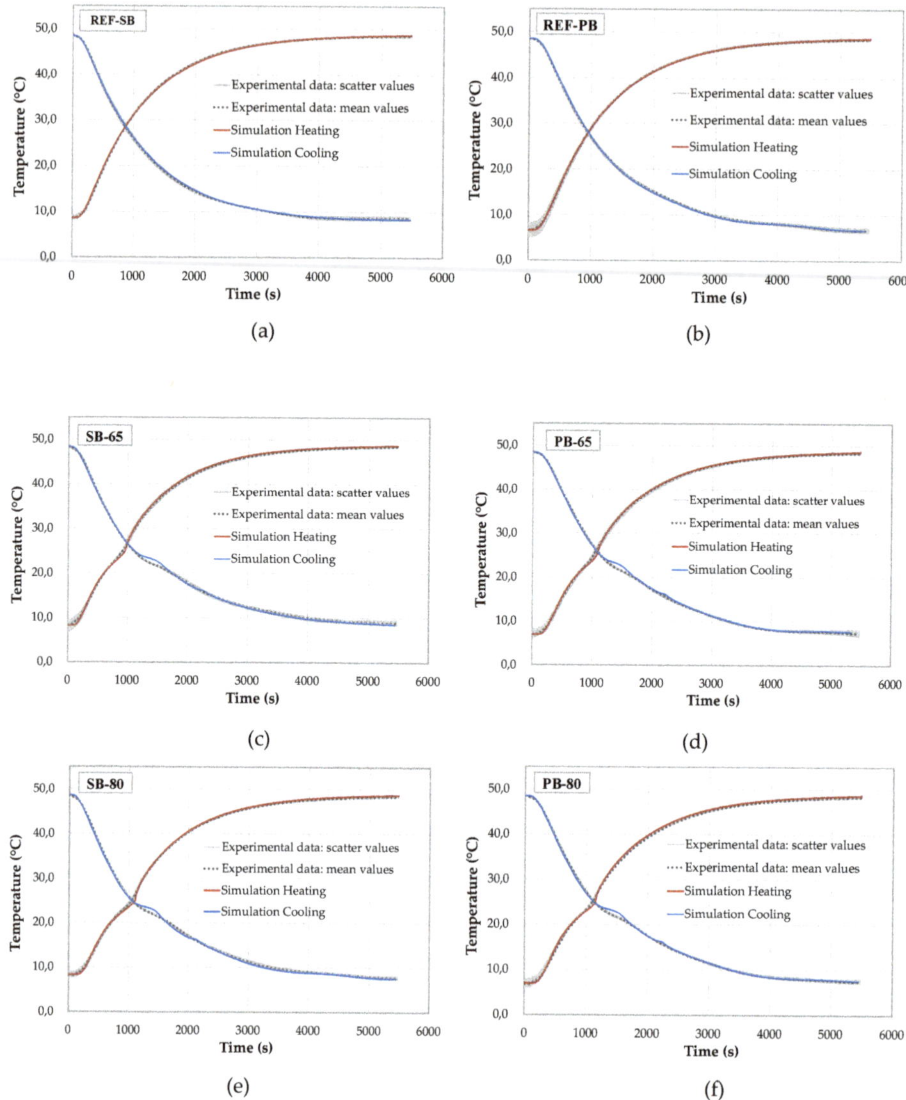

Figure 9. Experimental "DKK" vs. numerical results: (**a**,**c**,**e**) SB brick type mixtures and (**b**,**d**,**f**) PB brick type mixtures.

6. Conclusions

Based on the results shown in this paper, the following conclusions can be drawn:

- An enthalpy-based model, formulated for spherical coordinates and symmetry, was proposed for predicting the thermal energy storage in the tested PCM-RBA specimens.
- Thermal measurements, obtained from dynamic DSC and steady-state Hot Disk tests, were employed for calibrating the model of the numerical activities.

- A mixture theory, based on volume fractions deduced from 3D micro μX-ray computer tomography measurements, was used for generating the resulting meso-composite thermal parameters adopted in the numerical analysis. Particularly the $C_{eff}(T)$ curves have been based on this approach.
- The numerical simulations for the temperature evolution, compared with the experimental DKK results, showed accurate and consistent agreement. It may be important to highlight that these numerical results were just based on input parameters obtained from the experimental characterizations of the mortar components.
- No fitting adjustments, re-calibrations, or numerical adaptions were necessary for reaching good agreement between numerical to experimental data. This can confirm that on the one hand the experimental activities were performed in an accurate way, and on the other hand, that the numerical assumptions and procedures are very accurate to model TES responses in cementitious materials.

Future numerical developments, which follow this research, will include micro-to-meso scale analysis taking into considerations local effects like inclusions (PCM-RBAs), air bubbles, porosity, and interface effects. These further steps will lead to optimizing the "best" recipe for achieving the most performing energy-saving and sustainable cementitious composite.

Author Contributions: Conceptualization, C.M. and A.C.; methodology: materials, methods and processing C.M., A.K. and M.N.S.; data curation, C.M. and A.C.; numerical simulations and comparisons, C.M., A.C. and D.S.S.; writing—original draft preparation, C.M. and A.C.; writing—review and editing, everybody; supervision, A.C. and E.K.; project administration, E.K.; funding acquisition, E.K. All authors have read and agreed to the published version of the manuscript.

Funding: The second author acknowledge the Alexander von Humboldt-Foundation (www.humboldt-foundation.de/) for funding his position at the WiB – TU Darmstadt under the research grant ITA-1185040-HFST-P (2CENERGY project). The support to networking activities provided by the PoroPCM Project (part of the EIG CONCERT-Japan funding, http://concert-japan.eu/) is also gratefully acknowledged.

Conflicts of Interest: The authors declare no conflict of interest.

References

1. Pérez-Lombard, L.; Ortiz, J.; Pout, C. A review on buildings energy consumption information. *Energy Build.* **2008**, *40*, 394–398.
2. Energy Performance of Buildings. Available online: https://ec.europa.eu/energy/en/topics/energy-efficiency/energy-performance-of-buildings (accessed on 18 February 2020).
3. Garcia, J.; Kranzl, L. Ambition Levels of Nearly Zero Energy Buildings (nZEB) Definitions: An Approach for Cross-Country Comparison. *Buildings* **2018**, *8*, 143. [CrossRef]
4. Galvin, R.; Sunikka-Blank, M. Economic viability in thermal retrofit policies: Learning from ten years of experience in Germany. *Energy Policy* **2013**, *54*, 343–351. [CrossRef]
5. Voss, K.; Musall, E.; Lichtmeß, M. From low-energy to Net Zero-Energy Buildings: Status and perspectives. *J. Green Build.* **2011**, *6*, 46–57. [CrossRef]
6. Ekardt, F.; Heitmann, C. Probleme des EEWärmeG bei Neubauten. *Z. Neues Energ.* **2009**, *4*, 348.
7. Amoruso, G.; Donevska, N.; Skomedal, G. German and Norwegian policy approach to residential buildings' energy efficiency—A comparative assessment. *Energy Effic.* **2018**, *11*, 1375–1395. [CrossRef]
8. Cabeza, L.F. (Ed.) *Advances in Thermal Energy Storage Systems: Methods and Applications*; Woodhead Publishing Series I Energy: Number 66; Elsevier: Amsterdam, The Netherlands, 2014.
9. Sukontasukkul, P.; Uthaichotirat, P.; Sangpet, T.; Sisomphon, K.; Newlands, M.; Siripanichgorn, A.; Chindaprasirt, P. Thermal properties of lightweight concrete incorporating high contents of phase change materials. *Constr. Build. Mater.* **2019**, *207*, 431–439. [CrossRef]
10. Kenisarin, M.; Mahkamov, K. Passive thermal control in residential buildings using phase change materials. *Renew. Sustain. Energy Rev.* **2016**, *55*, 371–398. [CrossRef]
11. D'Alessandro, A.; Pisello, A.L.; Fabiani, C.; Ubertini, F.; Cabeza, L.F.; Cotana, F. Multifunctional smart concretes with novel phase change materials: Mechanical and thermo-energy investigation. *Appl. Energy* **2018**, *212*, 1448–1461. [CrossRef]

12. Bahrar, M.; Djamai, Z.I.; Mankibi, M.E.; Larbi, A.S.; Salvia, M. Numerical and experimental study on the use of microencapsulated phase change materials (PCMs) in textile reinforced concrete panels for energy storage. *Sustain. Cities Soc.* **2018**, *41*, 455–468. [CrossRef]
13. Silva, A.S.; Ghisi, E.; Lamberts, R. Performance evaluation of long-term thermal comfort indices in building simulation according to Ashrae Std 55. *Build. Environ.* **2016**, *102*, 95–115. [CrossRef]
14. Souayfane, F.; Fardoun, F.; Biwole, P.H. Phase change materials (PCM) for cooling applications in buildings: A review. *Energy Build.* **2016**, *129*, 396–431. [CrossRef]
15. Vélez, C.; Khayet, M.; de Zárate, J.O. Temperature-dependent thermal properties of solid/liquid phase change even-numbered n-alkanes: N-Hexadecane, n-octadecane and n-eicosane. *Appl. Energy* **2015**, *143*, 383–394. [CrossRef]
16. Mankel, C.; Caggiano, A.; Ukrainczyk, N.; Koenders, E. Thermal energy storage characterization of cement-based systems containing Microencapsulated-PCMs. *Constr. Build. Mater.* **2019**, *199*, 307–320. [CrossRef]
17. Caggiano, A.; Mankel, C.; Koenders, E. Reviewing theoretical and numerical models for PCM-embedded cementitious composites. *Buildings* **2019**, *9*, 3. [CrossRef]
18. Schicchi, D.S.; Caggiano, A.; Hunkel, M.; Koenders, E.A. Thermodynamically consistent multiscale formulation of a thermo-mechanical problem with phase transformations. *Contin. Mech. Thermodyn.* **2019**, *31*, 273–299. [CrossRef]
19. Rubinšteĭn, L.I. *The Stefan Problem*, 1st ed.; American Mathematical Society: Providence, RI, USA, 1971; ISBN 13: 978-1470428501.
20. Ceretani, A.N.; Salva, N.N.; Tarzia, D.A. An exact solution to a Stefan problem with variable thermal conductivity and a Robin boundary condition. *Nonlinear Anal. Real World App* **2018**, *40*, 243–259. [CrossRef]
21. Mirzaei, P.A.; Haghighat, F. Modeling of phase change materials for applications in whole building simulation. *Renew. Sustain. Energy Rev.* **2012**, *16*, 5355–5362. [CrossRef]
22. AL-Saadi, S.N.; Zhai, Z.J. Modeling phase change materials embedded in building enclosure: A review. *Renew. Sustain. Energy Rev.* **2013**, *21*, 659–673. [CrossRef]
23. Nedjar, B. An enthalpy-based finite element method for nonlinear heat problems involving phase change. *Comput. Struct.* **2002**, *80*, 9–21. [CrossRef]
24. Šavija, B.; Schlangen, E. Use of phase change materials (PCMs) to mitigate early age thermal cracking in concrete:Theoretical considerations. *Constr. Build. Mater.* **2016**, *126*, 332–344. [CrossRef]
25. Fachinotti, V.; Cardona, A.; Huespe, A. A fast convergent and accurate temperature model for phase-change heat conduction. *Int. J. Numer. Methods Eng.* **1999**, *44*, 1863–1884. [CrossRef]
26. Koenders, E.A.B.; Mankel, C.; Caggiano, A. Dynamische Kugel Kalorimetrie (DKK). German Patent 123-0069 AZ 2018/21, 18 October 2018.
27. *BS EN 196-1 Methods of Testing Cement—Part 1: Determination of Strength*; British Standards: London, UK, 2005.
28. Mankel, C.; Caggiano, A.; Koenders, E. Thermal energy storage characterization of cementitious composites made with recycled brick aggregates containing PCM. *Energy Build.* **2019**, *202*, 109395. [CrossRef]
29. RUBITHERM®RT. *Datasheet of RT25HC*; Rubitherm Technologies GmbH Company: Berlin, Germany, 2018.
30. Koenders, E.A.B.; Mankel, C. Wärmespeichergranulat aus Rezyklierten Gesteinskörnungen. Patent DE 102016123739, 2018.
31. Gschwander, S.; Haussmann, T.; Hagelstein, G.; Sole, A.; Diarce, G.; Hohenauer, W.; Lager, D.; Rathgeber, C.; Hennemann, P.; Lazaro, A.; et al. *Standard to determine the heat storage capacity of PCM using hf-DSC with constant heating/cooling rate (dynamic mode). A Technical Report of Subtask A2.1 of IEA-SHC 42/ECES Annex 29*; IEA Solar Heating and Cooling: Brussels Belgium, 2015.
32. Ford Versypt, A.N.D.; Braatz, R. Analysis of finite difference discretization schemes for diffusion in spheres with variable diffusivity. *Comput. Chem. Eng.* **2014**, *71*, 241–252. [CrossRef] [PubMed]
33. Dehn, F.; Thalheim, S.; König, A. Betontechnologische Maßnahmen gegen Brandeinwirkungen. In *Sicherheit durch Beton: Schutz vor Explosion, Brand und Risikostoffen: 13. Symposium Baustoffe und Bauwerkserhaltung, Karlsruher Institut fuer Technologie (KIT), 16 Maerz 2017*; KIT Scientific Publishing: Karlsruhe, Germany, 2017; p. 33.

Article

The Heat Conductivity Properties of Hemp–Lime Composite Material Used in Single-Family Buildings

Sławomir Pochwała [1,*], Damian Makiola [1], Stanisław Anweiler [2] and Michał Böhm [3]

[1] Department of Thermal Engineering and Industrial Facilities, Faculty of Mechanical Engineering, Opole University of Technology, Ulica Prószkowska 76, 45-758 Opole, Poland; dmakiola@gmail.com

[2] Department of Environmental Engineering, Faculty of Mechanical Engineering, Opole University of Technology, Ulica Prószkowska 76, 45-758 Opole, Poland; s.anweiler@po.edu.pl

[3] Department of Mechanics and Machine Design, Faculty of Mechanical Engineering, Opole University of Technology, Ulica Prószkowska 76, 45-758 Opole, Poland; m.bohm@po.edu.pl

* Correspondence: s.pochwala@po.edu.pl; Tel.: +48-77-449-8456

Received: 27 January 2020; Accepted: 21 February 2020; Published: 24 February 2020

Abstract: The main goal of the paper is to calculate the heat conductivity for three experimental hemp–lime composites used for structural construction purposes with the use of the experimental stand inside two compartments. Due to current construction trends, we are constantly searching for eco-friendly materials that have a low carbon footprint. This is the case of the analyzed material, and additional thermographic heat distribution inside the material during a fire resistance test proves that it is also a perfect insulation material, which could be applied in addition of popular isolating materials. This paper presents the results of certain hemp–lime composite studies and the potential for using hemp–lime composite for the structural construction industry. Hemp–lime composite heat transfer coefficient, fire resistance, and bulk density properties are compared to those of other commonly used construction materials. The obtained results show that the material together with supporting beams made of other biodegradable materials can be the perfect alternative for other commonly used construction materials.

Keywords: hemp–lime composite; thermal conductivity; low-energy buildings; specific energy absorption; natural fiber

---

1. Introduction

The idea of sustainable development in construction, i.e., one that limits the negative impact of buildings on the environment and is user-friendly, is gaining in popularity [1]. Building in an eco-friendly way is becoming common not only among office space developers, who apply for global Leadership in Energy and Environmental Design (LEED) or Building Research Establishment Environmental Assessment Method (BREEAM) certificates, but also among individuals thinking about building a year-round or recreational house [2]. Hemp lime is a material that has a chance to revolutionize the world's natural construction as it strongly fits into the trend of renewable resources. It has a negative carbon footprint because during its growth, hemp absorbs more carbon dioxide than is used later to make building materials [3,4]. Hemp lime has high thermal insulation, heat accumulation, vapour permeability, is non-flammable, can be used as a fertiliser after demolition and is 100% decomposable [5,6].

The objective of the work is the experimental identification of heat conductivity for three experimental mixtures of hemp–lime composite used for construction purposes analyzed for the use for construction of single-family buildings. Industrial hemp (*Cannabis sativa* L.) has a long history with human civilization and was often found near early nomadic settlements close to streams in well-manured areas [7]. Industrial hemp or *Cannabis sativa* L. is a quick-growing, annual herb with a multitude of uses covering a range of products derived from fiber or oilseed that have been known throughout history [8]. True hemp (industrial hemp) found common application during the 19th century. The sturdy fibers of the plant—well-known and valued for their strength—were used to make ropes for maritime shipping and other industry, as well as paper and textiles [9]. Hemp use has been suppressed in recent times, but due to its usefulness and the ecological advantages in harvesting hemp over cotton and trees, there has been a call for hemp use which is growing worldwide [10].

The hemp–lime composite (Figure 1) consists of water, hemp shiv, lime, and other additives to further improve its properties [5]. Mixing and compressing the components results in a light and appropriately strong material whose structure makes it possible to fill any space or—when applying special formwork—build a complete partition [11,12]. Hemp–lime composite may also be used to make floor tiles or roof and ceiling insulation. Hemp–lime composite shows very poor load-bearing strength and for that reason, it may only be used with specifically designed load-bearing structures or pillars, most typically made of softwood timber [13].

Figure 1. Hemp–lime composite mixture 1.

Hemp lime composite debuted in the Polish market just a few years ago and needs extensive testing. The material meets all the requirements for an environmentally-friendly product, a characteristic that plays an ever more significant role [1,14] Some older technologies seem to experience a strong comeback (such as thatched roofs and wooden or adobe homes) owing to their positive impact on human health and comfort and since they utilize natural raw materials [15].

Hemp lime composite owes its excellent thermal conductivity properties to high hemp shiv porosity [16]. Its properties attract even more people both in this country and abroad to launch investigations and promote hemp–lime composite applications in the construction of new or improving thermal insulation performance of the existing buildings [17–20].

Such composite materials can appear to have variations to their properties due to the existence of structural changes [21]. The properties of the composite are influenced by many factors such as the morphology of the filers, the orientation of the filers, porosity, degree of compaction, the distribution of the filers, and others like gluing [22,23]. Also, some new information regarding nanotubes and fibers in concrete and cementitious materials are worth mentioning [24,25].

Hemp lime composite leaves practically no carbon footprint as hemp shiv, from which it is made, absorbs more CO_2 from the atmosphere during its lifespan than the amount of CO_2 released by the manufacturing process. Research showed that one ton of dry hemp sequesters almost 325 kgs of CO_2 [3].

In hemp–lime composite production, lime is used primarily as a binder, but it also helps inhibit the growth of fungi and mold in the wall. The mixture uses special hydrated lime with an alkaline pH to assure an appropriate biological environment and high vapor permeability of the product. Lime also contributes to better thermal performance and fire resistance properties of hemp–lime composite [26,27].

Hemp lime composite has found application in several construction technologies, one of which includes formwork mounted on a wooden frame whose structure corresponds to the design layout of the building walls, floors, and the roof [28]. The fresh mixture is then sprayed into the formwork, compacted, and left to bind. Subsequently, the formwork is removed to allow the partition to dry. Figure 2 shows hemp–lime composite techniques: casting monolithic walls, prefabrication of the entire wall elements, spraying, and bricklaying [29].

Figure 2. Hemp lime composite techniques: (**a**) casting monolithic walls, (**b**) prefabrication of the entire wall elements, (**c**) spraying, and (**d**) bricklaying [21].

2. Materials and Methods

Three different hemp–lime composite mixtures utilizing various binders were studied. Table 1 presents the percentage shares of the components of the tested composites.

Hemp lime composite mixtures study when completed will help define the basic heat conductivity properties of the material, which remain unknown, but which are important for environmentalists and researchers alike. As water plays different roles during the setting and the curing of hemp–lime, and because we needed material with relatively low brittleness and relatively high strength, our experience showed that the chosen in Table 1 water/cement (W/C) ratio was optimal [30,31].

Table 1. Composite mixtures used for testing.

Binder	Composite 1	Composite 2	Composite 3
Hydraulic lime	100%	10%	-
Hydrated lime	-	90%	70%
Portland cement	-	-	30%

So far, most product information has been supplied by manufacturers who, provide composite performance characteristics, they rather do not specify the detailed composition of the mixtures they use. Thus, comparing the material becomes quite difficult considering the great number of factors affecting the product properties. Some basic factors determining the final properties of the product include the following.

- Hemp shiv type and particle size fractions.
- Type of binder.
- Mixture component proportions.
- Mixing and material application methods.

Despite its ever-growing popularity, the composite properties have not been standardized yet, and that is why it has become necessary to carry out as many tests as possible to establish both the result repeatability and the properties of the product. The objective of our tests consisted of investigating the following properties.

- Bulk density.
- Heat transfer coefficient.
- Fire resistance.

Bulk density is described as a property related to the internal structure of the material; also known as apparent density, it determines the number of properties, such as thermal conductivity, strength, weight, and others. A lime-hemp composite should have a density of 300–500 kg/m^3 to assure adequate strength and thermal resistance. Apparent density depends mainly on the materials used but also on the density of the compacted mixture. Six samples of all the composite types were tested following a 28-day long maturation process. All the samples were kept in 15 cm × 15 cm × 15 cm containers. Table 2 presents the results obtained once the samples had been removed from the containers: the obtained average sample size of each composite and its volume.

Table 2. Volume of the samples tested for three composite materials (Mixtures).

Mixture Used	A (cm)	B (cm)	C (cm)	Volume (m^3)
Mixture 1	15.26	14.90	12.53	0.002849
Mixture 2	15.22	14.92	13.59	0.003086
Mixture 3	15.04	14.69	13.01	0.002874

Each container enclosed the same amount of the mixture. Compared to the volume of the wet mixture found in the containers after it had been maturing for 28 days, Mixture 2 is based on hydrated lime with addition of hydraulic lime demonstrated the least shrinkage and volume reduction. Mixture 1 is made of hydraulic lime and only showed the largest volume shrinkage. Table 3 presents the bulk density of the samples following 28-day long maturation of each of the six prepared samples for three composites and their averaged values.

Table 3. Bulk density of the tested mixtures.

Sample No	Mixture 1	Mixture 2	Mixture 3
	Bulk Density of the Samples Following 28-Day Long Maturation (kg/m^3)		
1	331.28	322.64	370.06
2	329.79	319.56	357.63
3	357.84	338.36	339.14
4	316.40	304.26	337.23
5	308.33	314.99	346.47
6	352.53	354.73	348.17
Average value	333.00	326.00	350.00

The bulk density of the tested composites seems to be similar; however, Mixture No 3, which also had some amounts of Portland cement, showed a slightly higher density. The composite based on hydraulic lime with some added hydrated lime presented the lowest apparent density, probably due to the calcium oxide turning into calcium hydroxide and thus increasing its volume.

Note that the literature related to hemp–lime composites discusses a lot of other important properties of the material. The compressive strength is one of the more common tests performed, as we can see by the number of papers by Cazacu et al. [32], Kremensas et al. [33], Brzyski et al. [22,34], or Li et al. [35]. As we can see from the information presented in these papers, the compressive strength results obtained in those researches have different values, which vary up to 20% in some cases depending on the mixture type. Another important factor, besides eco-friendliness, seems to be in favor of the hemp–lime composite is its great acoustics properties. We can find many papers related to this issue mostly dealing with the materials acoustic absorption by Kinnane et al. [36], or by Gle et al. [37]. In the last three years, we can also see that a lot of scientists like Bourebrab et al. [38] or Heidari et al. [6] are dealing with the surface coating of hemp–lime composites to increase their resistance to humidity or life cycle.

The main goal of the paper was the calculation of the heat conductivity, which is also related to the value of the heat transfer coefficient of a material. The heat transfer coefficient "λ" expressed as [W/m*K] describes the insulation properties of the material: the lower the value, the better the insulation parameters of the tested material. To obtain experimental data, the mixtures have been tested inside the two-compartment heat box experimental stand shown in Figure 3.

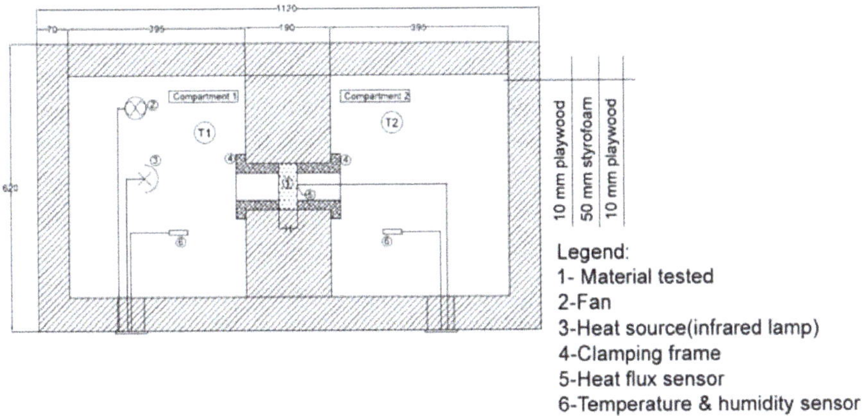

Figure 3. Two-compartment heat box experimental stand in cross section showing its two main compartments.

Both the inside and outside walls of the box are made of 1.0 cm thick plywood with a 5.0 cm thick Styrofoam insulation layer inside. The box features two compartments separated by a 19.0 cm thick insulated partition, which houses the sample to be tested. The sample is mounted with a clamping frame designed to reduce heat transfer through leaks in the edges caused by the nonuniform structure of the material. A 250W infrared lamp and a fan in Compartment 1 represent the heat source and provide regular air circulation. The study required a sustained 15–20 °C temperature differential between the two compartments be maintained for 2 h. The special insulation of Compartment 2 made it possible to reach the desired measurement stability. Testing a single sample continued for 2 h allowed the temperature and heat flux density to be stabilized. The materials used for testing of the heat conductivity had been weighed and measured to determine their density. Fire resistance of material means its durability, when exposed to high temperatures or flame, with some visual or structural changes if acceptable. The fire resistance test was designed to investigate hemp–lime composite behavior when exposed directly to an open flame and to analyze any ensuing structural changes by calculating potential mass loss. Testing was also performed on three other composite samples measuring 15 × 15 × 15 cm, with a Kemper gas burner of the manufacturer's maximum rated flame temperature of 1800 °C. The samples, set at a 10 cm distance away from the flame source, were tested at the room temperature of 22.6 °C for 10 min. The entire procedure was recorded with a TESTO 885-1 infrared camera in order to observe the temperature distribution within the material. Figure 4 shows the measuring station.

Figure 4. Hemp lime composite flame test, with the tested sample placed at a distance of 10 cm away from the flame source.

The primary energy needed by a home erected based on a single design, but in two different technologies, is compared:

- Traditional brick structure: Porotherm, Styrofoam, and wool;
- Natural materials: mainly hemp–lime composite, wattle and daub, and timber.

To this end, energy profiles of the analyzed materials had been prepared using ArCADia TermoCad software (PRO 7 version, Intersoft, Łódź, Poland). ArCADia-TERMOCAD is computer software available in several versions, it is one of the most popular programs on the Polish market designed for preparing energy performance certificates required for construction, modernization, and lease and sale transactions of buildings or premises, as well as for calculating the demand for heat and cooling of rooms and facilities. In ArCADia-TERMOCAD PRO version it is possible to perform energy audits, overhauls, and energy efficiency audits, e.g., for the purpose of obtaining modernization bonus. It can also be used for BREEAM certified calculations. Thanks to a rich database, the user of the program develop the necessary documents in accordance with the legal requirements applicable in Poland. ArCADia-TERMOCAD program in all versions has a built-in, fully functional graphics editor allowing to model the body of the building. The TERMOCADIA editor enables import of drawings in DWG format and import and export of ArCADia BIM system projects. Its main purpose is to perform building designs according to Building Information Modeling (BIM) technology assumptions. In addition to traditional architectural documentation, the program also performs a digital building model [39]. The homes featured an identical mechanical ventilation system capable of recovering roughly 60% of the heat, the same window frames and doors whose heat transfer coefficients complied with the 2020 design standards.

The calculations were performed for a single-family two-story home with a loft and a total floor area of ~220.0 m^2. Building energy performance certificates were issued based on the brick home design, while for our simulation we used the hemp–lime composite home that had been modified by changing its partition structure, i.e., using hemp–lime composite and other natural materials instead. The partitions were fabricated taking into account designers' and home contractors' new technology recommendations and guidelines. For our comparative analysis, we selected a new building made of commonly used construction materials as shown in Figure 5. The external walls of the home were made of 24.0 cm wide Porotherm ceramic blocks insulated with 20.0 cm thick Styrofoam (the picture shows the home without the facade). The home features Teriva ceilings and a timber roof truss with complete formwork that has been insulated with 30.0 cm thick Rockwool and covered with ceramic tiles. The "warm installation" method was used to install the balcony doors and windows. The building is heated with a gas-heated condensing boiler.

Figure 5. The analyzed building constructed with traditional materials.

3. Results

Testing was performed while the box was tightly closed. During the tests, the obtained value of the heat transfer coefficient was comparable to that found in the literature and reported by manufacturers. The calculated value of the coefficient was also used later on in the study to determine the power demand of a single-family home constructed with hemp–lime composite. Table 4 presents the results of six samples of the hemp–lime composite that showed the most favorable (lowest) heat transfer coefficient.

Table 4. Heat transfer coefficient "λ" calculations.

Sample No	Left Chamber Temperature (K)	Sample Left Wall Temperature (K)	Right Chamber Temperature (K)	Sample Right Wall Temperature (K)	Heat Flux Density (W/m^2)	Temperature Gradient (K)	Wall Thickness (m)	λ (W/m*K)
1	320.11	318.75	300.03	301.70	14.1	17.05	0.0442	0.037
2	339.00	333.86	301.07	303.02	23.4	30.84	0.0615	0.047
3	340.77	332.23	302.83	307.48	32.4	24.75	0.0343	0.045
4	333.50	333.29	300.18	304.98	41.0	28.31	0.0378	0.055
5	339.05	328.17	302.67	307.28	29.1	20.89	0.0386	0.054
6	331.25	326.53	304.24	307.31	21.9	19.22	0.0346	0.039

Our calculations indicate a traditionally constructed building requires 85.35 kWh/m^2 × year of nonrenewable primary energy. The energy characteristics provide a lot of vital information about the material and its environmental impact. Table 5 shows selected major highlights of the building energy performance characteristics.

Table 5. Energy consumption and the environmental impact of a traditional building.

Building Energy Characteristics Evaluation	
Energy characteristics indicators	Analyzed building
Annual usable energy demand	$EU = 52.37 \frac{kWh}{(m^2 \times year)}$
Annual final energy demand	$EK = 71.90 \frac{kWh}{(m^2 \times year)}$
Annual demand for nonrenewable primary energy	$EP = 85.36 \frac{kWh}{(m^2 \times year)}$
CO_2 emission unit	$ECO2 = 0.01497 \frac{1\ CO2}{(m^2 + year)}$
The percentage share of renewable energy Resources in annual final energy demand	URER = 0%

A hemp–lime composite building requires nonrenewable primary energy amounting to 44.81 (kWh/m^2 × year).

The thermogram shown in Figure 6 illustrates the temperature distribution within the composite; the picture proves the material has good insulation properties since it does not allow for heat to be transferred to the other side. Manufacturers, suppliers, and various websites promoting hemp–lime often claim fire resistance properties of hemp–lime, but, however, do not provide any actual data [40]. The literature review did not list any research papers pertaining to the fire properties of hemp–lime [41,42]. A clear standard for fire resistance tests is still missing for this type of material samples. The American Society for Testing and Materials (ASTM) did not have a restrictive standard for this type of material. But they are working on a new standard [43]. Nevertheless, fire-resistant tests, under very restrictive conditions, have been performed for the investigated composite sample. The temperature of the flame was 1800 °C during 10 min of testing time. After that time the surface of the sample shown in Figure 6 warmed up to a maximum of 165 °C.

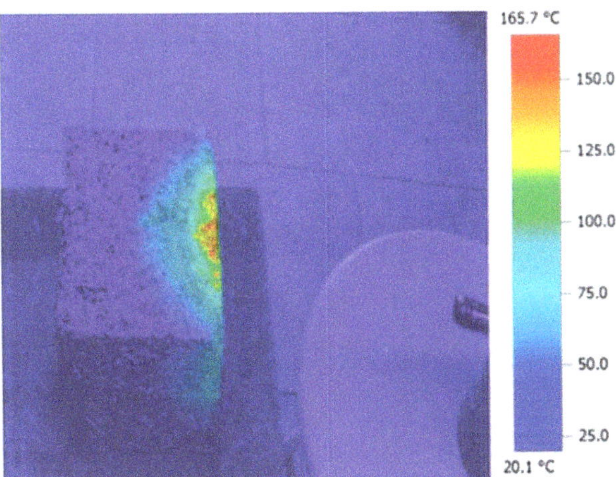

Figure 6. Sample thermal distribution following the flame test.

Following a 10 min-long flame test, the composite failed to ignite and showed no tendency to spread the fire. The only effect observed was limited to a certain glow and carbonization of the composite structure, which became considerably weakened around the area exposed to the flame and prone to crumbling. The exposure to flame and oxidization resulted in the composite sample losing some mass. The composite samples had all been carefully weighed using a lab-scale both prior to and following the flame test. Table 6 shows the weight loss results.

Table 6. Mass loss calculations following the flame test.

Sample No	Sample Mass before the Test (g)	Sample Mass after the Test (g)	Mass Loss after the Test (g)	Mass Loss Percentage (%)
1	1582.0	1568.4	13.6	0.9
2	1541.4	1533.8	7.6	0.5
3	1598.8	1589.4	9.4	0.6

The composite based on a hydrated and hydraulic lime binder showed the smallest mass loss of 0.5%, whereas Mixture No 1 lost the most mass compared with other tested mixtures. In summary, results of the hemp–lime composite flame test lead to the conclusion that the material is non-flammable and that using hydrated lime binder may improve the material fire resistance. Following exposure to direct flame, the material structure within the area affected by the high temperature had changed causing significant material weakening due to its increased looseness.

As various materials have been used, losses caused by partition permeability will vary considerably. Figure 7 shows total heat losses due to material permeability.

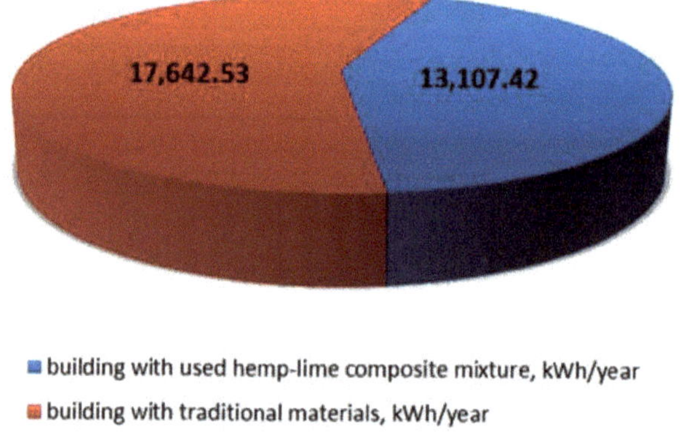

Figure 7. Comparison of total heat losses due to material permeability.

The building made of natural materials clearly shows much better heat insulation performance compared to a traditional building, owing to lower heat losses attributable to permeability which may reach as much as 4500 kWh per year.

4. Discussion

The proposed single-family building structure used hemp–lime composite and other natural materials only. As mentioned before, the same architectural design as the one developed for a building constructed in traditional technology was used. The building energy performance characteristics were then developed for the proposed design using the ArCADia TermoCad software. Commonly available materials and technologies were used to build the hemp–lime composite building. The average value of the coefficient for the composite with the lowest heat transfer coefficient was assumed in Table 7. Table 7 illustrates the structural partition layout required as an atypical construction material was used. For our simulations, we used the results of certain prior hemp–lime composite tests, mainly the heat transfer coefficient of 0.046 W/(m*K) and the 370 kg/m^3 material density.

Table 7. The main partition structure of a hemp–lime composite building.

No	External Wall	d (m)	λ (W/m·K)	R (m^2K/W)
		External partition		
1	Clay	0.015	0.850	0.018
2	Concentrated hempcrete	0.400	0.046	8.696
3	Wattle mat	0.010	0.070	0.143
4	Clay	0.015	0.850	0.018
-	Internal partition	-	-	UC = 0.11 $\frac{W}{m^2K}$
No	**Inside Ceiling**	**d (m)**	**λ (W/m·K)**	**R (m^2K/W)**
		Inside partition		
1	Oak fibers lengthwise	0.030	0.400	0.075
2	Pine and spruce fibers crosswise	0.025	0.160	0.156
3	Concentrated hempcrete	0.200	0.046	4.348
4	Pine and spruce fibers crosswise	0.025	0.160	0.156
-	Outside partition	-	-	UC = 0.20 $\frac{W}{m^2K}$
No	**Ground Floor**	**d (m)**	**λ (W/m·K)**	**R (m^2K/W)**
		Outside partition		
1	Granulated blast furnace slag, Keramzyt 700	0.300	0.200	1.500
2	Concentrated hempcrete	0.150	0.046	3.261
3	Sand-lime plaster	0.080	0.800	0.100
4	Oak fibers lengthwise	0.025	0.400	0.063
	Inside partition			UC = 0.20 $\frac{W}{m^2K}$
No	**Roof**	**d (m)**	**λ (W/m·K)**	**R (m^2K/W)**
		Outside partition		
1	Wattle slabs	0.350	0.070	5.000
2	Pine and spruce fibers crosswise	0.025	0.160	0.156
3	Hempcrete	0.150	0.044	3.409
4	Pine and spruce fibers crosswise	0.025	0.160	0.156
5	Straw slabs	0.010	0.080	0.125
6	Clay	0.030	0.850	0.035
-	Inside partition	-	-	UC = 0.11 $\frac{W}{m^2K}$

The data characterizing the tested material (hemp–lime) were entered into the library of ArCADia-TERMOCAD program, which enabled to calculate the coefficients of penetration of individual building partitions (walls, ceiling, floor, and roof) and, as a result, to determine the energy demand of the whole analyzed building. Next, the obtained results were compared with a building made of traditional building materials used in Poland. The obtained results are presented in Figure 7. The analysis allows to draw a conclusion do that the examined composite can be an alternative to traditional materials, and the calculated energy demand for objects with the same functional arrangement and dimensions is lower for the examined material, which confirms the advisability of using natural materials in single-family buildings.

5. Conclusions and Observations

The study was designed to analyze the potential of using natural construction materials for single-family buildings. To perform such an analysis, researchers developed their own natural material to take advantage of the hemp life cycle potential for sustainable construction industry. The most significant benefits of using hemp–lime composite include its potential for use as a non-combustible, renewable and natural raw material with a zero-carbon footprint, and good thermal insulation properties. The lime–hemp composite is very light thanks to its high porosity and low bulk density, which ranges from 300 to 400 kg/m^3. The parameters give hemp–lime composite very good thermal insulation properties with the thermal conductivity coefficient ranging from 0.038–0.055 (W/m*K), depending on how much the shiv has been compacted and what mixing method was applied. The heat transfer coefficient obtained during the calculations had the value of U_C = 0.11 (W/m^2*K) for external wall and 0.20 (W/m^2*K) for inside wall with the addition of the hemp–lime composite. The above value makes it possible to erect walls without any additional insulation needed: the hemp–lime composite is a technology that eliminates thermal bridges in the building. The material was also tested for its flame resistance; the sample did not ignite following a 10 min-long exposure to an open 1800 °C flame, because the material contains lime which not only improves its flame resistance but it also protects it from biological degradation corrosion or fungal deterioration [44]. The test results clearly show the use of hydrated lime binder may enhance the material fire resistance characteristics. Utilizing the established properties and parameters of hemp–lime composite, a simulation study was carried out for a home built with hemp–lime composite. A comparison of the power demand characteristics points to a conclusion that a building made of hemp–lime composite will use less energy of each kind, i.e., primary energy Ep, usable energy Eu, and final energy Ef. The only plus of a brick home is that construction materials are easily available at properly qualified contractors. Brick partitions hardly meet the required heat diffusion parameters and the production and application of such materials have been found harmful and detrimental both to the environment andhuman health. Several conclusions may be drawn based on our tests:

- Further development of conventional building materials is not critical for the construction industry since nature offers the best choices it is up to us to use them properly
- Knowledge of hemp–lime as a building material is still at the beginning of the process. In spite of the new research undertaken in this area, there are still no unified standards to ensure the appropriate parameters of a given composite
- A great variability of parameters, such as the conductivity, fire, weather, and biological resistance, of the hemp–lime composite is related to so many factors such as morphology of the fillers, the orientation of the filler particles, porosity, the method and the ratio of compaction, the distribution of the filers and many others
- The authors acknowledge the importance of these factors in terms of hemp–lime structure-related issues. This is a very wide range of interdisciplinary research the authors are in the process of preparing samples of the composites for further investigations.

Author Contributions: Conceptualization, S.P.; methodology, S.P.; software, D.M.; validation, S.A. and M.B.; formal analysis, S.P.; investigation, D.M.; resources, S.P.; data curation, D.M.; writing—original draft preparation, S.P.; writing—review and editing, S.P., S.A and M.B.; visualization, D.M.; supervision, S.P.; project administration, S.P.; funding acquisition, S.A. All authors have read and agreed to the published version of the manuscript.

Funding: This research received no external funding.

Conflicts of Interest: The authors declare no conflict of interest.

References

1. Bedlivá, H.; Isaacs, N. Hempcrete—An environmentally friendly material. *Adv. Mater. Res.* **2014**, *1041*, 83–86. [CrossRef]
2. Radogna, D.; Mastrolonardo, L.; Forlani, M.C. Hemp for a healthy and sustainable building in abruzzo. In *Advances in Intelligent Systems and Computing*; Springer International Publishing: Milan, Italy, 2018.
3. Jami, T.; Rawtani, D.; Agrawal, Y.K. Hemp concrete: Carbon-negative construction. *Emerg. Mater. Res.* **2016**, *5*, 240–247. [CrossRef]

4. Maalouf, C.; Ingrao, C.; Scrucca, F.; Moussa, T.; Bourdot, A.; Tricase, C.; Presciutti, A.; Asdrubali, F. An energy and carbon footprint assessment upon the usage of hemp-lime concrete and recycled-PET façades for office facilities in France and Italy. *J. Clean. Prod.* **2018**, *170*, 1640–1653. [CrossRef]
5. Mikulica, K.; Hela, R. Hempcrete—Cement Composite with Natural Fibres. *Adv. Mater. Res.* **2015**, *1124*, 130–134. [CrossRef]
6. Heidari, M.D.; Lawrence, M.; Blanchet, P.; Amor, B. Regionalised Life Cycle Assessment of Bio-Based Materials in Construction; the Case of Hemp Shiv Treated with Sol-Gel Coatings. *Materials* **2019**, *12*, 2987. [CrossRef]
7. Kaiser, C.; Cassady, C.; Ernst, M. Industrial Hemp Production. Center for Crop Diversification, University of Kentucky, September 2015. Available online: https://www.uky.edu/ccd/sites/www.uky.edu.ccd/files/hempproduction.pdf (accessed on 23 February 2020).
8. Young, E.M. Revival of Industrial Hemp: A Systematic Analysis of the Current Global Industry to Determine Limitations and Identify Future Potentials within the Concept Of Sustainability. Master's Thesis, Lund University, Lund, Sweden, 2005.
9. Allegret, S. The history of hemp. In *Hemp: Industrial Production and Uses*; CAB International: Surrey, UK, 2013; pp. 4–26.
10. Gibson, K. Hemp: A Substance of Hope. *J. Ind. Hemp.* **2006**, *10*, 75–83. [CrossRef]
11. Gołębiewski, M. Kompozyty konopno-wapienne (hempcrete). *Mater. Bud.* **2016**, *1*, 93–96. [CrossRef]
12. Elfordy, S.; Lucas, F.; Tancret, F.; Scudeller, Y.; Goudet, L. Mechanical and thermal properties of lime and hemp concrete ("hempcrete") manufactured by a projection process. *Constr. Build. Mater.* **2008**, *22*, 2116–2123. [CrossRef]
13. Woolley, T. Building physics, natural materials and policy issues. In *Low Impact Building*; Wiley: Hoboken, NJ, USA, 2013; pp. 148–186.
14. Amziane, S.; Arnaud, L.; Challamel, N. *Bio-Aggregate-Based Building Materials*; Wiley: Hoboken, NJ, USA, 2013.
15. Korjenic, A.; Petránek, V.; Zach, J.; Hroudová, J. Development and performance evaluation of natural thermal-insulation materials composed of renewable resources. *Energy Build.* **2011**, *43*, 2518–2523. [CrossRef]
16. Amziane, S.; Sonebi, M. Overview on biobased building material made with plant aggregate. *RILEM Tech. Lett.* **2016**, *1*, 31–38. [CrossRef]
17. Prabesh, K. Hempcrete Noise Barrier Wall for Highway Noise Insulation: Research & Construction. Bachelor's Thesis, Hame University of Applied Sciences, Hämeenlinna, Finland, December 2016.
18. Piot, A.; Béjat, T.; Jay, A.; Bessette, L.; Wurtz, E.; Barnes-Davin, L. Study of a hempcrete wall exposed to outdoor climate: Effects of the coating. *Constr. Build. Mater.* **2017**, *139*, 540–550. [CrossRef]
19. Arnaud, L.; Gourlay, E. Experimental study of parameters influencing mechanical properties of hemp concretes. *Constr. Build. Mater.* **2012**, *28*, 50–56. [CrossRef]
20. Arrigoni, A.; Pelosato, R.; Dotelli, G. Hempcrete from cradle to grave: The role of carbonatation in the material sustainability. In Proceedings of the International Conference on Sustainable Built Environment, Hamburg, Germany, 8–11 March 2016; Available online: https://re.public.polimi.it/retrieve/handle/11311/989311/253111/Arrigoni%20et%20al.%20with%20cover-%20SBE16Hamburg_ConferenceProceedings.pdf (accessed on 23 February 2020).
21. Bolcu, D.; Stănescu, M.M. The Influence of Non-Uniformities on the Mechanical Behavior of Hemp-Reinforced Composite Materials with a Dammar Matrix. *Materials* **2019**, *12*, 1232. [CrossRef]
22. Brzyski, P.; Barnat-Hunek, D.; Suchorab, Z.; Łagód, G. Composite Materials Based on Hemp and Flax for Low-Energy Buildings. *Materials* **2017**, *10*, 510. [CrossRef]
23. Viel, M.; Collet, F.; Pretot, S.; Lanos, C. Hemp-Straw Composites: Gluing Study and Multi-Physical Characterizations. *Materials* **2019**, *12*, 1199. [CrossRef]
24. Ramezanianpour, A.A.; Ghahari, S.A.; Khazaei, A. Feasibility study on production and sustainability of poly propylene fiber reinforced concrete ties based on a value engineering survey. In Proceedings of the Sustainable Construction Materials and Technologies, Kyoto, Japan, 18–21 August 2013.
25. Ghahari, S.A.; Ramezanianpour, A.M.; Esmaeili, M. An Accelerated Test Method of Simultaneous Carbonation and Chloride Ion Ingress: Durability of Silica Fume Concrete in Severe Environments. *Adv. Mater. Sci. Eng.* **2016**, *2016*, 1–12. [CrossRef]

26. Nguyen, T.T.; Picandet, V.; Amziane, S.; Baley, C. Influence of compactness and hemp hurd characteristics on the mechanical properties of lime and hemp concrete. *Eur. J. Environ. Civ. Eng.* **2009**, *13*, 1039–1050. [CrossRef]
27. Pietruszka, B.; Gołębiewski, M.; Lisowski, P. Characterization of Hemp-Lime Bio-Composite. In Proceedings of the IOP Conference Series: Earth and Environmental Science, Prague, Czech Republic, 2–4 July 2019.
28. Bevan, R.; Woolley, T. Constructing a Low Energy House From Hempcrete and Other Natural Materials. In Proceedings of the 11th International Conference on Non-conventional Material Technology (NOCMAT2009), Bath, UK, 6–9 September 2009.
29. Bevan, R.; Woolley, T. *Hemp Lime Construction: A Guide to Building. with Hemp lime Composes*; IHS BRE Press: London, UK, 2008.
30. Brocklebank, I. The lime spectrum. *Context* **2006**, *97*, 21–23. Available online: https://www.kalkforum.org/uploads/pdf/artikler/The_Lime_Spectrum_pdf.pdf (accessed on 23 February 2020).
31. Colinart, T.; Glouannec, P.; Chauvelon, P. Influence of the setting process and the formulation on the drying of hemp concrete. *Constr. Build. Mater.* **2012**, *30*, 372–380. [CrossRef]
32. Cazacu, C.; Muntean, R.; Gălățanu, T.; Taus, D. Hemp Lime Technology. *Bull. Transilv. Univ. Brașov* **2016**, *9*, 19.
33. Kremensas, A.; KAIRYTĖ, A.; Vaitkus, S.; Vėjelis, S.; Balčiūnas, G. Mechanical Performance of Biodegradable Thermoplastic Polymer-Based Biocomposite Boards from Hemp Shivs and Corn Starch for the Building Industry. *Materials* **2019**, *12*, 845. [CrossRef] [PubMed]
34. Brzyski, P.; Grudzińska, M.; Majerek, D. Analysis of the Occurrence of Thermal Bridges in Several Variants of Connections of the Wall and the Ground Floor in Construction Technology with the Use of a Hemp-lime Composite. *Materials* **2019**, *12*, 2392. [CrossRef] [PubMed]
35. Li, Z.; Wang, X.; Wang, L. Properties of hemp fibre reinforced concrete composites. *Compos. Part A Appl. Sci. Manuf.* **2006**, *37*, 497–505. [CrossRef]
36. Kinnane, O.; Reilly, A.; Grimes, J.; Pavia, S.; Walker, R. Acoustic absorption of hemp-lime construction. *Constr. Build. Mater.* **2016**, *122*, 674–682. [CrossRef]
37. Glé, P.; Gourdon, E.; Arnaud, L. Acoustical properties of materials made of vegetable particles with several scales of porosity. *Appl. Acoust.* **2011**, *72*, 249–259. [CrossRef]
38. Bourebrab, M.; Durand, G.G.; Taylor, A. Development of Highly Repellent Silica Particles for Protection of Hemp Shiv Used as Insulation Materials. *Materials* **2017**, *11*, 4. [CrossRef]
39. INTERsoft ArCADia-TERMOCAD PRO 7. Available online: https://www.intersoft.pl/cad/index.php?kup-program-cad=audyt-energetyczny-arcadia-termocad-pro-efektywnosc-energetyczna (accessed on 14 February 2020).
40. Gregor, L. Performance of Hempcrete Walls Subjected to a Standard Time-temperature Fire Curve. Master's Thesis, Victoria University Melbourne, Melbourne, Australia, August 2014.
41. Daly, P. Hemp lime bio-composite in construction: A study into the performance and application of hemp lime bio-composite as a construction material in Ireland. In Proceedings of the PLEA 2011—Architecture and Sustainable Development, 27th International Conference on Passive and Low Energy Architecture, Louvain-La-Neuve, Belgium, 13–15 July 2011; UCL Presses: Louvain, Belgium, 2011.
42. Amziane, S.; Arnaud, L.; Challamel, N. *Bio-Aggregate-Based Building Materials: Applications to Hemp Concretes*; Wiley-ISTE: Hoboken, NJ, USA, 2013; ISBN 9781848214040.
43. ASTM International New Test Methods for Evaluating the Appropriateness/Applicability of Current R-Value and Fire Resistance Test Methods to Testing the Insulative Properties of Hempcrete Insulation Samples. Available online: https://www.astm.org/DATABASE.CART/WORKITEMS/WK70549.htm (accessed on 14 February 2020).
44. Crawford, B.; Pakpour, S.; Kazemian, N.; Klironomos, J.; Stoeffler, K.; Rho, D.; Denault, J.; Milani, A.S. Effect of Fungal Deterioration on Physical and Mechanical Properties of Hemp and Flax Natural Fiber Composites. *Materials* **2017**, *10*, 1252. [CrossRef]

Article

Elucidation of Conduction Mechanism in Graphene Nanoplatelets (GNPs)/Cement Composite Using Dielectric Spectroscopy

Guido Goracci [1,2,*] and Jorge S. Dolado [2,3]

1 BASKRETE-Euskampus Fundazioa, Ed. Rectorado Barrio Sarriena s/n, 48940 Leioa, Spain
2 Centro de Física de Materiales, (CSIC-UPV/EHU)-Material Physics Centre (MPC), Paseo Manuel de Lardizabal 5, 20018 San Sebastián, Spain; jorge_dolado002@ehu.eus
3 Donostia International Physics Center (DIPC), Paseo Manuel Lardizábal 4, 20018 Donostia-San Sebastián, Spain
* Correspondence: guido_goracci@ehu.eus

Received: 3 December 2019; Accepted: 31 December 2019; Published: 8 January 2020

Abstract: Understanding the mechanisms that govern the conductive properties of multifunctional cement-materials is fundamental for the development of the new applications proposed to enhance the energy efficiency, safety and structural properties of smart buildings and infrastructures. Many fillers have been suggested to increase the electrical conduction in concretes; however, the processes involved are still not entirely known. In the present work, we investigated the effect of graphene nanoplatelets (1 wt% on the electrical properties of cement composites (OPC/GNPs). We found a decrease of the bulk resistivity in the composite associated to the enhancement of the charge transport properties in the sample. Moreover, the study of the dielectric properties suggests that the main contribution to conduction is given by water diffusion through the porous network resulting in ion conductivity. Finally, the results support that the increase of direct current in OPC/GNPs is due to pore refinement induced by graphene nanoplatelets.

Keywords: electrical conductive concrete; multifunctional composite; conductive filler; graphene nanoplatelets; dielectric properties; electric properties

---

1. Introduction

In recent years, the interest on the smart city concept to promote environmental sustainability through the implementation of new technologies has shown a large growth. New construction technologies have been investigated to enhance the energy efficiency, safety, and structural performance of buildings and infrastructure. In this framework, the development of multifunctional cement-based materials has a key role. Indeed, a strong effort is needed to design innovative concretes that could serve as structural material with tailored functional behavior to meet specific requirements.

Multifunctional cement-based materials have been proposed for several applications [1]. Regarding the infrastructure, much interest has been devoted to snow melting and de-icing systems with conductive concrete composites as heating elements [2–4], cathodic protection of steel reinforcement concrete to prevent corrosion damages [5–7] and traffic sensors with conductive concrete [8]. Cement-based composites have been proposed for grounding systems [9,10] as well as electromagnetic wave shielding [11,12]. Regarding the development of smart buildings, structural health monitoring systems are fundamental for the modern structures and concrete composite sensors are a good candidate due to the intrinsic compatibility with the cement matrix [1]. Finally, multifunctional cement-based materials have been suggested for the development of structural supercapacitors [13–15].

All the applications mentioned above base their efficiency on the conductive properties of the structural material. However, it is well known that concrete is characterized by an insulating behavior.

To overcome such a drawback, different conductive fillers have been indicated as good aggregates to achieve the design of conductive concrete [1,16]. Metal conductive admixtures have been proposed with the addition of steel fibers and micro fibers [17–19] and steel shaving [2]. Among the carbon admixtures, graphite [20–22], carbon fibers [3,18,19,23–26] and graphene [27,28] have been investigated for electrical conductive concretes. Finally, carbon nanomaterials gathered a lot of attention as they have, beyond high electrical conductivity, unique physical properties [16,29]. Among carbon nanomaterials employed in concrete composites we mention carbon black [30,31], carbon nanotubes [32–35] and nanofibers [36–40], as well as graphene nanoplatelets [41–45].

Performance of electrical conductive concrete depends clearly on the nature and amount of the filler. However, it has been demonstrated that water as well plays a crucial role in the conduction mechanism. Indeed, electrical resistivity depends on aging due to changes of pore water amount [18,40] and an ionic conduction in wet concretes has been observed due to the free water molecules [46]. Nonetheless, the conductive mechanism in cement-based materials is still not completely understood. Moreover, the studies on the effect of fillers on electrical properties focused the attention mainly on the formation of a conduction path.

With the aim of elucidating the processes involved in conduction in cement-based materials and, in particular, of investigating the role of water molecules and the indirect impact of the addition of fillers, we studied graphene nanoplatelets/cement composite by means of dielectric spectroscopy technique at different temperatures. Such a technique, commonly used to study conductivity in ceramic materials [47–52] and dielectric and electrical properties of porous systems [53–60] and polymer composite [61–68], is a very suitable tool due to its unique properties. In fact, due to its large frequency range, it is possible to investigate the impedance and dielectric response of the material on different time scale and, therefore, to obtain information on the different processes involved in the ionic and electronic conduction phenomena. First, due to the strong response of dielectric spectroscopy to water amount, thermal gravimetric analysis results are shown. Therefore, the impedance response of the specimens is examined to understand how graphene nanoplatelets affect the electrical properties. Finally, the dielectric response of the system is discussed to reveal the conduction mechanism and the role of the filler.

2. Materials and Methods

In this study, two samples were prepared: OPC paste (as reference) and OPC/Graphene nanoplatelets (GNPs) composite. To focus the attention on ion conduction, the porosity of the system was increased by using a water-to-cement ratio of w/c = 0.6 and curing the samples during seven days. Moreover we added only 1 wt% of GNPs to keep the sample below the percolation threshold suggested by literature [41] to minimize the electrical conduction that may contributes when a complete conductive path is formed. The cement used was CEM II/A-LL 42.5 R and GPL from GrapheneTech (Zaragoza, Spain) was used as filler. This product presents a specific surface area around 200 $m^2\ g^{-1}$, lateral size between 500–1000 nm and a carbon content above 97%. For sample preparation, first powders were mixed using a mechanical blender at low speed (350 rpm) for 1 min to obtain a uniform dispersion of GNPs in the OPC powder. Afterwards, ultrapure water was added and the solution was mixed at 750 rpm. Both sets were cast in cylindrical silicone molds with d = 4 cm and sealed. After 24 h, specimens were demolded and cured in water for 7 days. Finally, the cylinders were crashed into fine powder and kept overnight in desiccator with silica gel before measured.

Thermal gravimetric analyses were carried out using a TGA-500 (TA Instruments, New Castle, DE, USA) to investigate the water amount in the samples and verify if any phase transformation occurs when GNPs are added. All the measurements were conducted under high-purity nitrogen flow over a temperature of 303–1173 °C with a ramp rate of 5 K/min.

A broadband dielectric spectrometer, Novocontrol Alpha-A, (Novocontrol, Montabaur, Germany) was used to measure the complex dielectric permittivity, defined as $\varepsilon^*(\omega) = C^*(\omega)/C_0 = 1/(i\omega Z^*(\omega)C_0) = \varepsilon'(\omega) - i\varepsilon''(\omega)$ where C_0 is the capacitance of the free space, C^* is the complex capacitance function, Z^* is the complex impedance and $\omega = 2\pi f$ the angular frequency with f the applied electric field frequency. Data were collected over a broad frequency range, from 10^{-2} to 10^6 Hz. Samples were prepared by placing the sample powder between two parallel gold-plated electrodes of a diameter of d = 30 mm and thickness of about 0.7 mm. First, the sample was kept at room temperature inside the spectrometer during 10 min to overcome humidity signal. Therefore, isothermal scans were performed on heating every 5 degrees over the temperature range of 290–310 K. Temperature was controlled by a nitrogen gas flow with stability better than ±0.1 K.

3. Results and Discussion

3.1. Thermal Gravimetric Analysis

Thermal Gravimetric Analysis (TGA) and Differential Thermal Gravimetric (DTG) (TA instruments, New Castle, DE, USA) measurements allow to identify the different water population in the sample. These data are deeply relevant for the interpretation of the dielectric response of the material. TGA and DTG curves of OPC and OPC/GNPs samples are shown in Figure 1. Clearly, both TGA and DTG measurements share similar temperature dependence behavior. The first peak at ~98 °C in DTG curve is associated with evaporable water in C-S-H gel and ettringite [69,70]. At ~140 °C we observe a further decrease in weight than can be related to gypsum or amorphous carbon illuminate hydrate decompositions [71]. Moreover, a stiff decrease is observed in TGA curve in the range between 390 °C–460 °C. Such event is associated to the dehydroxylation of $Ca(OH)_2$ [72]. Finally, the peak at around ~680 °C is related to the decarbonation, together with possible solid-solid phase transformations [69].

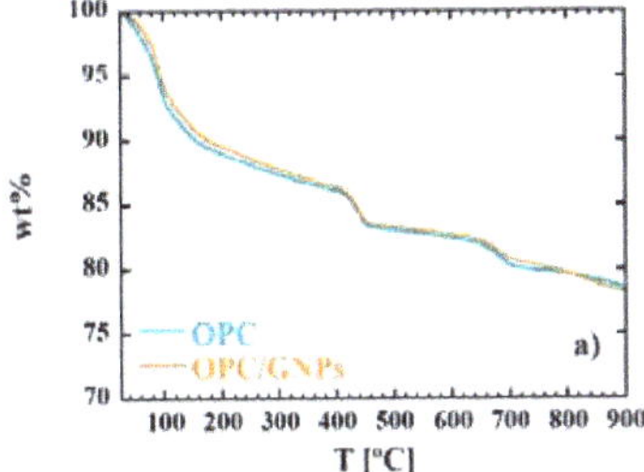

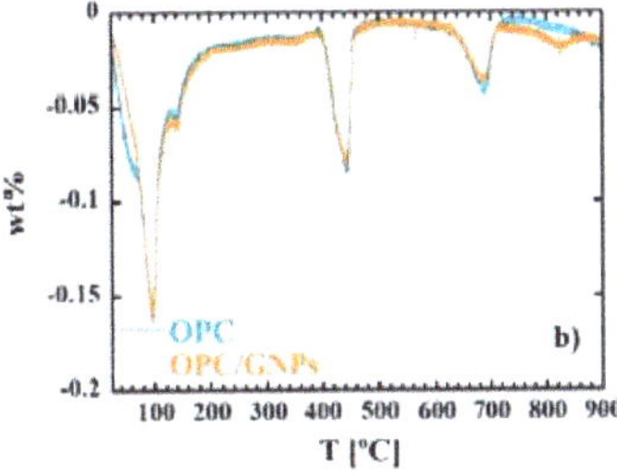

Figure 1. Thermal Gravimetric (TG) (**a**) and Differential Thermal Gravimetric (DTG) (**b**) curves for OPC and OPC/GNPs samples.

In OPC/GNPs sample, a further peak in DTG curve is observed at ~825 °C. In Table 1, the weight percent loss corresponding to water and portlandite is shown. Both samples are characterized by a similar amount of free water (~7 wt%) at a temperature lower than 105 °C) and physical bound water of hydrates (~7.5 wt%). Moreover, OPC and OPC/GNPs specimens contains the same amount of Portlandite (15 wt%).

Table 1. Free water, non-evaporable water cw% and portlandite content obtained by TGA analysis.

Sample	Free Water	Bound Water	$Ca(OH)_2$ (%)
OPC	7	7.2	15
OPC/GNPs	6.5	7.5	15

3.2. Impedance Response

The complex impedance function is defined as:

$$Z^* = Z' + iZ'' \quad (1)$$

where the real part and the imaginary part are defined as $Z' = R$ and $Z'' = 1/\omega C$, with R and C resistance and capacitance respectively. The plot in the complex Z″-Z′ plane, called Nyquist plot (Z″ vs. Z′), in the temperature range investigated (290–310 K) is shown in Figure 2. This representation allows to separate bulk properties from electrode polarization [73,74]. In the OPC sample, in the low frequency region (right side of the figure), a sloped line that can be related to cement-electrode interface contribution is observed [75–77]. At around 40 Hz, the line starts to diverge to a broad and asymmetric semi-circle corresponding to an overlap of polarization mechanisms in the bulk. From the intersection between interface and polarization contributions the bulk resistance value of 8.9 KΩ at 300 K was extracted. When graphene nanoplatelets are added to the cement paste some relevant changes in the Nyquist plot are observed. In fact, a stretching of the semi-arcs for frequencies higher than 40 Hz was clearly noticed and the bulk resistance at room temperature decreased to 3.9 KΩ indicating an enhance of charge transport in the composite sample. Such behavior is confirmed when the real part of the impedance of the reference is compared to that of the composite. In fact, the resistance of the OPC samples is, at low frequencies, almost three times higher than that measured for OPC/GNPs (see Figure 3a) and, even though such difference decreases with frequency, Z′ values of OPC are still two times larger than those observed in OPC/GNPs for f > 1 KHz. This effect has been already observed in alkali activated slag (AAS) composites with graphene and carbon nanotubes addition and it was related to the creation of conductive paths resulting in a reduction of electrical resistance [78–80]. However, due to the small amount of GNPs in our composite, a continuous conductive path is not formed as demonstrated by Bai et al. [41]. In Figure 3b, the capacitance of OPC and OPC/GNPs at 300 K is compared. In the low frequency region, the addition of graphene nanoplatelets increases the electrical capacitance of three times. However, as the frequency increased, the difference between the two samples almost disappeared. Finally, we observed an increment of C values as a function of temperature (Figure 4): Such effect appears stronger for f ≤ 1 KHz and, in particular, in the composite sample. Summarizing, an enhancement of electrical capacitance and charge transport properties is observed when 1 wt% of GNPs is added to the cement paste. As such an amount of filler is not sufficient to create a complete conductive path, a deeper analysis must be carried out to clarify how the fillers lead to the electrical improvements in the composite.

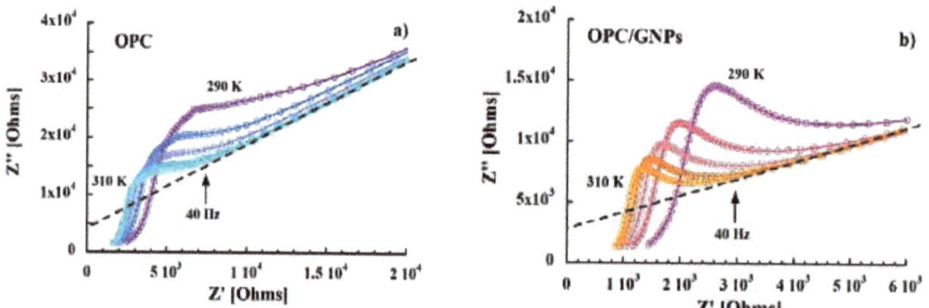

Figure 2. Nyquist plot of the impedance of OPC (**a**) and OPC/GNPs (**b**). Dashed line is the extrapolation to high frequencies of the cement-electrode interface contribution.

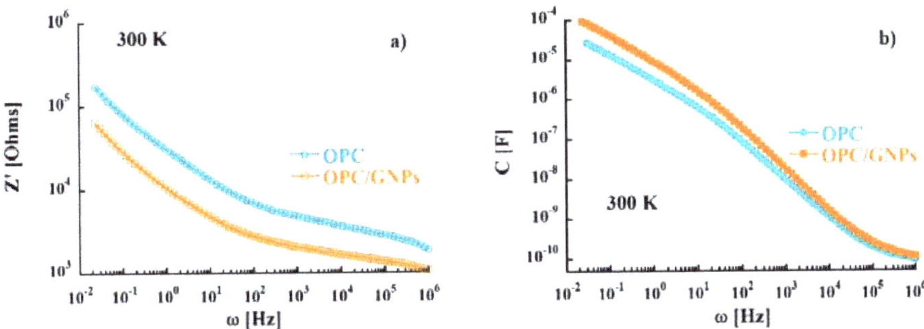

Figure 3. (**a**) Dependence of the real part of the complex impedance function as a function of frequency. (**b**) Frequency dependence of capacitance at 300 K.

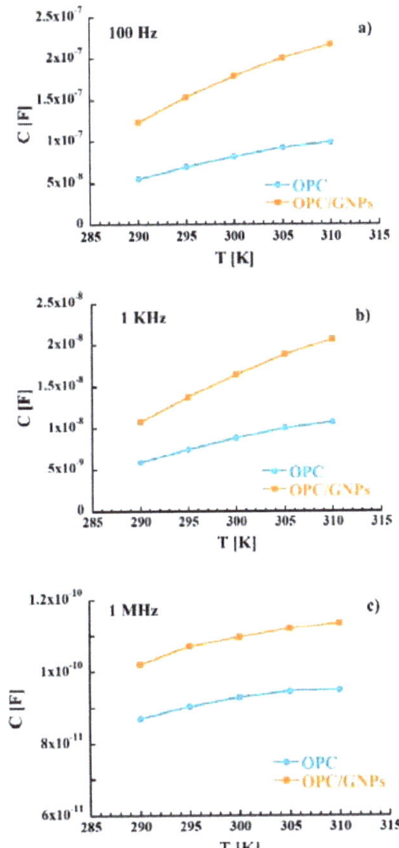

Figure 4. Temperature dependence of the capacitance as at 100 Hz (**a**), 1 KHz (**b**) and 1 MHz (**c**).

3.3. Dielectric Response

Dielectric Spectroscopy allows us to investigate molecules dynamics, charge transport and interface interactions through the study of dipole reorientations when an alternative electric field is applied. In fact, depending on the temperature and frequency range, different dipole relaxations are observed, reflecting the distinct nature of the processes and interactions with the environment.

Through the characterization of the energy activations and characteristic relaxation times the information on the origin of the processes can be extracted. Moreover, the complex conductivity function can be calculated from the dielectric loss according to the relation $\sigma^*(\omega) = \sigma'(\omega) + i\sigma''(\omega) = i\omega\varepsilon^*(\omega)$ and, hence, it is possible to investigate the mechanisms that mark the electrical properties of the system. Figure 5 shows the variation of the real part of the complex conductivity $\sigma'(\omega)$ in the temperature range 290–310 K. The presence of GNPs leads to an increase of $\sigma'(\omega)$ values in the whole frequency range. In both samples, the same frequency pattern is observed: (1) frequency dependent conductivity at high frequencies (f > 10^5 Hz) indicating an alternating current (ac) dominating contribution (2) an almost flat region related to the direct current (dc) (10^1 to 10^5 Hz) (3) a drastic drop at frequencies <100 Hz due to electrode polarization effects. Data were analyzed by using:

$$\sigma'(\omega) = \sigma_{DC} + A\omega^n \tag{2}$$

where σ_{DC} is the dc conductivity, A is the pre-exponential factor and n is the exponential factor with values between 0 and 1 [81]. The resulting parameters are listed in Table 2. As expected, σ_{DC} increases with temperature in both reference sample and composite. Moreover, the n parameter values of 0.17 tend low compared to those obtained for ionic conductors, $0.5 < n < 1$, [82] and they are independent from both temperature and graphene nanoplatelets addition. In contrast, the values of σ_{DC} are affected by GNPs: indeed, at room temperature they increase 4 times with respect to those measured in the reference sample at room temperature, and this effect grows with temperature.

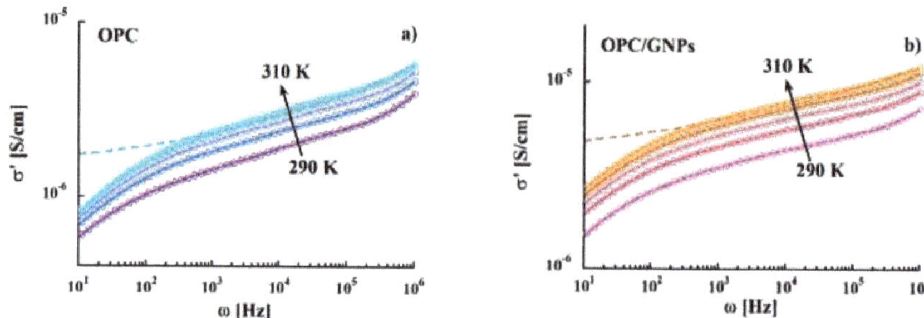

Figure 5. Electrical conductivity $\sigma'(\omega)$ as a function of frequency in the temperature range 290–310 K of OPC (**a**) and OPC/GNPs (**b**).

Table 2. Parameters obtained from the Arrhenius equation applied to data in Figure 5.

	OPC			OPC/GNPs		
T(K)	σ_{DC} (S/cm)	A	n	σ_{DC} (S/cm)	A	n
290	3.9×10^{-7}	3.11×10^{-7}	0.17	1.8×10^{-6}	5×10^{-7}	0.17
295	6.0×10^{-7}	3.6×10^{-7}	0.17	2.4×10^{-6}	6.1×10^{-7}	0.17
300	7.0×10^{-7}	4.1×10^{-7}	0.17	2.9×10^{-6}	6.9×10^{-7}	0.17
305	7.6×10^{-7}	4.6×10^{-7}	0.17	3.2×10^{-6}	7.7×10^{-7}	0.17
310	8.5×10^{-7}	4.7×10^{-7}	0.17	3.6×10^{-6}	8.3×10^{-7}	0.17

Additionally, the temperature dependence of conductivity was analyzed considering the dc conductivity as a thermally activated process with activation energy calculated according to the Arrhenius relation [83]:

$$\sigma_{DC} = \sigma_0 \exp(-\frac{E_A}{k_B T}) \tag{3}$$

where σ_0 is the pre-exponential factor associated with the charge carrier mobility and density of states, E_A is the activation energy, k_B is the Boltzmann constant and T is the temperature. The obtained activation energies, shown in Table 3, are low compared to those found for hopping conductivity (typically about 0.7–0.9 eV) [84]. Finally, we observed a decrease of the activation energy as we add graphene nanoplatelets. Figure 6 compares the variation of the real and imaginary part of the complex permittivity function of the samples with frequency at different temperatures. Regarding the real part ε' (Figure 6a,c) a strong dispersion in the low frequency region is observed, followed by an almost frequency-independent behavior above 10 Hz. The decrease of ε' can be attributed to electrode polarization and Maxwell-Wagner effect [85]. In the composite sample, we found higher values of ε' at room temperature for $f < 10$ Hz. Moreover, we observed a severe temperature dependence of the dielectric constant that reach values almost one order of magnitude higher than in the reference sample. Regarding the imaginary part of the complex permittivity (Figure 6b,d), spectra of both specimens are characterized by a strong dispersion, as observed in ε'.

Table 3. Activation Energy (E_A) and pre-Exponential Factor (log [σ_0]), which were obtained from the Arrhenius equation.

Sample	Log [σ_0] (s)	E_A (eV)
OPC	−1.5	0.28
OPC/GNPs	−1.2	0.26

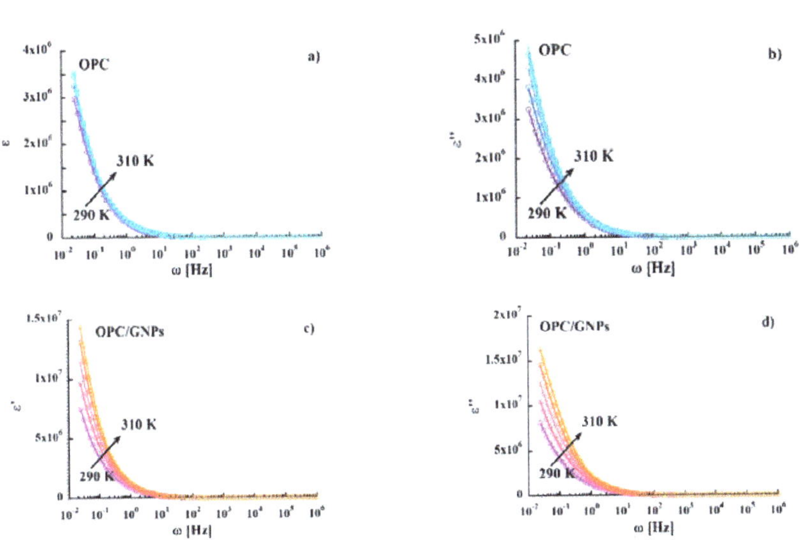

Figure 6. Frequency dependence of real and imaginary part of permittivity in OPC, (**a**,**b**), and in OPC/GNPs, (**c**,**d**), in the temperature range 290–310 K.

Hence, in order to investigate the relaxation processes occurring in the samples, we focused our attention on the loss tangent [85] defined as:

$$tg\delta(\omega) = \frac{\varepsilon(\omega)''}{\varepsilon(\omega)'} \tag{4}$$

In Figure 7a the spectra of the loss tangent of the reference, measured at different temperatures, are shown. In the whole temperature range the presence of two peaks is noticed. The most intense has the maximum centered at around 30 KHz and it does not show a strong dependence on temperature of the maximum position. On the other hand, the less intense peak at lower frequency is characterized by a clear dependence on the temperature of the peak maximum. Spectra of the loss tangent of the OPC/GNPs sample are shown in Figure 7b. The peak in the high frequency region is more intense than that found in OPC sample, even though the temperature dependence appears to be similar. On the other hand, the low frequency peak shows a different behavior when temperature is increased: The position of the maximum does not change with temperature, while the intensity increases.

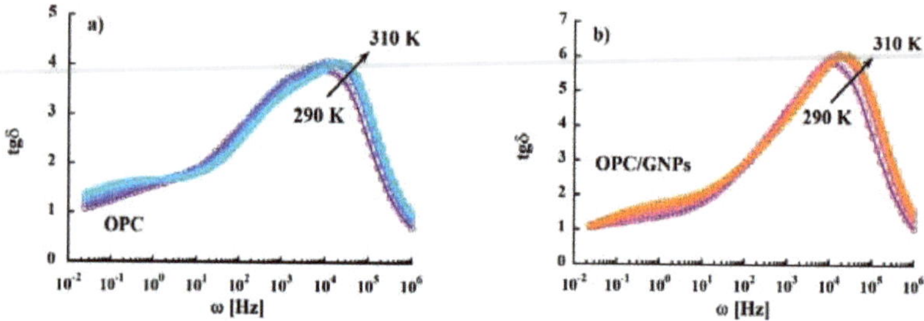

Figure 7. Variation of loss tangent in OPC (**a**) and OPC/GNPs (**b**) as a function of frequency in the temperature range 290–310 K.

Regarding the origin of those contributions, the peak at lower frequency appears where polarization effects were found both in the impedance response and in conductivity. Therefore, we believe that this contribution is given by electrode polarization or Maxwell-Wagner effect instead of some relaxation process. On the contrary, the intense peak of loss tangent appears in the frequency region where direct current is observed. A contribution with a similar weak temperature dependence was found at room temperature and at the same frequency range also in other porous materials [86–88]. In these works, the peak was related to the percolation of charge carriers through the porous network. To confirm this origin, we analyzed deeper the dielectric loss tangent spectra. In Figure 8 the relaxation times extracted from the position of the maximum of the peak are shown as a function of the inverse of temperature. In the composite specimen, a slight decrease of the relaxation times is observed. Furthermore, we observed a linear temperature dependence of log (τ_0) typical of a simply thermal activated process.

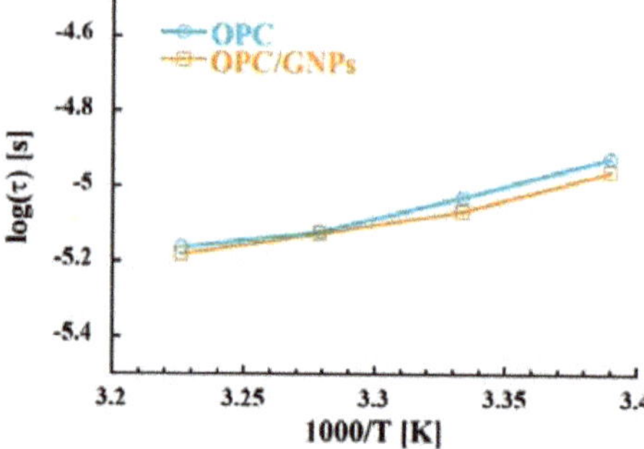

Figure 8. Temperature dependence of relaxation times extracted by the position of the peak maximum in the loss tangent.

Therefore, with the aim of obtaining information on the nature of the relaxation processes, the temperature dependence of τ values has been fitted with an Arrhenius law:

$$\tau = \tau_0 \exp\left(-\frac{E_A}{k_B T}\right) \quad (5)$$

where τ_0 is the pre-exponential factor the characteristic relaxation times of the process, E_A is the activation energy, k_B is the Boltzmann constant and T is the temperature. The resulting activation energy and pre-exponential factor parameters are shown in Table 4. First, we observed that the values of the pre-factor are larger than those typically observed for molecular vibrations ($\tau_0 \approx 10^{-14}$–10^{-12} s). Moreover, the activation energies are similar to those found for the self-diffusion of ions through water in argillaceous rock porous network [89]. It is worth noting that the values of the activation energy calculated by the dielectric loss tangent correspond to those obtained by the direct current analysis indicating a correlation between these quantities. Therefore, we can associate the intense peak observed in dielectric loss tangent to the percolation of water molecules contributing to ion conductivity.

Table 4. Activation Energy (E_A) and pre-Exponential Factor (log [τ_0 (s)]) values obtained from the Arrhenius equation applied to data in Figure 8.

Sample	Log [τ_0 (s)]	E_A (eV)
OPC	−10	0.30
OPC/GNPs	-9.5	0.26

Finally, such a scenario is supported by modulus formalism analysis with complex modulus defined as:

$$M^{*} = M' + iM'' = \frac{1}{\varepsilon^{*}} \tag{6}$$

In Figure 9 the peaks of Z″, M″ and tgδ (normalized to the maximum value) are shown. We observed a significant mismatch between the position of the peak of the maximum in the imaginary part of electric modulus and the imaginary part of impedance. Such condition indicates a short-range nature of the motion of the charge carriers instead of long-range hopping mechanism [90].

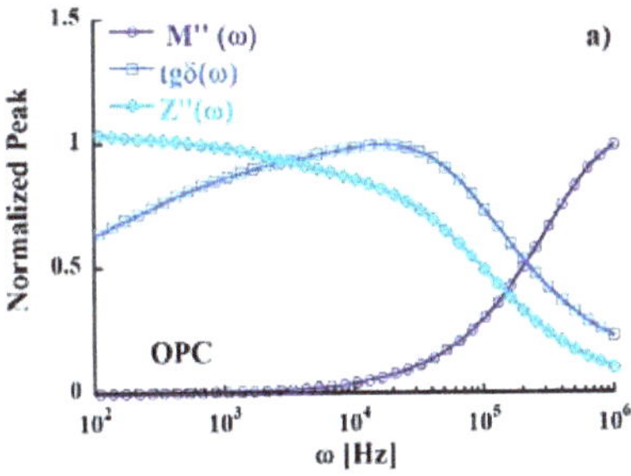

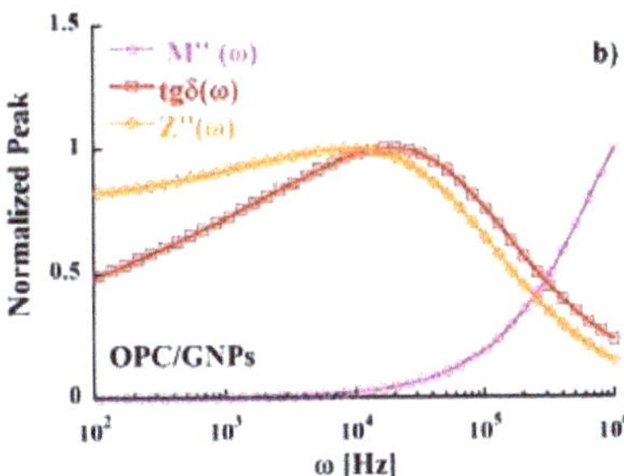

Figure 9. Comparison of tgδ, M″ and Z″ as a function of frequency at 300 K in OPC (**a**) and OPC/GNPs (**b**).

Trying to elucidate how the presence of graphene nanoplatelets affect conductivity in cement paste, we summarize the main results of this work: First we observed an enhancement of the electrical properties in the composite by means of impedance response study, while dielectric response investigation proved that the main mechanism of charge transport is water percolation. At first sight, an increase of conductivity might be related to an increase of charge carriers in the composite, that is more water molecules in the specimen with GNPs. However, TG and DTG measurements showed that the samples have similar free water and bound water amount. Therefore, it is not possible to associate a raise of conductivity to an increase of charge carriers caused by GNPs presence. Consequently, the addition of graphene nanoplatelets must lead to some topological changes in the cement paste. This interpretation is supported by previous experimental works. In fact, it has been shown how the addition of graphene nanoplatelets lead to a refinement of the pore structure [43,91]. A small amount of GNPs reduces the percentage of macropores, even though the total porosity is slightly affected suggesting a broader distribution of small gel pores in the composite. Moreover, during the hydration process, C-S-H particles bonded to graphene flakes act as a nucleation site and promote the growth of C-S-H gel with higher crystallinity degree [92]. Under this scenario, we can explain the increase of ionic conductivity in relation with the enhancement of ordered C-S-H gel that promote a smoother diffusion of water molecules through the porous network. This would also explain the decrease of the energy activation of the relaxation process associated to water percolation and the slight smaller values of its relaxation times in OPC/GNPs sample.

4. Conclusions

With the aim of understanding the processes involved in conduction in cement-based materials, OPC cement paste and OPC/GNPs composite were studied by dielectric spectroscopy over a broad frequency range (10^{-2}–10^{6} Hz) and at different temperatures. At low frequency, the enhancement of charge transport properties, reflected by the decrease of bulk resistivity and Z', is observed once graphene nanoplatelets are added to the cement paste. Moreover, the electrical capacitance shows higher values in the composite over all the frequency range investigated. Finally, the effect of GNPs on the dielectric response of the system was also investigated. Regarding conductivity, larger values of direct current was found in OPC/GNPs. Information on the origin σ_{DC} was extracted by analyzing its temperature dependence: E_A and n parameters calculated for both samples suggest that the main contribution to direct current is given by ion diffusion. Moreover, at around ~10^4 Hz, an intense peak is found in the loss tangent spectra. Previous dielectric studies on porous systems found a similar process related to water percolation through the pore network. The relaxation times associated to the maximum of this contribution were studied as a function of temperature: the calculated activation energy is close to that found for σ_{DC} in the same frequency window and, therefore, these processes are correlated. Hence, we associated the intense peak observed in dielectric loss tangent to the percolation of water molecules contributing to ion conductivity. With respect to the role of the filler in the conduction enhancement, an increase of conductivity in the composite cannot be associated with an increase of charge carriers because TG and DTG measurements revealed that GNPs do not affect the water populations in the pore network. As graphene leads to a refinement of the pore, we propose that the enhancement of conductivity is mainly given by the ordered C-S-H gel structure growth around GNPs promoting the water diffusion and, therefore, resulting in an increase of conduction. It is clear that our results are calling for new experiments devoted to exploring the subtle changes provoked by GNP into the cementitious pore-network to favor the water percolation. In that sense, the combination of experiments and techniques like SANS [93,94], H-NMR [95,96], porosity measurements [97] can provide valuable help in understanding the conductivity of cementitious composites.

Author Contributions: Conceptualization, G.G. and J.S.D.; Experiments and Data analysis, G.G.; Writing—Original draft preparation, G.G. and J.S.D.; Writing—Review and Editing, G.G. and J.S.D.; Supervision, J.S.D. All authors have read and agreed to the published version of the manuscript.

Funding: This work is partially supported by the Gobierno Vasco-UPV/EHU project IT1246-19 and the Spanish Ministry of Science, Innovation and Universities projects PCI2019-103657 and RTI2018-098554-B-I00. The project is co-funded by EUSKAMPUS FUNDAZIOA.

Conflicts of Interest: The authors declare no conflict of interest.

References

1. El-Dieb, A.S.; El-Ghareeb, M.A.; Abdel-Rahman, M.A.H.; Nasr, E.S.A. Multifunctional electrically conductive concrete using different fillers. *J. Build. Eng.* **2018**, *15*, 61–69. [CrossRef]
2. Yehia, S.A.; Tua, C.Y. Thin conductive concrete overlay for bridge deck deicing and anti-icing. *Transp. Res. Rec.* **2000**, *1698*, 45–53. [CrossRef]
3. Zhao, H.; Wu, Z.; Wang, S.; Zheng, J.; Che, G. Concrete pavement deicing with carbon fiber heating wires. *Cold Reg. Sci. Technol.* **2011**, *65*, 413–420. [CrossRef]

4. Wu, J.; Liu, J.; Yang, F. Three-phase composite conductive concrete for pavement deicing. *Constr. Build. Mater.* **2015**, *75*, 129–135. [CrossRef]
5. Bertolini, L.; Bolzoni, F.; Pastore, T.; Pedeferri, P. Effectiveness of a conductive cementitious mortar anode for cathodic protection of steel in concrete. *Cem. Concr. Res.* **2004**, *34*, 681–694. [CrossRef]
6. Jing, X.; Wu, Y. Electrochemical studies on the performance of conductive overlay material in cathodic protection of reinforced concrete. *Constr. Build. Mater.* **2011**, *25*, 2655–2662. [CrossRef]
7. Yao, W.; Xu, J. Current distribution in reinforced concrete cathodic protection system. *Tongji Daxue Xuebao/J. Tongji Univ.* **2009**, *37*, 1014–1018.
8. Han, B.; Kwon, E.; Yu, X. Self-sensing CNT/cement composite for traffic monitoring Strength improvement additives for cement bitumen emulsion mixture View project A self-sensing carbon nanotube/cement composite for traffic monitoring. *Nanotechnology* **2009**, *20*, 445501. [CrossRef]
9. Qin, Z.; Wang, Y.; Mao, X.; Xie, X. Development of graphite electrically conductive concrete and application in grounding engineering. *New Build. Mater* **2009**, *11*, 46–48.
10. Chung, D.D.L. Electrically conductive cement-based materials. *Adv. Cem. Res.* **2004**, *16*, 169–176. [CrossRef]
11. Guan, H.; Liu, S.; Duan, Y.; Cheng, J. *Cement Based Electromagnetic Shielding and Absorbing Building Materials*; Elsevier: Amsterdam, The Netherlands, 2006.
12. Wen, S.; Chung, D.D.L. Double percolation in the electrical conduction in carbon fiber reinforced cement-based materials. *Carbon N. Y.* **2007**, *45*, 263–267. [CrossRef]
13. Chung, D.D.L. Development, design and applications of structural capacitors. *Appl. Energy* **2018**, *231*, 89–101. [CrossRef]
14. Zhang, J.; Xu, J.; Zhang, D. A structural supercapacitor based on graphene and hardened cement paste. *J. Electrochem. Soc.* **2016**, *163*, E83–E87. [CrossRef]
15. Xu, J.; Zhang, D. Multifunctional structural supercapacitor based on graphene and geopolymer. *Electrochim. Acta* **2017**, *224*, 105–112. [CrossRef]
16. Wang, L.; Aslani, F. A review on material design, performance, and practical application of electrically conductive cementitious composites. *Constr. Build. Mater.* **2019**, *229*, 116892. [CrossRef]
17. Berrocal, C.G.; Hornbostel, K.; Geiker, M.R.; Löfgren, I.; Lundgren, K.; Bekas, D.G. Electrical resistivity measurements in steel fibre reinforced cementitious materials. *Cem. Concr. Compos.* **2018**, *89*, 216–229. [CrossRef]
18. Banthia, N.; Djeridane, S.; Pigeon, M. Electrical resistivity of carbon and steel micro-fiber reinforced cements. *Cem. Concr. Res.* **1992**, *22*, 804–814. [CrossRef]
19. Wen, S.; Chung, D.D.L. A comparative study of steel- and carbon-fibre cement as piezoresistive strain sensors. *Adv. Cem. Res.* **2003**, *15*, 119–128. [CrossRef]
20. Lunak, M.; Kusak, I.; Chobola, Z. Carbon Admixtures Influence on the Electrical Properties of Slag Mortars Focusing on Alternating Conductivity and Permittivity. *Procedia Eng.* **2016**, *151*, 236–240. [CrossRef]
21. He, Y.; Lu, L.; Jin, S.; Hu, S. Conductive aggregate prepared using graphite and clay and its use in conductive mortar. *Constr. Build. Mater.* **2014**, *53*, 131–137. [CrossRef]
22. Wang, H.; Yang, J.; Liao, H.; Chen, X. Electrical and mechanical properties of asphalt concrete containing conductive fibers and fillers. *Constr. Build. Mater.* **2016**, *122*, 184–190. [CrossRef]
23. Chen, B.; Wu, K.; Yao, W. Conductivity of carbon fiber reinforced cement-based composites. *Cem. Concr. Compos.* **2004**, *26*, 291–297. [CrossRef]
24. Fu, X.; Chung, D.D.L. Carbon fiber reinforced mortar as an electrical contact material for cathodic protection. *Cem. Concr. Res.* **1995**, *25*, 689–694. [CrossRef]
25. Hou, Z.; Li, Z.; Wang, J. Electrical conductivity of the carbon fiber conductive concrete. *J. Wuhan Univ. Technol. Mater. Sci. Ed.* **2007**, *22*, 346–349. [CrossRef]
26. Donnini, J.; Bellezze, T.; Corinaldesi, V. Mechanical, electrical and self-sensing properties of cementitious mortars containing short carbon fibers. *J. Build. Eng.* **2018**, *20*, 8–14. [CrossRef]
27. Peyvandi, A.; Soroushian, P.; Balachandra, A.M.; Sobolev, K. Enhancement of the durability characteristics of concrete nanocomposite pipes with modified graphite nanoplatelets. *Constr. Build. Mater.* **2013**, *47*, 111–117. [CrossRef]
28. Alkhateb, H.; Al-Ostaz, A.; Cheng, A.H.D.; Li, X. Materials genome for graphene-cement nanocomposites. *J. Nanomech. Micromech.* **2013**, *3*, 67–77. [CrossRef]
29. Raut, P.; Swanson, N.; Kulkarni, A.; Pugh, C.; Jana, S.C. Exploiting arene-perfluoroarene interactions for dispersion of carbon black in rubber compounds. *Polymer* **2018**, *148*, 247–258. [CrossRef]

30. Ding, Y.; Chen, Z.; Han, Z.; Zhang, Y.; Pacheco-Torgal, F. Nano-carbon black and carbon fiber as conductive materials for the diagnosing of the damage of concrete beam. *Constr. Build. Mater.* **2013**, *43*, 233–241. [CrossRef]
31. Monteiro, A.O.; Cachim, P.B.; Costa, P.M.F.J. Electrical Properties of Cement-based Composites Containing Carbon Black Particles. *Mater. Today Proc.* **2015**, *2*, 193–199. [CrossRef]
32. Yu, X.; Kwon, E. A carbon nanotube/cement composite with piezoresistive properties. *Smart Mater. Struct.* **2009**, *18*. [CrossRef]
33. Saafi, M.; Andrew, K.; Tang, P.L.; McGhon, D.; Taylor, S.; Rahman, M.; Yang, S.; Zhou, X. Multifunctional properties of carbon nanotube/fly ash geopolymeric nanocomposites. *Constr. Build. Mater.* **2013**, *49*, 46–55. [CrossRef]
34. Kusak, I.; Lunak, M.; Rovnanik, P. Electric Conductivity Changes in Geopolymer Samples with Added Carbon Nanotubes. *Procedia Eng.* **2016**, *151*, 157–161. [CrossRef]
35. Kim, H.K.; Nam, I.W.; Lee, H.K. Enhanced effect of carbon nanotube on mechanical and electrical properties of cement composites by incorporation of silica fume. *Compos. Struct.* **2014**, *107*, 60–69. [CrossRef]
36. Gao, D.; Sturm, M.; Mo, Y.L. Electrical resistance of carbon-nanofiber concrete. *Smart Mater. Struct.* **2009**, *18*. [CrossRef]
37. Konsta-Gdoutos, M.S.; Aza, C.A. Self sensing carbon nanotube (CNT) and nanofiber (CNF) cementitious composites for real time damage assessment in smart structures. *Cem. Concr. Compos.* **2014**, *53*, 162–169. [CrossRef]
38. Azhari, F.; Banthia, N. Cement-based sensors with carbon fibers and carbon nanotubes for piezoresistive sensing. *Cem. Concr. Compos.* **2012**, *34*, 866–873. [CrossRef]
39. Materazzi, A.L.; Ubertini, F.; D'Alessandro, A. Carbon nanotube cement-based transducers for dynamic sensing of strain. *Cem. Concr. Compos.* **2013**, *37*, 2–11. [CrossRef]
40. Yoo, D.Y.; You, I.; Lee, S.J. Electrical properties of cement-based composites with carbon nanotubes, graphene, and graphite nanofibers. *Sensors* **2017**, *17*, 1064. [CrossRef]
41. Bai, S.; Jiang, L.; Jiang, Y.; Jin, M.; Jiang, S.; Tao, D. Research on electrical conductivity of graphene/cement composites. *Adv. Cem. Res.* **2018**, 1–8. [CrossRef]
42. Sun, S.; Han, B.; Jiang, S.; Yu, X.; Wang, Y.; Li, H.; Ou, J. Nano graphite platelets-enabled piezoresistive cementitious composites for structural health monitoring. *Constr. Build. Mater.* **2017**, *136*, 314–328. [CrossRef]
43. Du, H.; Pang, S.D. Enhancement of barrier properties of cement mortar with graphene nanoplatelet. *Cem. Concr. Res.* **2015**, *76*, 10–19. [CrossRef]
44. Jing, G.; Ye, Z.; Lu, X.; Hou, P. Effect of graphene nanoplatelets on hydration behaviour of Portland cement by thermal analysis. *Adv. Cem. Res.* **2017**, *29*, 63–70. [CrossRef]
45. Haddad, A.S.; Chung, D.D.L. Decreasing the electric permittivity of cement by graphite particle incorporation. *Carbon N. Y.* **2017**, *122*, 702–709. [CrossRef]
46. Wen, S.; Chung, D.D.L. The role of electronic and ionic conduction in the electrical conductivity of carbon fiber reinforced cement. *Carbon N. Y.* **2006**, *44*, 2130–2138. [CrossRef]
47. Molak, A.; Paluch, M.; Pawlus, S.; Klimontko, J.; Ujma, Z.; Gruszka, I. Electric modulus approach to the analysis of electric relaxation in highly conducting (Na0.75Bi0.25)(Mn0.25Nb 0.75)O3 ceramics. *J. Phys. D Appl. Phys.* **2005**, *38*, 1450–1460. [CrossRef]
48. Raevski, I.P.; Prosandeev, S.A.; Bogatin, A.S.; Malitskaya, M.A.; Jastrabik, L. High dielectric permittivity in AFe1/2B1/2O3 nonferroelectric perovskite ceramics (A=Ba, Sr, Ca; B=Nb, Ta, Sb). *J. Appl. Phys.* **2003**, *93*, 4130–4136. [CrossRef]
49. Karthik, C.; Varma, K.B.R. Dielectric and AC conductivity behavior of BaBi2Nb2O9 ceramics. *J. Phys. Chem. Solids* **2006**, *67*, 2437–2441. [CrossRef]
50. Acharya, T.; Choudhary, R.N.P. Structural, dielectric and impedance characteristics of $CoTiO_3$. *Mater. Chem. Phys.* **2016**, *177*, 131–139. [CrossRef]
51. Lin, Y.Q.; Chen, X.M.; Liu, X.Q. Relaxor-like dielectric behavior in La2NiMnO6 double perovskite ceramics. *Solid State Commun.* **2009**, *149*, 784–787. [CrossRef]
52. Wang, W.G.; Li, X.Y. Impedance and dielectric relaxation spectroscopy studies on the calcium modified Na0.5Bi0.44Ca0.06TiO2.97 ceramics. *AIP Adv.* **2017**, *7*. [CrossRef]
53. Cerveny, S.; Arrese-Igor, S.; Dolado, J.S.; Gaitero, J.J.; Alegra, A.; Colmenero, J. Effect of hydration on the dielectric properties of C-S-H gel. *J. Chem. Phys.* **2011**, *134*. [CrossRef] [PubMed]

54. Monasterio, M.; Jansson, H.; Gaitero, J.J.; Dolado, J.S.; Cerveny, S. Cause of the fragile-to-strong transition observed in water confined in C-S-H gel. *J. Chem. Phys.* **2013**, *139*. [CrossRef] [PubMed]
55. Goracci, G.; Monasterio, M.; Jansson, H.; Cerveny, S. Dynamics of nano-confined water in Portland cement-Comparison with synthetic C-S-H gel and other silicate materials. *Sci. Rep.* **2017**, *7*, 1–10. [CrossRef]
56. Swenson, J.; Jansson, H.; Bergman, R. Relaxation processes in supercooled confined water and implications for protein dynamics. *Phys. Rev. Lett.* **2006**, *96*, 1–4. [CrossRef]
57. Ryabov, Y.; Gutina, A.; Arkhipov, V.; Feldman, Y. Dielectric Relaxation of Water Absorbed in Porous Glass. *J. Phys. Chem. B* **2001**, *105*, 1845–1850. [CrossRef]
58. Gutina, A.; Antropova, T.; Rysiakiewicz-Pasek, E.; Virnik, K.; Feldman, Y. Dielectric relaxation in porous glasses. *Microporous Mesoporous Mater.* **2003**, *58*, 237–254. [CrossRef]
59. Vasilyeva, M.A.; Gusev, Y.A.; Shtyrlin, V.G.; Gutina, A.G.; Puzenko, A.; Ishai, P.B.; Feldman, Y. Dielectric relaxation of water in clay minerals. *Clays Clay Miner.* **2014**, *62*, 62–73. [CrossRef]
60. Feldman, Y.; Puzenko, A.; Ryabov, Y. Dielectric relaxation phenomena in complex materials. *Fractals Diffus. Relax. Disord. Complex Syst. A Spec. Vol. Adv. Chem. Phys.* **2006**, *133*, 125.
61. Tsangaris, G.M.; Psarras, G.C.; Kouloumbi, N. Electric modulus and interfacial polarization in composite polymeric systems. *J. Mater. Sci.* **1998**, *33*, 2027–2037. [CrossRef]
62. Psarras, G.C. Hopping conductivity in polymer matrix-metal particles composites. *Compos. Part A Appl. Sci. Manuf.* **2006**, *37*, 1545–1553. [CrossRef]
63. Sanida, A.; Stavropoulos, S.G.; Speliotis, T.; Psarras, G.C. Development, characterization, energy storage and interface dielectric properties in SrFe12O19/ epoxy nanocomposites. *Polymer* **2017**, *120*, 73–81. [CrossRef]
64. Dang, Z.M.; Nan, C.W.; Xie, D.; Zhang, Y.H.; Tjong, S.C. Dielectric behavior and dependence of percolation threshold on the conductivity of fillers in polymer-semiconductor composites. *Appl. Phys. Lett.* **2004**, *85*, 97–99. [CrossRef]
65. Zhu, M.; Huang, X.; Yang, K.; Zhai, X.; Zhang, J.; He, J.; Jiang, P. Energy storage in ferroelectric polymer nanocomposites filled with core-shell structured polymer@BaTiO3 nanoparticles: Understanding the role of polymer shells in the interfacial regions. *ACS Appl. Mater. Interfaces* **2014**, *6*, 19644–19654. [CrossRef]
66. Palomba, M.; Carotenuto, G.; Longo, A.; Sorrentino, A.; Di Bartolomeo, A.; Iemmo, L.; Urban, F.; Giubileo, F.; Barucca, G.; Rovere, M.; et al. Thermoresistive properties of graphite platelet films supported by different substrate. *Materials* **2019**, *12*, 3638. [CrossRef]
67. Gorrasi, G.; Bugatti, V.; Milone, C.; Mastronardo, E.; Piperopoulos, E.; Iemmo, L.; Di Bartolomeo, A. Effect of temperature and morphology on the electrical properties of PET/conductive nanofillers composites. *Compos. Part B Eng.* **2018**, *135*, 149–154. [CrossRef]
68. Khan, S.; Lorenzelli, L. Recent advances of conductive nanocomposites in printed and flexible electronics. *Smart Mater. Struct.* **2017**, *26*, 083001. [CrossRef]
69. Sha, W.; O'Neill, E.A.; Guo, Z. Differential scanning calorimetry study of ordinary Portland cement. *Cem. Concr. Res.* **1999**, *29*, 1487–1489. [CrossRef]
70. Sha, W.; Pereira, G.B. Differential scanning calorimetry study of ordinary Portland cement paste containing metakaolin and theoretical approach of metakaolin activity. *Cem. Concr. Compos.* **2001**, *23*, 455–461. [CrossRef]
71. Alarcon-Ruiz, L.; Platret, G.; Massieu, E.; Ehrlacher, A. The use of thermal analysis in assessing the effect of temperature on a cement paste. *Cem. Concr. Res.* **2005**, *35*, 609–613. [CrossRef]
72. Esteves, L.P. On the hydration of water-entrained cement-silica systems: Combined SEM, XRD and thermal analysis in cement pastes. *Thermochim. Acta* **2011**, *518*, 27–35. [CrossRef]
73. Kyritsis, A.; Siakantari, M.; Vassilikou-Dova, A.; Pissis, P.; Varotsos, P. Dielectric and electrical properties of polycrystalline rocks at various hydration levels. *IEEE Trans. Dielectr. Electr. Insul.* **2000**, *7*, 493–497. [CrossRef]
74. Schwan, H.P. Electrod polarization impedance and measurements in biological materials. *Ann. N. Y. Acad. Sci.* **1968**, *148*, 191–209. [CrossRef] [PubMed]
75. Cabeza, M.; Merino, P.; Miranda, A.; Nóvoa, X.R.; Sanchez, I. Impedance spectroscopy study of hardened Portland cement paste. *Cem. Concr. Res.* **2002**, *32*, 881–891. [CrossRef]
76. Gu, P.; Xie, P.; Fu, Y.; Beaudoin, J.J. AC impedance phenomena in hydrating cement systems: Frequency dispersion angle and pore size distribution. *Cem. Concr. Res.* **1994**, *24*, 86–88. [CrossRef]
77. Keddam, M.; Takenouti, H.; Nóvoa, X.R.; Andrade, C.; Alonso, C. Impedance measurements on cement paste. *Cem. Concr. Res.* **1997**, *27*, 1191–1201. [CrossRef]

78. Lunak, M.; Kusak, I. Modern Electrical Measurement of Alkali Activated Slag Mortars with Increased Electrical Conductivity. *Appl. Mech. Mater.* **2016**, *861*, 64–71. [CrossRef]
79. Kusak, I.; Lunak, M.; Chobola, Z. Monitoring of concrete hydration by electrical measurement methods. *Procedia Eng.* **2016**, *51*, 271–276. [CrossRef]
80. Kusak, I.; Lunak, M.; Mikova, M.; Rovnanik, P. Influence of carbon admixtures to the electrical conductivity of slag mortars. *Solid State Phenom.* **2017**, *258*, 465–468. [CrossRef]
81. Jonscher, A.K. Dielectric relaxation in solids. *J. Phys. D Appl. Phys.* **1999**, *32*, R57–R70. [CrossRef]
82. Mauritz, K.A.; Yun, H. Dielectric Relaxation Studies of Ion Motions in Electrolyte-Containing Perfluorosulfonate Ionomers. 3. ZnSO4 and CaCl2 Systems. *Macromolecules* **1989**, *22*, 220–225. [CrossRef]
83. Dyre, J.C. The random free-energy barrier model for ac conduction in disordered solids. *J. Appl. Phys.* **1988**, *64*, 2456–2468. [CrossRef]
84. MacOvez, R.; Zachariah, M.; Romanini, M.; Zygouri, P.; Gournis, D.; Tamarit, J.L. Hopping conductivity and polarization effects in a fullerene derivative salt. *J. Phys. Chem. C* **2014**, *118*, 12170–12175. [CrossRef]
85. Kremer, F.; Schönhals, A. *Broadband Dielectric Spectroscopy*; Springer: Berlin/Heidelberg, Germany, 2003; ISBN 9783642628092.
86. Axelrod, E.; Givant, A.; Shappir, J.; Feldman, Y.; Sa'ar, A. Dielectric relaxation and porosity determination of porous silicon. *J. Non. Cryst. Solids* **2002**, *305*, 235–242. [CrossRef]
87. Øye, G.; Axelrod, E.; Feldman, Y.; Sjöblom, J.; Stöcker, M. Dielectric properties and Fourier transform IR analysis of MCM-48, Al-MCM-48 and Ti-MCM-48 mesoporous materials. *Colloid Polym. Sci.* **2000**, *278*, 517–523. [CrossRef]
88. Gutina, A.; Axelrod, E.; Puzenko, A.; Rysiakiewicz-Pasek, E.; Kozlovich, N.; Feldman, Y. Dielectric relaxation of porous glasses. *J. Non Cryst. Solids* **1998**, *235–237*, 302–307. [CrossRef]
89. Van Loon, L.R.; Müller, W.; Iijima, K. Activation energies of the self-diffusion of HTO, 22Na+ and 36Cl- in a highly compacted argillaceous rock (Opalinus Clay). *Appl. Geochem.* **2005**, *20*, 961–972. [CrossRef]
90. Gerhardt, R. Impedance and Dielectric Revisited: Distinguishing From Long-Range Conductivity. *J. Phys. Chem. Solids* **1994**, *55*, 1491–1506. [CrossRef]
91. Du, H.; Gao, H.J.; Pang, S.D. Improvement in concrete resistance against water and chloride ingress by adding graphene nanoplatelet. *Cem. Concr. Res.* **2016**, *83*, 114–123. [CrossRef]
92. Dimov, D.; Amit, I.; Gorrie, O.; Barnes, M.D.; Townsend, N.J.; Neves, A.I.S.; Withers, F.; Russo, S.; Craciun, M.F. Ultrahigh Performance Nanoengineered Graphene—Concrete Composites for Multifunctional Applications. *Adv. Funct. Mater.* **2018**, *28*. [CrossRef]
93. Jennings, H.M.; Thomas, J.J.; Gevrenov, J.S.; Constantinides, G.; Ulm, F.J. A multi-technique investigation of the nanoporosity of cement paste. *Cem. Concr. Res.* **2007**, *37*, 329–336. [CrossRef]
94. Allen, A.J.; Windsor, C.G.; Rainey, V.; Pearson, D.; Double, D.D.; Alford, N.M. A Small-Angle Neutron-Scattering Study of Cement Porosities. *J. Phys. D Appl. Phys.* **1982**, *15*, 1817–1833. [CrossRef]
95. Valori, A.; McDonald, P.J.; Scrivener, K.L. The morphology of C-S-H: Lessons from 1H nuclear magnetic resonance relaxometry. *Cem. Concr. Res.* **2013**, *49*, 65–81. [CrossRef]
96. Mogami, Y.; Yamazaki, S.; Matsuno, S.; Matsui, K.; Noda, Y.; Takegoshi, K. Hydrogen cluster/network in tobermorite as studied by multiple-quantum spin counting 1H NMR. *Cem. Concr. Res.* **2014**, *66*, 115–120. [CrossRef]
97. Raut, P.; Liang, W.; Chen, Y.M.; Zhu, Y.; Jana, S.C. Syndiotactic Polystyrene-Based Ionogel Membranes for High Temperature Electrochemical Applications. *ACS Appl. Mater. Interfaces* **2017**, *9*, 30933–30942. [CrossRef]

 *materials*

Article

Heat Transfer Through Insulating Glass Units Subjected to Climatic Loads

Zbigniew Respondek

Faculty of Civil Engineering, Czestochowa University of Technology, 42-201 Częstochowa, Poland; zbigniew.respondek@pcz.pl

Received: 19 December 2019; Accepted: 6 January 2020; Published: 8 January 2020

Abstract: One of the structural elements used in the construction of insulating glass units (IGUs) are tight gaps filled with gas, the purpose of which is to improve the thermal properties of glazing in buildings. Natural changes in weather parameters: atmospheric pressure, temperature, and wind influence the gas pressure changes in the gaps and, consequently, the resultant loads and deflections of the component glass panes of a unit. In low temperature conditions and when the atmospheric pressure increases, the component glass panes may have a concave form of deflection, so that the thickness of the gaps in such loaded glazing may be less than its nominal thickness. The paper analyses the effect of reducing this thickness in winter conditions on the design heat loss through insulating glass units. For this purpose, deflections of glass in sample units were determined and on this basis the thickness of the gaps under operating conditions was estimated. Next, the thermal transmittance and density of heat-flow rate determined for gaps of nominal thickness and of thickness reduced under load were compared. It was shown that taking into account the influence of climatic loads may, under certain conditions, result in an increase in the calculated heat loss through IGUs. This happens when the gaps do not transfer heat by convection, i.e., in a linear range of changes in thermal transmittance. For example, for currently manufactured triple-glazed IGUs in conditions of "mild winter", the calculated heat losses can increase to 5%, and for double-glazed IGUs with 10–14 mm gaps this ratio is about 4.6%. In other cases—e.g., large thickness of the gaps in a unit, large reduction in outside temperature—convention appears in the gaps. Then reducing the thickness of the gaps does not worsen the thermal insulation of the glazing. This effect should be taken into account when designing IGUs. It was also found that the wind load does not significantly affect the thickness of the gaps.

Keywords: glass in building; insulating glass units; heat loss in buildings; climatic loads

---

1. Introduction

Generally used in the building industry as a filling of windows or glass facades, insulating glass units (IGUs) consist of two or more component glass panes, connected at the edges with a glass spacer. The space between the component glass panes forms a tight gap filled with gas. In order to improve the thermal performance of the building partition constructed this way, the gap is filled with gas with lower thermal conductivity than air, most often argon. Further improvement of the performance is achieved by the use of component glass panes with a low-E coating—such a coating must be located on the side of the gap because it corrodes quickly when exposed to weather conditions. The tightness of the gap in insulating glass units is therefore a necessary factor to maintain good thermal insulation of transparent glazing [1,2].

Tight gaps, however, determine some specific properties of IGUs in the context of environmental loads transfer and the associated deformation of structural elements. The gap is filled with gas in the production process of the unit, therefore the gas in the gap has some initial parameters of pressure, temperature, and volume. Under operating conditions, an insulating glass unit is exposed to climatic loads which generate loads and deflections of the component glass panes due to the pressure difference between the gap and the environment. For example, an increase in atmospheric pressure or a decrease in the gas temperature in the gap results in a concave form of deflection of the panes (Figure 1a) and the opposite changes of these parameters in a convex form (Figure 1b). The magnitude of the under or overpressure in the gap depends not only on the value of climatic loads, but also on the structure of IGU. In general, it increases with reduced IGU dimensions (width × length), increased thickness of the gas gap, and increased thickness of the component glass plates. How the pressure difference affects the deflections in IGU will be presented later in this article.

In the case of wind exposure (Figure 1c), the tightness of the gap has a positive effect on the load distribution in an IGU. Due to changes in gas pressure in the gaps, the external load is partly transferred to the other panes of the unit.

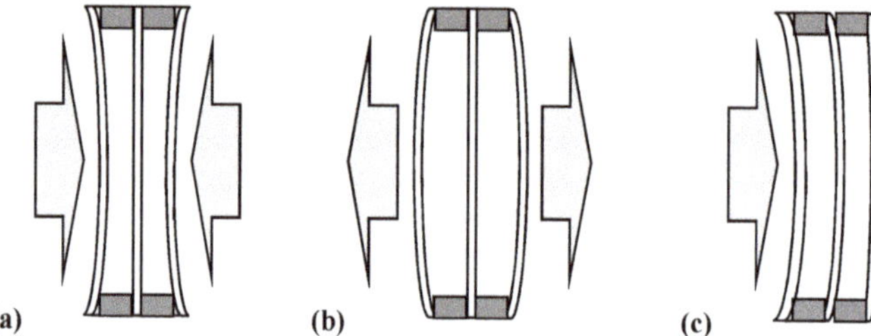

Figure 1. Typical deflections of insulating glass units: (**a**) concave form of deflection, (**b**) convex form of deflection, (**c**) deflection characteristic of wind load.

The deflections of the glass described above result in deformation of the image viewed in the light reflected from the glass in windows or on glass facades (Figure 2). It is important that under conditions of low air temperatures, i.e., during the heating season, insulating glass units tend to take the concave form of deflection. The result is a reduction in the thickness of the gas space—especially in the central part of the glazing, where the component glass panes are closest to each other—which makes it possible to reduce the thermal insulation of the IGU.

Figure 2. Visible distorted reflection of the image of the neighboring building from both insulating glass unit (IGU) component glass panes indicates the concave form of deflection of the unit.

The aim of the analysis carried out in the paper was to determine the effect of taking into account the reduction in thickness of gas-filled gaps in insulating glass units in winter conditions on the calculated heat losses through these partitions. The analysis was made for example for double- and triple-glazed IGUs. A detailed numerical quantification of this phenomenon was carried out for various IGU constructions.

In the literature, studies describing previous research in this area can be found: Barnier and Bourret [3] analyzed the effect of plate curvature in IGUs on the thermal transmittance (U_g-value). The authors determined the U_g-value for IGUs with variable gap thickness (limited by the surfaces of deflected panes), considering the average gap thickness in the loaded IGU as reliable. The authors stated that this assumption becomes reasonable when plate curvature is small and it is certainly acceptable in the conduction regime, where the convective movement is not significant. This article provides the results of sample calculations for double- and triple-glazed units under winter conditions. It was found that taking into account the plate curvature increases the calculated U_g-value from 4.4% to 5.8%. Calculations were also made to account for changes in weather conditions (typical meteorological year) for Montreal and Toulouse. The results indicate that U_g may vary up to 5% above and 10% below the yearly average.

Hart et al. [4] analyzed the U_g-value calculated from real deflections of double and triple-glazed units, measured in summer and winter at several locations in the USA. It was found that a 20 °C temperature difference reduces thermal performance by 4.6% for double-glazed IGUs and by 3.6% for triple-glazed IGUs.

Penkova et al. [5] presented examples of numerical analysis and experimental research regarding both parameters related to heat flow and climate loads. However, no detailed analysis of the change in thermal transmittance related to the deflections in the IGUs was carried out.

Thermal imaging photographs illustrating a decrease in thermal insulation in the central part of the glazing were published as well [6]. Examples are presented where the temperature in the central part of the glazing is 1–3 °C higher than the average on its surface (in images from the outside). An example of an IGU in which the component panes came into contact due to climatic loads is also presented.

2. Methodology for the Calculation of Static Quantities in IGUs

The methods of calculation of static quantities in double-glazed IGUs loaded with climatic factors are described in the literature. Mention may be made here of analytical models presented in papers [7–10] and numerical models allowing for consideration of the possibility of non-linear deflections of component glass [11,12]. The results of calculations presented in this paper were obtained using the author's analytical model proposed in the article [13], which allows to calculate the load and deflection of component glass panes in units with any number of tight gaps.

The basis for the calculation of static quantities in IGUs is the assumption that the gas in the gaps meets the ideal gas equation:

$$\frac{p_0 \cdot v_0}{T_0} = \frac{p_{op} \cdot v_{op}}{T_{op}} = \text{const}, \tag{1}$$

where:

p_0, T_0, v_0—initial gap gas parameters: pressure [kPa], temperature [K], volume [m^3], obtained in the production process,

p_{op}, T_{op}, v_{op}—operating parameters—analogously.

It is also assumed that the glass panes are simply supported at the edges and that the linear dependence of deflection w [m] of the component glass pane on its resultant surface load q [kN/m^2] is assumed. The latter assumption is considered to be sufficiently accurate if the deflection is not greater than the thickness of the glass [14]. The deflection function of a simply supported single pane of the a [m] width and b [m] length, subjected to the q [kN/m^2] load, placed centrally in the x-y coordinate system, can be recorded as [15]:

$$w(x,y) = \frac{4qa^4}{\pi^5 D} \sum_{i=1,3,5\ldots} \frac{(-1)^{(i-1)/2}}{i^5} \cos\frac{i\pi x}{a} \cdot \left(1 - \frac{\beta_i \cdot \text{th}\beta_i + 2}{2 \cdot \text{ch}\beta_i} \cdot \text{ch}\frac{i\pi y}{a} + \frac{1}{2 \cdot \text{ch}\beta_i} \cdot \frac{i\pi y}{a} \cdot \text{sh}\frac{i\pi y}{a}\right), \tag{2}$$

with

$$\beta_i = \frac{i\pi b}{2a}, \tag{3}$$

D [kNm] is the flexural rigidity of glass pane:

$$D = \frac{E \cdot d^3}{12 \cdot (1 - \mu^2)}, \tag{4}$$

where:

d—is the glass pane thickness [m],
E—is the Young's modulus of glass [kPa],
μ—is the Poisson's ratio [-].

Change in gap volume Δv [m^3] resulting from the deflection of one of the limiting glass pane may be determined by integration of Equation (2):

$$\Delta v = \int_{-b/2}^{b/2} \int_{-a/2}^{a/2} w(x, y) dx dy, \tag{5}$$

$$\Delta v = \frac{4qa^6}{\pi^7 D} \sum_{i=1,3,5\ldots} \frac{(-1)^{(i-1)/2}}{i^7} \cdot \frac{\sin \frac{i \cdot \pi}{2}}{(\mathrm{ch}\beta_i)^2} \cdot (4 \cdot \beta_i + 2 \cdot \beta_i \cdot \mathrm{ch}(2 \cdot \beta_i) - 3 \cdot \mathrm{sh}(2 \cdot \beta_i)), \tag{6}$$

After the relevant calculations have been made:

$$\Delta v = \alpha\prime_{\mathrm{v}} \cdot \frac{q \cdot a^6}{D} = \alpha_{\mathrm{v}} \cdot q, \tag{7}$$

where:

α'_{v}—is the dimensionless coefficient dependent on the b/a ratio (Table 1) [-],
α_{v}—is the proportionality factor, [m^5/kN].

Any change in climatic conditions (atmospheric pressure, temperature, wind) results in a change in the gas pressure in the gaps which affects the resultant operating load of each of the component glass panes. For each gap of an IGU it is possible to formulate the equation of state:

$$p_0 \cdot v_0 \cdot T_{\mathrm{op}} = p_{\mathrm{op}} \cdot \left(v_0 + \sum \Delta v\right) \cdot T_0, \tag{8}$$

where:

$\Sigma\Delta v$—is the change in gap volume caused by deflection of both panes limiting it [m^3].

As already mentioned in the article, double- and triple-glazed IGUs were analyzed. In the remaining part, the parameters of the individual component glass panes and gaps were marked with appropriate indices (Figure 3). It is also assumed that loads and deflections are positive if they face the interior, i.e., from left to right as in Figure 3.

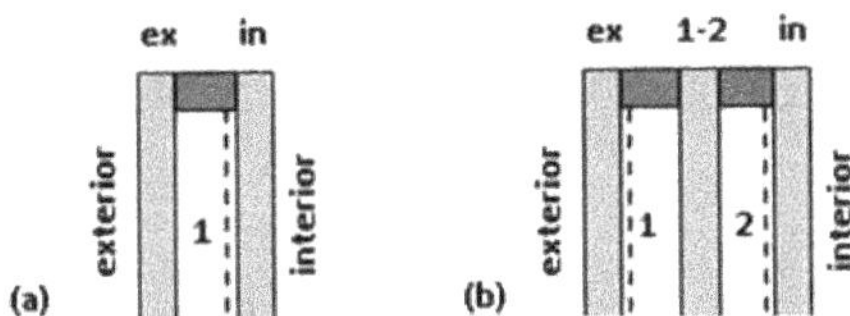

Figure 3. Index designations of IGU elements and location of low-emission coatings (dashed line): (**a**) double-glazed IGU, (**b**) triple-glazed IGU.

Taking into account the adopted markings and conventions, Equation (8) for a double-glazed IGU can be presented in the form:

$$\frac{p_0 \cdot v_{01} \cdot T_{op1}}{T_0} = p_{op1} \cdot \left[\left(p_{op1} - c_{ex} \right) \cdot \alpha_{v,ex} + \left(p_{op1} - c_{in} \right) \cdot \alpha_{v,in} \right], \tag{9}$$

with

$$c_{ex} = p_a + q_{z,ex},\ c_{in} = p_a - q_{z,in}, \tag{10}$$

where:

p_a—current atmospheric pressure [kPa],
$q_{z,ex}$, $q_{z,in}$—load per area from outer factors, primarily wind [kN/m^2], almost always $q_{z,in} = 0$.

After the relevant transitions have been made:

$$B \cdot p_{op1}^2 - A \cdot p_{op1} - \frac{p_0 \cdot v_{01} \cdot T_{op1}}{T_0} = 0, \tag{11}$$

with

$$A = c_{ex} \cdot \alpha_{v,ex} + c_{in} \cdot \alpha_{v,in} - v_{01}, \tag{12}$$

$$B = \alpha_{v,ex} + \alpha_{v,in}. \tag{13}$$

Equation (11) has one solution giving non-negative results:

$$p_{op1} = \frac{A}{2 \cdot B} + \sqrt{\left(\frac{A}{2 \cdot B} \right)^2 + \frac{p_0 \cdot v_{01} \cdot T_{op1}}{B \cdot T_0}}. \tag{14}$$

In the case of a triple-glazed unit, a system of quadratic equations should be solved:

$$\begin{cases} p_{op1} \cdot \left[v_{01} + \left(p_{op1} - c_{ex} \right) \cdot \alpha_{v,ex} + \left(p_{op1} - p_{op2} \right) \cdot \alpha_{v,1-2} \right] - \frac{p_0 \cdot v_{01} \cdot T_{op1}}{T_0} = 0 \\ p_{op2} \cdot \left[v_{02} + \left(p_{op2} - p_{op1} \right) \cdot \alpha_{v,1-2} + \left(p_{op2} - c_{in} \right) \cdot \alpha_{v,in} \right] - \frac{p_0 \cdot v_{02} \cdot T_{op2}}{T_0} = 0 \end{cases} \tag{15}$$

This system has no analytical solution, but it can be solved numerically by iteration.

After calculating the operating pressure p_{op} for each of the gaps, the resultant loading q for each of the component glass panes can be determined:

- for a double-glazed IGU

$$q_{ex} = c_{ex} - p_{op1},\ q_{in} = p_{op1} - c_{in}, \tag{16}$$

- for a triple-glazed IGU

$$q_{ex} = c_{ex} - p_{op1},\ q_{1-2} = p_{op1} - p_{op2},\ q_{in} = p_{op2} - c_{in} \tag{17}$$

Deflection w_c [mm] in the center of the glass pane can be determined by the formula:

$$w_c = \alpha'_w \cdot \frac{q \cdot a^4}{D} \cdot 1000, \tag{18}$$

where:

α'_w—is the dimensionless coefficient dependent on the b/a ratio (Table 1) [-].

Table 1. Coefficients for calculating volume change and deflection for simply supported glass pane.

b/a	1.0	1.1	1.2	1.3	1.4	1.5
α'_v	0.001703	0.002246	0.002848	0.003499	0.004189	0.004912
α'_w	0.004062	0.004869	0.005651	0.006392	0.007085	0.007724
b/a	1.6	1.7	1.8	1.9	2.0	3.0
α'_v	0.005659	0.006427	0.00721	0.008004	0.008808	0.017055
α'_w	0.008308	0.008838	0.009316	0.009745	0.010129	0.012233

However, the average deflection of the component glass panes w_m [mm] was determined from the formula:

$$w_m = \frac{\Delta v}{a \cdot b} \cdot 1000. \tag{19}$$

3. Materials and Methods

Thermal transmittance U_g [W/(m^2·K)] of IGUs was calculated on the basis of the methodology described in standard [16], and heat losses were expressed by density of heat-flow rate Φ [W/m^2] from the formula:

$$\Phi = U_g \cdot (t_i - t_e), \tag{20}$$

where:

t_i, t_e—are the internal and external air temperature [°C].

The heat flow through an insulating glass unit is complex—through conduction, convection, and radiation. The thermal resistance of gas-filled gaps R_s [(m^2·K)/W] has the greatest influence on the U-value. For each gap:

$$R_s = \frac{1}{h_g + h_r}, \tag{21}$$

with

$$h_g = \frac{\lambda_g \cdot Nu}{s}, \tag{22}$$

$$h_r = \frac{4 \cdot \sigma \cdot T_m^3}{\frac{1}{\varepsilon_{sur1}} + \frac{1}{\varepsilon_{sur2}} - 1}, \tag{23}$$

where:

h_r—is the thermal conductance by radiation [W/(m^2·K)],
h_g—is the thermal conductance of gas (by conduction and convection) [W/(m^2·K)].
λ_g—is the thermal conductivity of gas [W/(m·K)],
s—is the gas gap thickness [m],
Nu—is the Nusselt number [-],
σ—is the Boltzmann constant 5.6693×10^{-8} W/(m^2·K^4)
T_m—is the average temperature of both surfaces delimiting the gap [K],
ε_{sur1}, ε_{sur2}—are the emissivity of surfaces delimiting the gap [-].

It is particularly important whether convection occurs in the gaps. In the case of narrow gaps ($Nu < 1$) it is assumed that convection does not occur—thermal insulation of the gap increases linearly with its thickness. If a certain gap thickness limit (for $Nu = 1$) is exceeded, the effect of convection is taken into account. In this non-linear range (for $Nu > 1$) thermal insulation of the IGU does not improve. The value of this thickness limit depends on many factors (see also [17,18]), first of all on:

- the type of gas; the calculation was based on the use of argon,
- location in the structure; the calculations assume a horizontal position, in units situated horizontally or diagonally convection increases.
- increasing the temperature difference on the surfaces of the glass panes limiting the gap affects the increase in convection,
- convection also increases when the average gas temperature in the gap increases.

The thermal resistance of the gaps is primarily influenced by the use of low-emission glass. Glass without coating has a standard coefficient of emission of $\varepsilon = 0.837$. Application of low-emission coating reduces the emissivity of the plate surface, which results in a significant reduction of heat transfer by radiation. Currently, in Central and Northern Europe, IGUs are most often produced, in which each gap is adjacent to one coated surface and one without coating (Figure 3). This solution is most often used in units currently produced in Central and Northern Europe. The values $\varepsilon_{sur1} = 0.837$ and $\varepsilon_{sur2} = 0.04$ were used in the calculations.

Glass conducts heat well, therefore the thickness of the glass panes has no significant effect on the U_g-value. Physical parameters of argon were adopted on the basis of the standard [17].

Of course, thermal insulation is also affected by thermal surface resistance at the external side (R_e [(m^2·K)/W]) and at the internal side of a window (R_i [(m^2·K)/W]). They depend primarily on the positioning of the window in the structure and the velocity of air (a short analysis on this subject is presented in Chapter 5). The calculations assumed $R_i = 0.13$ (m^2·K)/W (vertical position) and $R_e = 0.04$ (m^2·K)/W (for wind velocity $V = 4$ m/s). These are often accepted comparative values, also in the standard [16].

The calculations according to the adopted model require the use of numerical methods, because we encounter several interdependent values here. For example, the temperature values of gas and glass surfaces depend on the temperature distribution in the cross-section of the IGU. This distribution depends on the resulting thermal resistance values. The results of calculations were obtained by iteration after building the appropriate spreadsheet, assuming the steady state of heat transfer.

Figure 4 shows the effect of gap thickness s [mm] on the design U_g-value for double- and triple-glazed IGUs, with glass thickness $d = 4$ mm, assuming $t_i = 20$ °C and in two variants of the outside air temperature $t_e = 0$ °C and $t_e = -20$ °C. The dashed line was used to determine the limits of gap thickness at which $Nu = 1$.

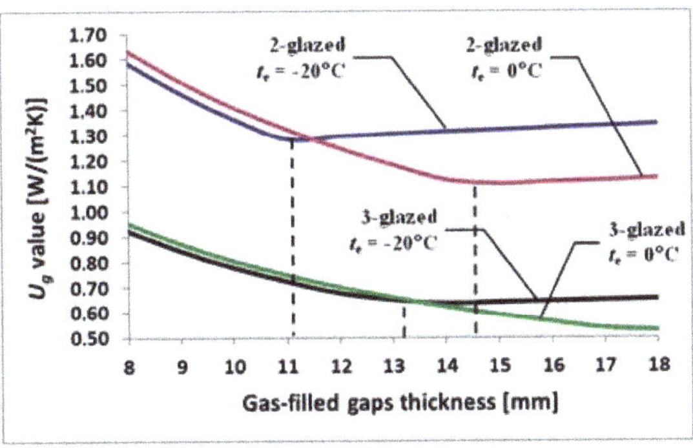

Figure 4. The dependence of the U_g-values of IGUs on the thickness of the gaps.

Figure 4 shows that at low temperatures t_e the limit thickness decreases. It can also be stated that in the case of triple-glazed IGUs, the difference in temperature in the gap is smaller, and the thickness of the boundary increases.

4. IGUs Under Pressure and Temperature Changes—Presentation and Discussion of Test Results

An analysis of the influence of climate loads on heat loss through IGUs under winter conditions was carried out for sample units with dimensions 0.7 × 1.4 m. Glass material parameters were adopted according to the standard [19]: $E = 70$ GPa, $\mu = 0.2$.

It was also assumed that the following initial parameters were obtained in the argon-filled gaps during the production process $T_0 = 20$ °C = 293.15 K, $p_0 = 100$ kPa. In these conditions, the component glass panes are flat.

Two variants of the temperature drop load were used.

Variant 1. Reduced temperature conditions: $t_i = 20$ °C, $t_e = -20$ °C; the gas temperature in each gap was calculated for each case based on the temperature distribution in the particular IGU: for a double-glazed IGU $T_{op1} = -2.37$ to -2.25 °C, for triple-glazed IGU $T_{op1} = -10.09$ to -9.66 °C, $T_{op2} = 7.60$ to 7.97 °C.

Variant 2. Conditions for a "mild winter": $t_i = 20$ °C, $t_e = 0$ °C; gas temperature: for a double-glazed IGU $T_{op1} = 8.80$ to 9.01 °C, for triple-glazed IGU $T_{op1} = 4.97$ to 5.08 °C, $T_{op2} = 13.66$ to 14.10 °C.

First, the effect of varying glass thickness on the gap width in the loaded set was investigated. IGUs with 16 and 12 mm gap thickness were analyzed in various combinations of 3, 4, and 6 mm thick panes. It was assumed that IGUs are only loaded with the temperature drop, as in variant 1, i.e., the current atmospheric pressure $p_a = p_0 = 100$ kPa. The results of the calculations are presented in Table 2.

The resultant loading q_{ex} and q_{in} (absolute value) illustrates the underpressure in the gaps in relation to atmospheric pressure. The parameter $q_{1\text{-}2}$ is the difference in operating pressure between the gaps in a triple-glazed IGU. From Equations (18) and (19) the extreme deflection (in the center of the pane) w_c and the average deflection w_m (the w_m values are given between parentheses) were calculated for each pane.

On the basis of these deflections, the minimum gap thickness in the center of the IGU s_c [mm] and the average gap thickness s_m [mm] were calculated.

On the basis of the calculations presented in Table 2, it was found that the calculated values of s_c and s_m for analyzed IGUs with gaps of the same nominal thickness do not differ much from each other. This is despite the fact that the deflection of component glass panes varies considerably. The effect of gas interactions in tight gaps can be seen here. Rigid panes are less susceptible to deflection, but the external load is less compensated for by the gas pressure in the gap. After changing the thickness of all the panes in a unit, the load changes, but the deflections are similar. Therefore, when one of the glass panes changes to a stiffer one, the absolute load values of component glass panes increase, although their algebraic sum for each IGU is equal to 0. After such conversion, the less rigid panes deflect more because they are exposed to a higher loading—for this reason, a loaded IGU has approximately constant volume of gaps, despite the change in thickness of the component panes.

To identify the extent of the phenomenon described above, another example was solved. Figure 5 shows the influence of IGU width (at a constant ratio $b/a = 2$) on the maximum deflection of component panes w_c in double-glazed units at 3, 4 i 6 mm thick panes and 16 mm thick gap. It was assumed that IGU is loaded only by a change in atmospheric pressure by $\Delta p = p_a - p_o = 3$ kPa. This means that the current atmospheric pressure is $p_a = 103$ kPa. It can be added here that the results of calculations of static quantities are not very sensitive to the value of p_o, and to a significant extent to Δp. This means that if we assumed, for example, $p_o = 950$ kPa i $p_a = 980$ kPa, the results would be almost identical.

Table 2. Static quantities and gap thicknesses in IGUs—under reduced temperature conditions (Variant 1).

Structure of IGU [mm]	Resultant Loading q [kN/m^2]			Deflection w_c (w_m) [mm]			Resultant Thickness of Gap s [mm]			
							gap 1		gap 2	
	ex	1-2	in	ex	1-2	in	s_{c1}	s_{m1}	s_{c2}	s_{m2}
d_{ex}-s_1-d_{in}	Double-glazed units									
4-16-4	0.218	-	−0.218	1.36 (0.59)	-	−1.36 (−0.59)	13.28	14.82	-	-
6-16-4	0.330	-	−0.330	0.61 (0.27)	-	−2.08 (−0.90)	13.31	14.83	-	-
4-12-4	0.164	-	−0.164	1.03 (0.45)	-	−1.03 (−0.45)	9.94	11.10	-	-
6-12-4	0.251	-	−0.251	0.47 (0.20)	-	−1.57 (−0.68)	9.96	11.12	-	-
3-12-3	0.070	-	−0.070	1.04 (0.45)	-	−1.04 (−0.45)	9.92	11.10	-	-
d_{ex}-s_1-$d_{1\text{-}2}$-s_2-d_{in}	Triple-glazed units									
4-16-4-16-4	0.463	−0.117	−0.346	2.90 (1.26)	−0.73 (−0.32)	−2.16 (−0.94)	12.37	14.42	14.57	15.38
6-16-4-16-4	0.840	−0.311	−0.529	1.56 (0.68)	−1.94 (−0.85)	−3.34 (−1.44)	12.50	14.47	14.63	15.41
4-12-4-12-4	0.345	−0.087	−0.258	2.16 (0.99)	−0.54 (−0.24)	−1.61 (−0.70)	9.30	10.77	10.93	11.54
6-12-4-12-4	0.631	−0.233	−0.398	1.17 (0.51)	−1.46 (−0.63)	−2.49 (−1.08)	9.37	10.86	10.97	11.55
6-12-3-12-4	0.540	−0.112	−0.429	1.00 (0.43)	−1.66 (−0.72)	−2.68 (−1.17)	9.34	10.85	10.98	11.55
3-12-3-12-3	0.149	−0.037	−0.112	2.21 (0.96)	−0.55 (−0.24)	−1.66 (−0.72)	9.24	10.80	10.89	11.52

Figure 5 shows that that greater diversity of deflections for units of different component panes thickness occurs in the case of smaller IGU sizes. Then, however, the deflection values are smaller and it can be expected that changes in the gap thickness are also small.

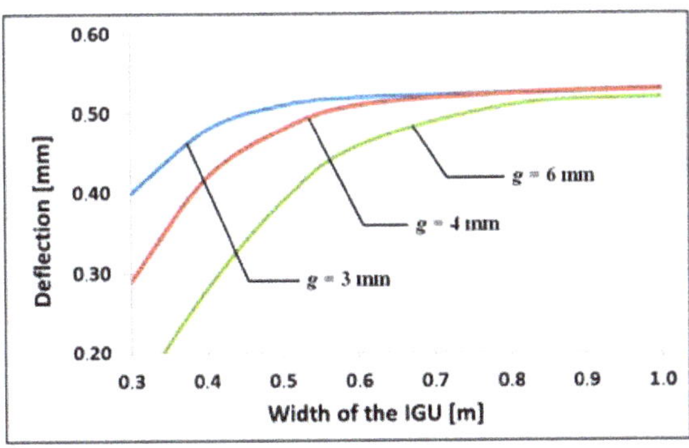

Figure 5. Dependence of the deflection of component glass panes w_c on the width of the IGU (atmospheric pressure increase of $\Delta p = 3$ kPa).

In the context of the above, further analysis was carried out for IGUs with the same thickness of component glass panes $d = 4$ mm and different gap thicknesses s were assumed. Table 3 presents calculations of w_c, w_m, s_c, and s_m values for units loaded with temperature change as in variant 1 and simultaneously operating loading with external atmospheric pressure increase of $\Delta p = 3$ kPa. These are particularly unfavorable operating conditions in the context of reducing the thickness of the gaps. An analogous calculation was carried out for variant 2 (Table 4).

Based on the above data, Table 5 (for variant 1) and Table 6 (for variant 2) present the results of calculations of the thermal transmittance U and the density of heat-flow rate Φ [W/m^2]:

- U_g, Φ_g—describe heat loss without taking into account the curvature of the panes, calculated for the nominal thickness of the gaps,
- U_c, Φ_c—describe possible local heat loss near the IGU center, i.e., where the distance between the panes is the smallest, calculated for the thickness of the gaps s_c,
- U_m, Φ_m—describe the average heat loss through the IGU, calculated for the thickness of the gaps s_m.

Finally, the percentage increase in the calculated quantities is also presented ($\Delta\Phi_c$, $\Delta\Phi_m$) for units of nominal gap thickness.

Table 3. Static quantities and gap thicknesses in IGUs under reduced temperature conditions (Variant 1) and atmospheric pressure increase by $\Delta p = 3$ kPa.

Structure of IGU [mm]	Resultant Loading q [kN/m^2]			Deflection w_c (w_m) [mm]			Resultant Thickness of Gap s [mm]			
	ex	1-2	in	ex	1-2	in	gap 1		gap 2	
							s_{c1}	s_{m1}	s_{c2}	s_{m2}
d_{ex}-s_1-d_{in}				Double-glazed units						
4-16-4	0.295	-	−0.295	1.84 (0.81)	-	−1.84 (−0.81)	12.32	14.38	-	-
4-14-4	0.259	-	−0.259	1.62 (0.70)	-	−1.62 (−0.70)	10.76	12.60	-	-
4-12-4	0.233	-	−0.233	1.39 (0.61)	-	−1.39 (−0.61)	9.22	10.78	-	-
4-10-4	0.187	-	−0.187	1.17 (0.51)	-	−1.17 (−0.51)	7.66	8.98	-	-
d_{ex}-s_1-$d_{1\text{-}2}$-s_2-d_{in}				Triple-glazed units						
4-16-4-16-4	0.613	−0.114	−0.500	3.83 (1.67)	−0.71 (−0.31)	−3.13 (−1.36)	11.46	14.02	13.58	14.95
4-14-4-14-4	0.540	−0.100	−0.440	3.38 (1.47)	−0.63 (−0.27)	−2.75 (−1.20)	9.99	12.26	11.88	13.07
4-12-4-12-4	0.460	−0.085	−0.375	2.87 (1.25)	−0.53 (−0.23)	−2.35 (−1.02)	8.60	10.52	10.18	11.21
4-10-4-10-4	0.387	−0.070	−0.317	2.42 (1.05)	−0.44 (−0.19)	−1.98 (−0.86)	7.14	8.76	8.46	9.33

Table 4. Static quantities and gap thicknesses in IGUs under conditions for a "mild winter" (Variant 2) and atmospheric pressure increase by Δp = 3 kPa.

Structure of IGU [mm]	Resultant Loading q [kN/m^2]			Deflection w_c (w_m) [mm]			Resultant Thickness of Gap s [mm]			
							gap 1		gap 2	
	ex	1-2	in	ex	1-2	in	s_{c1}	s_{m1}	s_{c2}	s_{m2}
d_{ex}-s_1-d_{in}				Double-glazed units						
4-16-4	0.188	-	−0.188	1.18 (0.51)	-	−1.18 (−0.51)	13.64	14.98	-	-
4-14-4	0.165	-	−0.165	1.03 (0.45)	-	−1.03 (−0.45)	11.94	13.10	-	-
4-12-4	0.143	-	−0.143	0.89 (0.40)	-	−0.89 (−0.40)	10.22	11.20	-	-
4-10-4	0.120	-	−0.120	0.75 (0.33)	-	−0.75 (−0.33)	8.50	9.34	-	-
d_{ex}-s_1-$d_{1\text{-}2}$-s_2-d_{in}				Triple-glazed units						
4-16-4-16-4	0.384	−0.058	−0.326	2.41 (1.04)	−0.36 (−0.16)	−2.04 (−0.89)	13.23	14.80	14.32	15.27
4-14-4-14-4	0.339	−0.050	−0.289	2.12 (0.92)	−0.31 (−0.14)	−1.81 (−0.79)	11.57	12.94	12.50	13.35
4-12-4-12-4	0.293	−0.042	−0.251	1.83 (0.80)	−0.26 (−0.11)	−1.56 (−0.68)	9.91	11.09	10.70	11.43
4-10-4-10-4	0.246	−0.035	−0.212	1.54 (0.67)	−0.22 (−0.09)	−1.32 (−0.57)	8.24	9.24	8.90	9.52

Table 5. Quantities describing heat losses by IGUs under conditions of reduced temperature (Variant 1) and atmospheric pressure increase by Δp = 3 kPa.

Gas Gap Thickness [mm]	Thermal Transmittance [W/(m^2·K)]			Density of Heat-Flow Rate Φ [W/m^2]			$\Delta\Phi_c$ [%]	$\Delta\Phi_m$ [%]
	U	U_c	U_m	Φ	Φ_c	Φ_m		
				Double-glazed units				
16	1.330	1.299	1.317	53.20	51.96	52.68	−2.3	−1.0
14	1.314	1.298	1.302	52.56	51.92	52.08	−1.2	−0.9
12	1.296	1.441	1.297	51.84	57.64	51.88	11.2	−0.1
10	1.364	1.630	1.467	54.56	65.20	58.68	19.5	7.6
				Triple-glazed units				
16	0.643	0.653	0.636	25.72	26.12	25.44	1.6	−1.1
14	0.634	0.727	0.648	25.36	29.08	25.92	14.7	2.2
12	0.675	0.817	0.730	27.00	32.68	29.20	21.0	8.1
10	0.776	0.941	0.839	31.04	37.64	33.56	21.3	8.1

Table 6. Quantities describing heat losses by IGUs under conditions for a "mild winter" (Variant 2) and atmospheric pressure increase by Δp = 3 kPa.

Gas Gap Thickness [mm]	Thermal Transmittance [W/(m²·K)]			Density of Heat-Flow Rate Φ [W/m²]			$\Delta\Phi_c$ [%]	$\Delta\Phi_m$ [%]
	U	U_c	U_m	Φ	Φ_c	Φ_m		
Double-glazed units								
16	1.113	1.142	1.107	22.26	22.84	22.14	2.6	−0.5
14	1.122	1.250	1.174	22.44	25.00	23.48	11.4	4.6
12	1.246	1.388	1.305	24.92	27.76	26.10	11.4	4.7
10	1.408	1.567	1.473	28.16	31.34	29.46	11.3	4.6
Triple-glazed units								
16	0.563	0.631	0.590	11.26	12.62	11.80	12.1	4.8
14	0.623	0.699	0.653	12.46	13.98	13.06	12.2	4.8
12	0.700	0.786	0.735	14.00	15.72	14.70	12.3	5.0
10	0.804	0.903	0.844	16.08	18.06	16.88	12.3	5.0

The data presented in Tables 5 and 6 and Figure 4 indicate that the reduction in the thickness of the gaps of insulating glass units due to their deflection under a drop in gas temperature and a rise in atmospheric pressure may result in an increase in design heat losses in relation to the calculations without taking into account the curvature of the panes. The increase in heat loss occurs in the linear range of the U_g-value change, i.e., when the conditions inside the gap lead to $Nu < 1$. It is different when the U_g-value changes in the non-linear range ($Nu > 1$). Heat losses do not increase. Then, the reduction of gap thickness can lead to a slight decrease in the calculated Nu value, which translates into a slight reduction in the calculated heat losses.

In this context, it is preferable to design IGUs such that it has $Nu > 1$ with a certain margin based on glazing deflections. However, this task should be approached with great caution, taking into account local climate conditions. It is necessary to check if the thickening of the gaps between the panes will not lead to excessive overpressure during the summer, due to the heating of gas in the gaps.

One more feature of the described phenomenon should be noted. In the linear range of changes in the U_g-value, the indices $\Delta\Phi_c$ and $\Delta\Phi_m$ almost do not depend on the thickness of the gaps. This is due to the fact, as additional calculations have shown, that the relationship between the thickness of IGU gaps and static quantities (resultant loading of component glass and their deflections) is also linear.

For many years, double-glazed IGUs dominated the market. Currently, due to the need to save energy, in Central and Northern Europe, triple-glazed IGU 4-16-4-16-4 is the most commonly produced and sold glazing for windows. Figure 6 presents an analysis illustrating the dependence of the percentage change in the calculated heat loss $\Delta\Phi_m$ for these units on their width (at a constant ratio b/a = 2), under different external temperature conditions t_e. Simultaneous pressure increase Δp = 3 kPa was assumed. Other data was used as in previous examples.

It was found that the described effect is important for the currently sold glazing in "mild winter" conditions, i.e., when the outside temperatures fluctuate within between −5 °C and 5 °C. For IGU width above 0.7m, the ratio $\Delta\Phi_m$ changes from 3.9% to 5.0%. These values are characteristic of the average temperature during the winter months in many places around the world.

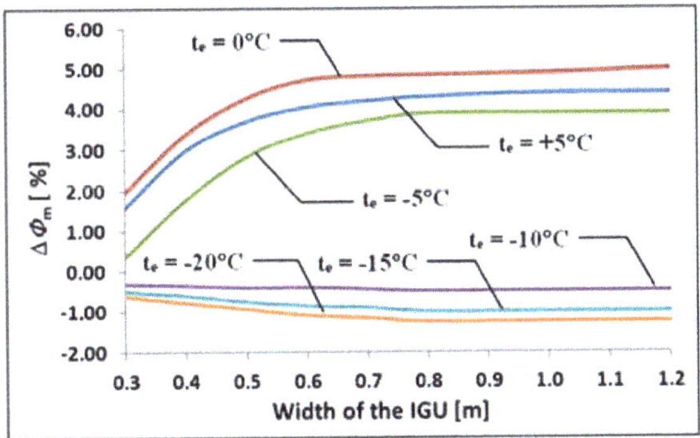

Figure 6. Dependence of the $\Delta\Phi_m$ index on the width of the IGU and the outside temperature.

5. Notes on IGUs Wind Load

Wind pressure or suction are also factors that cause deflection of the component glass panes in an IGU. As already mentioned, the wind velocity pressure acts directly only on the outer pane, but due to the change in the gas pressure in the gaps, the resultant load is distributed over all the panes of the unit. Table 7 shows the resultant loads and deflections in sample unit's surface loaded with 0.3 kN/m^2, which approximately corresponds [20] to a pressure of wind with velocity V of approx. 80 km/h (22.2 m/s).

Table 7. Static quantities in IGUs loaded with wind pressure of 0.3 kN/m^2.

Structure of IGU [mm]	Resultant Loading q [kN/m^2]			Deflection w_c [mm]		
	ex	1–2	in	ex	1–2	in
d_{ex}-s_1-d_{in}			Double-glazed units			
4-16-4	0.154	-	0.146	0.96	-	0.91
8-16-4	0.268	-	0.032	0.21	-	0.20
4-16-8	0.047	-	0.253	0.29	-	0.20
d_{ex}-s_1-$d_{1\text{-}2}$-s_2-d_{in}			Triple-glazed units			
4-16-4-16-4	0.109	0.098	0.092	0.68	0.61	0.58
8-14-4-14-4	0.247	0.028	0.026	0.19	0.17	0.16
4-12-4-12-8	0.053	0.039	0.208	0.33	0.24	0.16

Table 7 demonstrates that in the majority of units the deflections of component glass have similar values. Greater variations may occur when thicker panes are used, but the deflection values are small. It can therefore be concluded that the change in the thickness of the gaps due to wind load is small and has no noticeable effect on heat loss by IGUs.

Wind velocity has an indirect effect on heat loss. It is a factor influencing external thermal surface resistance on the outside, which translates into the U_g-value. Graphic illustration of this effect is shown in Figure 7. The calculations were made for units with gap thickness of 16 mm. It can be noted that in the case of triple-glazed IGUs, the effect of wind velocity is negligible.

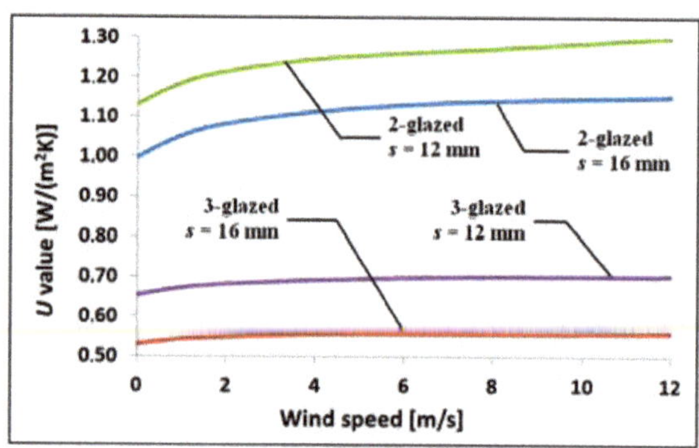

Figure 7. Influence of wind velocity on the U_g-value of sample insulating glass units.

6. Conclusions

One of the factors influencing thermal transmittance U_g of insulating glass units is the thickness of gas-filled tight gaps. It is assumed in the calculation procedures that this thickness is not dependent on temporary changes in climatic factors. The thickness is variable under real operating conditions. In winter conditions in particular, IGU component glass panes take a concave form of deflection, which reduces the thickness of the gaps. This effect increases if the atmospheric pressure increases at the same time.

Based on the example calculations carried out, it has been shown that the increase in the calculated heat losses associated with the reduction of the gap thickness occurs when the conditions in the gap lead to $Nu < 1$, i.e., when the thermal transmittance of the gas layer is linearly dependent on its thickness. Heat losses can then increase to about 4.6% for double-glazed IGUs and to about 5% for triple-glazed ones, for external air temperature $t_e = 0$ °C. These values almost do not depend on the nominal thickness of the gaps, which results from the linear dependence of static quantities in an IGU on this thickness. Under certain conditions, heat losses calculated according to standard procedures may therefore be underestimated.

It is different in the non-linear range of the U_g-value change ($Nu > 1$), i.e., when the outside temperature drops significantly or the gaps are thick enough. The thermal performance of glazing does not deteriorate. It is therefore advantageous to design IGUs so that $Nu > 1$, but it is necessary to take into account local climatic conditions and analyze loads that may also occur during the summer period.

In the case of the most commonly sold triple-glazed units 4-16-4-16-4 heat losses may be underestimated when the outside temperatures fluctuate between −5 °C and 5 °C. For large IGU dimensions, the $\Delta\Phi_m$ index totals then from 3.9% to 5.0%.

It was also shown that the effect of wind load on gap thickness change is negligible in the context of heat loss estimation.

Funding: This research received no external funding.

Conflicts of Interest: The author declares no conflict of interest.

Nomenclature

A	auxiliary parameter [m^3]
a	width (of glass pane) [m]
B	auxiliary parameter [m^5/kN]
b	length (of glass pane) [m]
c	auxiliary parameter [kPa]
D	flexural rigidity (of glass pane) [kNm]
d	thickness (of glass pane) [m] or [mm]

E	Young's modulus [kPa] or [GPa]
h	thermal conductance [W/(m^2·K)]
i	consecutive natural number
Nu	Nusselt number
p	pressure [kPa]
q	load per area, [kN/m^2]
R	thermal resistance [(m^2·K)/W]
s	thickness (of gas gap) [mm]
T	temperature (of gas in the gap) [K]
t	temperature (of air) [K] or [°C]
U	thermal transmittance [W/(m^2·K)]
V	wind velocity [km/h]
v	volume (of the gap) [m^3]
w	deflection [mm]
$w(x,y)$	function of deflection, [m]
x-y	coordinate system
Greek letters	
α	proportionality factor, [m^5/kN]
α'	dimensionless coefficient [-]
β_i	auxiliary parameter [-]
Δp	pressure change [kPa]
ΔT	temperature difference [K]
Δv	volume change [m^3]
$\Delta\Phi$	percentage increase in density of heat-flow rate [%]
ε	surface emissivity [-]
λ	thermal conductivity [W/(m·K)]
μ	Poisson's ratio [-]
π	number "pi"
σ	Boltzmann constant [W/(m^2·K^4)]
Φ	density of heat-flow rate [W/m^2]
Subscripts and markings	
0	initial gas parameters
1, 2	specific gas-filled gap
1-2	glass pane (between gaps)
c	center (of glass pane)
a	atmospheric
e	external
ex	exterior glass pane
g	regarding gas or regarding IGU (at U_g, Φ_g and $\Delta\Phi_g$)
i	internal
in	interior glass pane
m	mean, average
op	operating gas parameters
r	radiative
s	regarding gas gap
sur1, sur2	regarding surfaces
v	regarding volume
w	regarding deflection
z	outside

References

1. Huang, S.; Wang, Z.; Xu, J.; Lu, D.; Yuan, T. Determination of optical constants of functional layer of online Low-E glass based on the Drude theory. *Thin Solid Films* **2008**, *516*, 3179–3183. [CrossRef]

2. Solovyev, A.A.; Rabotkin, S.V.; Kovsharov, N.F. Polymer films with multilayer low-E coatings. *Mater. Sci. Semicond. Process.* **2015**, *38*, 373–380. [CrossRef]
3. Bernier, M.; Bourret, B. Effects of Glass Plate Curvature on the U-Factor of Sealed Insulated Glazing Units. *ASHRAE Trans.* **1997**, *103*, 4038.
4. Hart, R.; Goudey, H.; Arasteh, D.; Curcija, D.C. Thermal performance impacts of center-of-glass deflections in installed insulating glazing units. *Energy Build.* **2012**, *54*, 453–460. [CrossRef]
5. Penkova, N.; Krumov, K.; Zashkova, L.; Kassabov, I. Heat transfer and climatic loads at insulating glass units in window systems. *Int. J. Adv. Sci. Eng. Technol.* **2017**, *5*, 22–28.
6. Derwiński, W. Okna w kamerze termowizyjnej. *Świat Szkła* **2014**, *12*, 24–26.
7. Solvason, K.R. *Pressures and Stresses in Sealed Double Glazing Units*; Technical Paper No. 423; Division of Building Research, National Research Council Canada: Ottawa, ON, Canada, 1974.
8. Feldmeier, F. Klimabelastung und Lastverteilung bei Mehrscheiben-Isolierglas. *Stahlbau* **2006**, *75*, 467–478. [CrossRef]
9. Curcija, C.; Vidanovic, S. *Predicting Thermal Transmittance of IGU Subject to Deflection*; Lawrence Berkeley National Laboratory, Environmental Energy Technologies Division: Berkeley, CA, USA, 2012.
10. Feldman, M.; Kaspar, R.; Abeln, B.; Gessler, A.; Langosch, K.; Beyer, J.; Schneider, J.; Schula, S.; Siebert, G.; Haese, A.; et al. *Guidance for European Structural Design of Glass Components*; Support to the Implementation, Harmonization and Further Development of the Eurocodes; Luxembourg Publications Office of the European Union: Luxembourg, 2014.
11. Stratiy, P. Numerical-and-Analytical Method of Estimation Insulated Glass Unit Deformations Caused by Climate Loads. In *International Scientific Conference Energy Management of Municipal Transportation Facilities and Transport EMMFT 2017*; Murgul, V., Popovic, Z., Eds.; Advances in Intelligent Systems and Computing 2017; Springer International Publishing: New York, NY, USA, 2017; Volume 692, pp. 970–979.
12. Velchev, D.; Ivanov, I.V. A finite element for insulating glass units. In *Challenging Glass 4 & COST Action TU0905 Final Conference 2014*; Louter, C., Bos, F., Belis, J., Lebet, J.P., Eds.; Taylor & Francis Group: London, UK, 2014; pp. 311–318.
13. Respondek, Z. Rozkład obciążeń środowiskowych w wielokomorowej szybie zespolonej. *Constr. Optim. Energy Potential* **2017**, *1*, 105–110. [CrossRef]
14. Klindt, L.B.; Klein, W. *Glas als Baustoff: Eigenschaften, Anwendung, Bemessung*; Verlagsgesellschaft R. Müller: Köln-Braunsfeld, Germany, 1997.
15. Timishenko, S.; Woinowsky-Krieger, S. *Theory of Plates and Shells*; McGraw-Hill Book Company: New York, NY, USA; Toronto, ON, Canada; London, UK, 1959.
16. EN 673:2011. *Glass in Building—Determination of Thermal Transmittance (U Value)—Calculation Method*; European Committee for Standardization (CEN): Brussels, Belgium, 2011.
17. Aydın, O. Conjugate heat transfer analysis of double pane windows. *Build. Environ.* **2006**, *41*, 109–116. [CrossRef]
18. Arıcı, M.; Kan, M. An investigation of flow and conjugate heat transfer in multiple pane windows with respect to gap width, emissivity and gas filling. *Renew. Energy* **2015**, *75*, 249–256. [CrossRef]
19. EN 572-1:2004. *Glass in Building—Basic Soda Lime Silicate Glass Products—Part 1: Definitions and General Physical and Mechanical Properties*; British Standards Institution: London, UK, 2004.
20. Dynamic Pressure. Available online: https://en.wikipedia.org/wiki/Dynamic_pressure (accessed on 7 September 2019).

Article

Bio-Waste Thermal Insulation Panel for Sustainable Building Construction in Steady and Unsteady-State Conditions

Miloš Pavelek * and Tereza Adamová

Department of Wood Processing and Biomaterials, Faculty of Forestry and Wood Sciences, Czech University of Life Sciences, Kamýcká 129, 165 00 Prague 6-Suchdol, Czech Republic; adamovat@fld.czu.cz

* Correspondence: pavelek@fld.czu.cz; Tel.: +420-224-383-480

Received: 3 June 2019; Accepted: 20 June 2019; Published: 22 June 2019

Abstract: Apart from being used as an oil stock for bio-fuels production, an annual crop plant Brassica napus, thought to be an agro-waste, and used either as an animal feed, soil fertilizer or biomass for combustion and thermal energy production. Alternatively, as a bio-based and locally bio-sourced cellulosic material, it could be used as a thermal insulation in sustainable building fabrication, likewise woodchips, a bio-waste from the wood industry. In this study, the above-mentioned bio-waste materials' thermal properties were identified using a sandwich panel from medium density fibreboard (MDF) and wood studs. Premanufactured panels have been filled in with randomly oriented short-cut rapeseed and with short-cut woodchips. A modified guarded hot box method was used to designate steady and un-steady state thermo-physical parameters of such insulation panels. The examined bio-waste materials absorbed thermal fluctuations of the exterior environment and kept the indoor building environment at constant temperature regardless of such fluctuations. The ability of bio-based sandwich panels to store heat energy was found to be similar to mineral wool. Additionally, VOC (volatile organic compound) emissions of tested materials were identified using gas chromatography-mass spectrometry (GC-MS) combined with headspace solid-phase microextraction (HS-SPME) to declare materials' harmlessness to indoor environmental quality and human wellbeing. In conclusion, bio-based short-cut materials proved to be a viable environmentally friendly and energy efficient alternative to conventionally used thermal insulations.

Keywords: energy sustainability; thermal insulation; rapeseed; woodchips; bio-waste; thermal transmittance; un/steady conditions; VOC emission

---

1. Introduction

In the last decade, an increasing all-society interest in green materials, technologies, and services can be witnessed—partly due to natural resources exploitation, partly due to the constantly growing human liability to natural wealth and expanding consciousness of climate change. A significant amount of research is being carried out to help replace raw chemicals coming from oil feedstock with renewable ones, following the aim to cut down the carbon footprint and ecological human imprint on the environment. These chemical agents and their consequently derived compounds are often sourced from biological commodities, coming from industrial plants production [1]. As a consequence of the emerging technologies and sustainability implementation in technology practice, there has been considerable interest in agro-waste sourced from the industrial crops plantation, especially from oily-seeds plants that are a raw material for bio-diesel synthesis [2]. So far, animal feeds, soil fertilizers, and pellets to be incinerated for heat energy production are the main fields of application of the crop-stalks agro-waste, even though botanical fibers are praised for their low density, specific mechanical properties, biodegradability, low carbon footprint, renewability and affordability [3]. Likewise, there have been considerable traditions over generations in using fibers from plant sources in various applications, especially in textiles and affordable housing construction, straw reinforced mud bricks [4], rammed earth [5] and reed roofs [6] being examples.

A considerable amount of energy launched by the civilized cultures serves, apart from being used as a source in transportation, as heating facilities in constructions (almost 1/3 to 1/2 of the contribution to CO_2 emissions) [7]. Not only within the operation, but during the procedure of manufacturing, while installing and throughout the final demolition, the energy embodied in the building materials represents the total energy consumption of a building [8]. Low density, high specific heat capacity and low thermal conductivity represented by a coefficient of < 0.1 W m^{-1} K^{-1} is demanded from a material to qualify as a building thermal insulation [9]. So far, wooden products [10], bamboo [11], straw bales [12], and other industrial crops like, for example: sunflower [13], corn cobs [14], hemp [15] and cotton stalks [16], or less common fibers like Ichu [17] have served as a heat building insulation in structures.

Straw bales are perhaps the most explored naturally available and sustainable thermal insulation and construction material. Their thermal conductivity was reviewed during experimental measurements as a function of density and straw-fiber orientation towards the heat flow. Direction specified, perpendicular thermal conductivity was 0.045–0.056 W m^{-1} K^{-1}; then, in a parallel direction, it was 0.056–0.08 W m^{-1} K^{-1}, showing increasing values depending on the density of a straw bale [18]. Furthermore, specific thermal conductivities of bio-based and bio-sourced materials: bagasse (0.046–0.055 W m^{-1} K^{-1}), kenaf (0.034–0.043 W m^{-1} K^{-1}), and pineapple-leaf fibers (0.035–0.042 W m^{-1} K^{-1}) are not far from those of conventional man-made materials like mineral wool (0.033–0.040 W m^{-1} K^{-1}) and expanded polystyrene (0.031–0.038 W m^{-1} K^{-1}) [7]. Thermal conductivities demonstrated by some of the unconventional materials were lower than 0.1 W m^{-1} K^{-1}; in case banana and polypropylene (PP) commingled yarn, the highest thermal conductivity results (0.157–0.182 W m^{-1} K^{-1}) were reported. Due to anatomical structure of the hollow plant fibers, their thermal conductivities and heat capacities are significantly influenced by the sample's exposure towards heat flow from the source [18].

Sustainable insulation materials are accessible in several forms—bales, composite boards [19], and sandwich panels [20]; some of them are binder-less [21], some of them are glued together with sustainable plasters or polymer adhesives [22], and some of them may be constructed as a load bearing structures. To determine thermal conductivity of homogeneous materials, several experimental techniques are used—for example: steady state hot-plate, transient hot-bridge, hot-disk, and photo-thermal methods [23]. A hot box apparatus usually determines the thermal performance of complex heterogeneous structural elements built-up of diverse materials. It consists of two enclosed chambers (hot and cold) kept at constant temperatures and separated by the specimen (e.g., wooden panel with agro-waste thermal insulation). Due to the temperature gradient, heat flux between the two chambers through the specimen is measured and thermal resistance of the structure is determined [24].

The most prominent oily-seeds plant in Europe nowadays is rapeseed (Brassica napus), an annual plant which grows up to one meter in height. The stem consists of fibers 0.7–2 mm long and with a density of 1550 kg m^{-3}, containing 40–50% of cellulose, 25–30% of hemicelluloses and 17–21% of lignin as main chemical constituents [25]. Annual worldwide harvest of rapeseeds is about 12.6 millions of tonnes [26]. Simultaneously, another potentially applicable, and nowadays also accessible, bio-based material is wood [27], especially woodchips [28]. The yearly worldwide production of woodchips is 66.9 millions of tonnes [29]. Combining the use of industrial agro-waste from biofuels production with the requirements on energy savings in building industry, a possibility arises for short-cut rapeseed stalks and short-cut woodchips to be dried and used as a thermal envelope insulation in sustainable buildings construction.

In order to proclaim the advantages of these unconventional bio-waste thermal insulations (such as low priced, the local economy benefits because of the use of local resources, low energy consumption in production and in-site fastening, do-it-yourself affordable housing projects and simple and ecological end of life disposal), the thermal properties of the overall insulation panel (short-cut rapeseed and short-cut woodchips) were measured with a hot box apparatus. Thermal properties like thermal transmittance of the panel were characterized and the response of the structure towards periodic heating/cooling cycles was determined. The panel's ability to periodically store and dissipate heat was also determined, in order to show the material efficiency in retaining indoor air at constant conditions (despite the fact that the outside temperatures vary).

Lastly, the study deals with the indoor air quality while using the short-cut rapeseed and short-cut woodchips as a thermal insulation. Volatile organic compound (VOC) emissions from the tested materials were monitored to further negotiate the potential negative effect of VOCs on human health [30]. The GC-MS analysis was carried out and the specific VOCs were listed. To further support the use of bio-based materials and to contribute to society wellbeing in a sustainable future, every work or research that demonstrates that the long-term performance of sustainable materials, especially in comparison with commonly used conventional materials, is valuable.

2. Materials and Methods

2.1. Structure of Raw Materials

Scanning optical microscopy using a binocular magnifier was performed to picture the microstructure of raw materials—rapeseed (Brassica napus) and woodchips from coniferous trees (softwood; bark was deselected) bought from a local supplier. Figure 1a shows the structure of a transversal and a longitudinal section and a surface of the rapeseed stem. A specific oval shape of raw rapeseed stem can be seen in the transversal section.

A1 detail is focused on a rapeseed pith, a2 detail shows the interface between the pith and the bark (stem outer cortex). A variable shape of woodchips is depicted in Figure 1b.

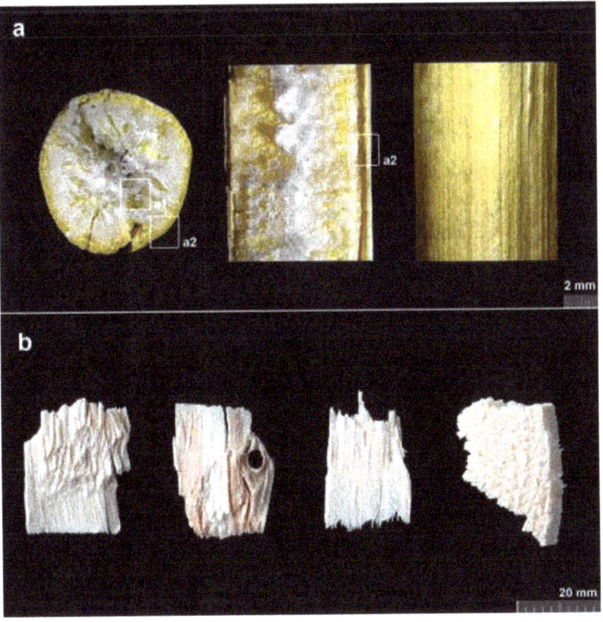

Figure 1. Images of raw materials' structure from binocular magnifier using scanning optical microscopy: (**a**) rapeseed (Brassica napus) stem; **a1**—rapeseed pith, **a2**—pith and bark interface; (**b**) variability of woodchips shapes.

2.2. Insulation Panel Structure—Core

For thermal properties' examinations, materials bought from a local supplier were shredded one-stage in a hammer mill to short-cut shape and placed into a premanufactured panel. Furthermore, sieve analysis was carried out and additional material characteristics, such as moisture content and VOC emission, were measured.

2.2.1. Moisture Content

For moisture content (u) determination, 10 samples of short-cut rapeseed and 10 samples of short-cut woodchips were placed in aluminous pan (each containing 100 g of tested material) and dried for 6 h at 105 °C ± 2 °C in a laboratory conditioning chamber. Afterwards, the absolute moisture content was determined as a percent weight difference between as received (m_w) and dry sample (m_d) and therefore calculated following the equation:

$$u(\%) = \frac{m_w - m_d}{m_d} \times 100 \quad (1)$$

The moisture content of short-cut rapeseed was 8.1 ± 1.6% and 6.9 ± 0.5% of short-cut woodchips.

2.2.2. Fraction Distribution—Sieve Analysis

Sieve analysis was performed to determine the tested short-cut rapeseed and short-cut woodchips fraction. Three randomly collected 100 g samples of each material were tested for fraction size. A screening machine with a laboratory metal sieve according to ISO 3310-1 [31] was employed.

Table 1 reports short-cut rapeseed fractions distribution obtained from sieve analysis. The fraction 0–8 mm was used as a core layer of tested insulating pane.

Table 1. Short-cut rapeseed fraction sizes from a sieve analysis.

Fraction (mm)	<0.25	0.25–0.5	0.5–0.8	0.8–1.6	1.6–2	2–3.15	3.15–8
Fraction representation (%)	1.2	2.8	4.8	39.4	20.1	23.1	8.6

The sieve analysis was carried out to observe the abundance of individual fraction width-sections. Although being processed (chopped/cut), short-cut rapeseed preserves the airy cellular morphology responsible for thermophysical properties (Figure 1a). Fractions <0.25 may be considered as a powder that has, contrary to short-cuts, rather isotropic thermophysical properties.

Table 2 demonstrates short-cut woodchips' fraction distribution obtained from sieve analysis. The most abundant fraction was 0.8–1.6 mm.

Table 2. Fractions of short-cut woodchips obtained from sieve analysis.

Fraction (mm)	<0.25	0.25–0.5	0.5–0.8	0.8–1.6	1.6–2	2–3.15	3.15–8
Fraction representation (%)	4.5	7.7	10.3	91.9	7.0	1.1	0

2.2.3. VOC Emission Samples

Dried-up short-cut rapeseed and short-cut woodchips samples as well as a mineral wool sample were placed into headspace vials (Figure 2). The vials were sealed with magnetic vial caps and were stored for 24 h airtight in a desiccator while preserving constant indoor air temperature (23 °C), corresponding with hot chamber steady state testing conditions.

Figure 2. Samples closed in vials ready for GC-MS analysis: (**a**) short-cut rapeseed; (**b**) short-cut woodchips; (**c**) mineral wool.

2.3. *Insulation Panel Structure—Shell*

An MDF (medium density fiberboard) envelope and a 150 mm thick bio-based insulation core sandwich panel of external dimensions 1700 × 1700 × 174 mm^3—length × height × thickness (Figure 3) was manufactured at the Faculty of Forestry and Wood Sciences, City, Prague to be subjected to thermal loading under steady-state and unsteady-state thermal conditions in a modified guarded hot box.

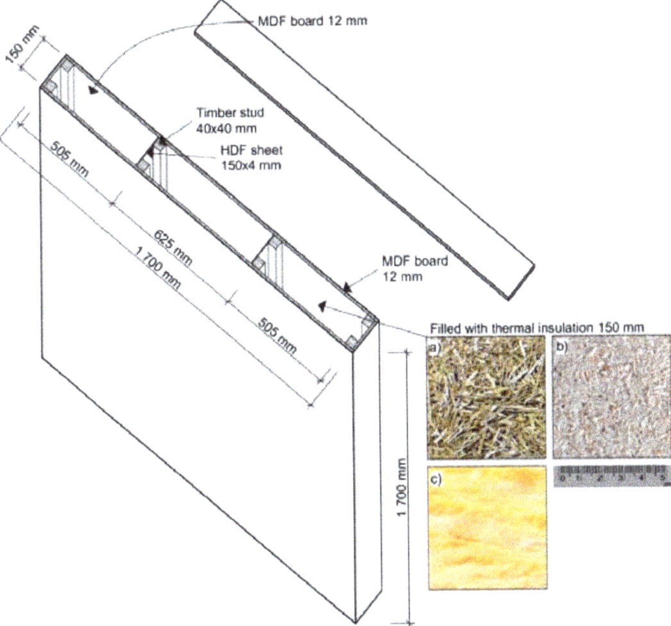

Figure 3. Structure of the insulating sandwich panel (external panel dimensions 1700 × 1700 × 174 mm) filled with: (**a**) short-cut rapeseed; (**b**) short-cut woodchips; (**c**) mineral wool.

The thermal insulation was loosely laid and randomly oriented—no binder was used. Bottom and upper MDF sealings were used to slightly press the insulation filler. The MDFs from Egger Ltd. (Hradec Králové, Czech Republic) were used because of their homogeneous cross-section over the entire thickness (12 mm), providing uniform thermal conductivity ($\lambda = 0.14$ W m^{-1} K^{-1}; $\varrho = 600$–650 kg m^{-3}). To ensure structural stability of a panel, the inner cavity was reinforced with joists combining pine studs ($40 \times 40 \times 1700$ mm^3) and HDF (high density fibreboard, $\lambda = 0.17$ W m^{-1} K^{-1}) of 4 mm thickness. The insulation core thickness was given by the joists' height (150 mm). All wooden elements, MDF and HDF boards were screwed together with 1.5×15 mm^2 and 3.0×30 mm^2 screws.

The bulk density of an insulation material was calculated as a ratio between the total weight of an insulating filler and the volume of a panel cavity. The bulk density of the short-cut rapeseed insulation panel was 110 kg m^{-3} and, in case of short-cut woodchips insulation panel, it was 205 kg m^{-3}. The panel of the same construction was filled with a conventionally used mineral wool thermal insulation.

2.4. Modified Guarded Hot Box Design

Thermal properties of tested samples in steady-state and unsteady-state conditions were determined in the modified guarded hot box. The guarded hot box method according to EN ISO 8990 [32] was slightly improved and adjusted in order to reduce heat loss through the hot box envelope. The hot box was composed of two chambers. A hot chamber was supplied with the heating system to maintain high temperatures; on the contrary, a cold chamber was supplied with the cooling system to maintain low temperatures. The tested sample was placed in between.

The difference between a guarded hot box, designed according to the standardized method, and the modified guarded hot box used in this experiment is illustrated in Figure 4. The box was located in a laboratory with controlled temperature and humidity (HVAC). The temperature on the hot side of the experimental box was kept constant at 24 °C, consistent with the ambient temperature.

This setting leads to the minimization of heat losses that potentially occurs through the hot chamber walls—heat flow φ3 (Figure 4b). This enables a use of the Hot Box constructed as a Calibrated Hot Box according to EN ISO 8990 [32]. No calibration is needed to minimize the system heat losses through the hot chamber walls. The second most significant advantage of this solution is a minimization of a three-dimensional heat transfer through the sample in a flanking area of the metering chamber—heat flow φ2 (Figure 4a) as standardized in the case of the Guarded Hot Box method.

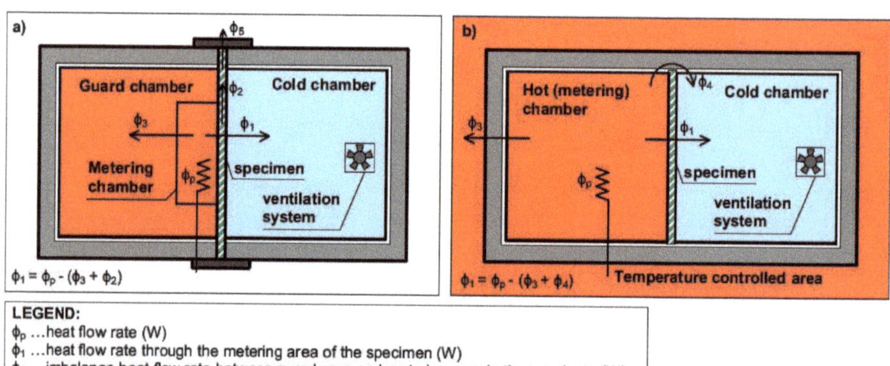

LEGEND:
ϕ_p ...heat flow rate (W)
ϕ_1 ...heat flow rate through the metering area of the specimen (W)
ϕ_2 ...imbalance heat flow rate between guard area and metering area in the specimen (W)
ϕ_3 ...heat flow rate through the metering box walls (W)
ϕ_4 ...flanking heat flow rate through the specimen frame (W)
ϕ_5 ...heat flow rate at the edge of the specimen (W)

Figure 4. Scheme of the difference between (**a**) a Guarded Hot Box (according to EN ISO 8990 [32]) and the Modified Guarded Hot Box used in this study (**b**).

The modified guarded hot box eliminates heat loss through the chamber wall; therefore, one can assume that the energy supply to the hot chamber flows only through the sample. The gap between the sample and the hot box was filled with an additional mineral wool insulation. The entire sample, as well as the hot box wall perimeter, was sealed with an airtight tape, causing the heat loss to be insignificant. The heat loss and airtightness were checked with a thermal camera after 12 h of sample conditioning.

For this study, a seven-day temperature cycle was selected with the temperature setting of Table 3. The preset temperature program was based on real climate conditions recorded at the local meteorological station, Prague–Suchdol, Czech Republic, in winter 2017.

Table 3. Description of temperature changes in the cold chamber of the Hot Box during the unsteady-state test.

Day	**0**	**1**	**2**	**3**	**4**	**5**	**6**	**7**	**0**	**1**
Time	16:00	16:00	9:00	9:00	9:00	9:00	16:00	9:00	16:00	Repeat
Temperature	+6 °C	−13 °C	+ 6 °C	−13 °C	+6 °C	−13 °C	−6 °C	−13 °C	+6 °C	Repeat
Hours	24	17	24	24	24	7	17	31	24	Repeat

Each sample was tested under identical conditions with room temperature (22–24 °C) in the hot chamber. The ambient temperature of 22–24 °C was maintained by a stable high performance air conditioning unit (deviation 0.2 °C). The temperature in the cold chamber varied between +6 °C and −13 °C. The current state of the art allows the design of the climate chambers to be precisely controlled and programmed with a temperature program.

The modified guarded hot box construction together with the specific sample position is depicted in Figure 5A. At the beginning of the experiment, a hot chamber was opened and the sample was placed in a hot box to separate the chambers. The temperature difference between the chambers determined the heat flow through the metering area of the sample—φ_1.

There was a heating system with a maximum power of 500 W in the hot chamber which was regulated by a PID (proportional–integral–derivative) panel controller with an additional thermocouple in the hot chamber. A Power Analyser Rohde & Schwarz HMC 8015 (Rohde & Schwarz, München, Germany) measured the heat flow, as an electric power flows straight to the heating system placed inside the hot chamber. The air temperature was measured using the Data Acquisition Base with humidity and temperature sensors. A set of temperature surface sensors was placed on each side of the measured sample according to EN ISO 8990. The temperature sensors' location is shown in Figure 5C. Air temperatures and humidity were measured on the cold and hot side 200 mm from the specimen surface (Figure 5B). All of the monitored parameters (air temperature, surface temperature and heat flow transmitted through the sample) were transferred to the computer.

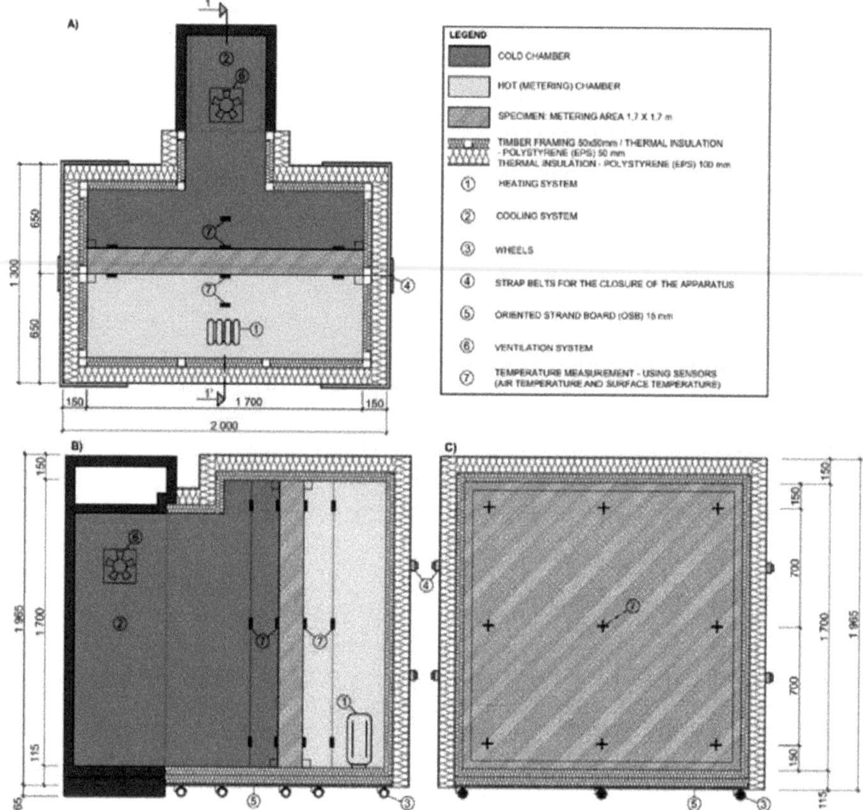

Figure 5. Modified guarded hot box design: (**A**) hot box horizontal section; (**B**) hot box vertical section 1-1′; (**C**) frontal view from the hot/cold side.

2.5. Calculation Procedure in the Steady-State Test

The total thermal transmittance Ut (W m^{-2} K^{-1}) of the experimental panel was calculated as the ratio of thermal energy φ_1 (W) transmitted through the sample area A (m^2) perpendicular to the heat flow. $T_{ai} - T_{ae}$ was the difference in air temperatures between the hot and cold sides of the sample (in K):

$$U_t = \frac{\phi_1}{A(T_{ai} - T_{ae})} \tag{2}$$

The total thermal resistance Rt (m^2 K W^{-1}) of the experimental panel was calculated as the inverted value of the total thermal transmittance. Referring to Figure 4, the heat flow through the sample (φ_1 in watts) was determined as input power (φ_p in watts). Heat flows φ_3 and φ_4 in watts have been neglected.

The input heat flow in the hot chamber was calculated from the electric heater power that was powered and controlled by the PID (proportional-integral-derivative) panel controller with an additional thermocouple in the hot chamber. The electrical output from the PID controller was measured using a Rohde & Schwarz HMC 8015 power analyzer. The metering area of the test panel was 1.7 × 1.7 m^2. The thermal conductivity calculation of the short-cut rapeseed/short-cut woodchips was based on the thermal resistance calculation R (m^2 K W^{-1}):

$$R_t = (\underbrace{R_{si} + \frac{d_{MDF}}{\lambda_{MDF}} + \frac{d_{\frac{rapeseed}{woodchips}}}{\lambda_{\frac{rapeseed}{woodchips}}} + \frac{d_{MDF}}{\lambda_{MDF}}}_{R} + R_{se}) \tag{3}$$

where d (m) represents the thickness of the material (rapeseed/woodchips/MDF), R_{si} is the internal surface thermal resistance (m^2 K W^{-1}), R_{se} is the external surface thermal resistance (m^2 K W^{-1}) and λ (W m^{-1} K^{-1}) is the thermal conductivity of the material (rapeseed/woodchips/MDF):

$$\lambda_{rapeseed/woodchips} = \frac{d_{rapeseed/woodchips}}{\underbrace{\left(\frac{(T_{si} - T_{se})A}{\phi_1}\right)}_{R} - \frac{2d_{MDF}}{\lambda_{MDF}}} \tag{4}$$

$T_{si} - T_{se}$ represents the difference in surface temperatures between the hot and cold side of the sample (K), A was the surface of the panel (m^2) and φ_1 was the heat flow in the sample (W). The thermal conductivity of the MDF boards λ_{MDF} (0.14 W m^{-1} K^{-1}) was given by the manufacturer.

2.6. Experiment Design—Unsteady-State Test

All tested panels were subjected to a one-week dynamic thermal loading. The temperature in the cold chamber (T_{ae}) fluctuated between +6 °C and −13 °C. The hot chamber temperature (T_{ai}) was continuously maintained at 24 °C and continuously measured (every min) as an indication of the structure's reaction to temperature changes. The total heat flow was depending on time, as the total energy transmitted through the panel $E_{searched}$ (Wh), was calculated using $U_{searched}$ for every min according to the following equation:

$$E_{searched} = U_{searched} \cdot A \cdot (T_{ai} - T_{ae}) \cdot t \tag{5}$$

where $U_{searched}$ is the thermal transmittance (W m^{-2} K^{-1}) given by the calculation using a solver function, A is the surface of the panel (m^2), $T_{ai} - T_{ae}$ is the difference between hot air temperature and cold air temperature of the panel (K), and t is the time (h):

$$Q = \sum_{j=1}^{n} (E_{searched,j} - E_{experimental,j})^2 \tag{6}$$

where $E_{experimental}$ is the total energy transmitted through the sample (Wh). To define the Q, a variable $U_{searched}$ to minimize the difference between $E_{experimental}$ and $E_{searched}$, the solver function was used. The thermal conductivity of the insulating core was calculated from the knowledge of the total $U_{searched}$ of the entire sandwich panel (respectively its inverse values of thermal resistance) and the thermal properties of the MDF envelope of the tested panel (Equation (4)). A similar calculation was used in Burrati et al. [33]. The alternative approach used previously by Pavelek et al. [24] and Trgala et al. [34] brought the opportunity to measure the total energy (Wh) transmitted through the sample and to find appropriate U value using dynamic conditions. Heat capacity, thermal response to real weather conditions and the influence of water and vapor content can be taken into account more suitably in the testing method compared to current steady-state conditions. More accurate calculations of the total annual heat loss due to transmission can be assured by the use of long-term real climate temperatures collected at 1 min resolution.

2.7. Extraction of Volatiles, GC–MS Analysis and Data Processing

Short-cut rapeseed, short-cut woodchips and mineral wool samples were analysed for their volatile content using gas chromatography coupled to mass spectrometry (GC-MS). To avoid instrumental sensitivity changes, samples were measured in one sequence. For volatile organic compound collection, solid-phase microextraction fiber with a divinylbenzen/carboxen/polydimethylsiloxan (DVB/CAR/PDMS 50/30 μm) coating from Supelco (Supelco Inc., Bellefonte, PA, USA) was employed. Vials were incubated for 10 min to increase volatiles emission from the sample and then volatiles were collected onto a fiber stationary phase for the next 10 min, both at 40 °C.

GC-MS was applied for VOC separation and identification. Basic measurements were performed using Quadrupole Shimadzu GC-MS QP2010 SE-Ultra (Kyoto, Japan), applying an SLB-5MS capillary column (30 m, 0.25 mm i.d., 0.25 μm film thickness) from Supelco. The injection was performed at 250 °C, while the transfer line was kept at 280 °C. The temperature program was as follows: 40 °C for 1 min and then with grad 5 °C min^{-1} to 250 °C and held for 2 min. Total run time was 45 min. Helium was used as a carrier gas at a flow rate of 1 ml min^{-1}.

In order to not focus only on a few compounds, the mass analyser was operated in a SCAN mode (scan speed 2000 ns, range 30–400 *m/z*. Identification of chemical compounds was based on mass spectral similarity with the in-built NIST MS library (NIST, Gaithersburg, MD, USA; 2017 released version). A group of approximately fifteen main volatile chemical compounds was identified through a literature review to be followed in all of the samples for their comparison. A group of key compounds for each material was defined. Reported intensities are areas of unique mass—the specific mass of compounds' mass spectrum, which were not coeluting with another compounds signal at a signal's retention time.

3. Results and Discussion

3.1. Panel U-Value Calculation from Steady-State Conditions

Experimental measurements were carried out after conditioning of each test sample in a closed hot box. Temperatures were measured at least four hours after steady state was reached, i.e., the temperature fluctuations in the range up to 1%. The temperature in the cold chamber—T_{ae} (°C)—was set to −13 °C and temperature in the hot chamber—T_{ai} (°C)—was continuously maintained at 24 °C. Detailed parameters from three experimental measurements, including the thermal transmittance and thermal resistance, are given in Table 4. The average thermal transmittance of the whole sandwich insulating panel filled with short-cut rapeseed was 0.308 ± 0.019 W m^{-2} K^{-1} and the average thermal resistance was 3.255 ± 0.217 m^2 K W^{-1}. The average thermal conductivity of short-cut rapeseed determined under steady-state conditions from all three measurements was 0.048 ± 0.003 W m^{-1} K^{-1}.

Table 4. Thermal transmittance and thermal resistance of sandwich panel filled with short-cut rapeseed insulation under steady-state conditions.

Test	Time (h)	φ_1 (W)	d (m)	A (m^2)	T_{ae} (°C)	T_{se} (°C)	T_{si} (°C)	T_{ai} (°C)	R (m^2 K W^{-1})	U (W m^{-2} K^{-1})
A1	4 h	27.81	0.174	2.89	−11.64	−8.21	21.03	22.64	3.562	0.280
A2	4 h	31.23	0.174	2.89	−10.84	−8.23	20.77	22.58	3.093	0.323
A3	4 h	30.46	0.174	2.89	−11.05	−8.26	20.07	21.74	3.111	0.321

All of the surface temperatures were constant during all three tests, with a maximum difference of 0.2 °C on the cold side and 0.1 °C difference on the hot side. The hot air temperature fluctuations were approximately ± 0.05 °C and ± 0.2 °C for cold air temperature (Figure 6). The laboratory ambient temperature was monitored continuously at 24 ± 0.5 °C.

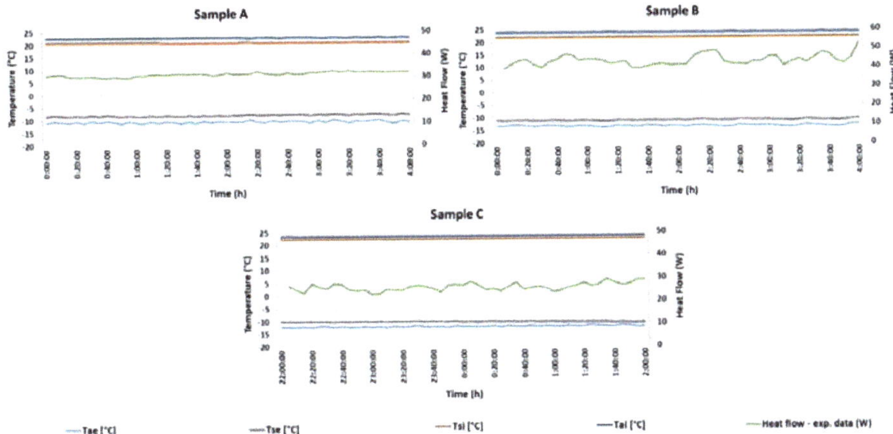

Figure 6. Heat flow and temperature measurements using the steady-state method: (**A**) short-cut rapeseed (test A2 in Table 4); (**B**) short-cut woodchips (test B2 in Table 5); (**C**) mineral wool (test C1 in Table 6); T_{ae} = air temperature in the cold chamber, T_{ai} = air temperature in the hot chamber, T_{se} = surface temperature of sample in the cold chamber, and T_{si} = surface temperature of sample in the hot chamber.

The average thermal transmittance of the whole sandwich insulating panel filled with short-cut woodchips was 0.403 ± 0.010 W m^{-2} K^{-1} and the average thermal resistance was 2.484 ± 0.060 m^2 K W^{-1}. The average thermal conductivity of short-cut woodchips determined under steady-state conditions from all three measurements was 0.065 ± 0.002 W m^{-1} K^{-1}. Detailed parameters from three experimental measurements are given in Table 5.

Table 5. Thermal transmittance and thermal resistance of sandwich panel filled with short-cut woodchips insulation under steady-state conditions.

Test	Time (h)	φ_1 (W)	d (m)	A (m^2)	T_{ae} (°C)	T_{se} (°C)	T_{si} (°C)	T_{ai} (°C)	R (m^2 K W^{-1})	U (W m^{-2} K^{-1})
B1	4 h	42.10	0.174	2.89	−13.41	−11.11	21.94	23.89	2.561	0.391
B2	4 h	43.44	0.174	2.89	−13.31	−11.11	21.92	23.89	2.475	0.404
B3	4 h	44.61	0.174	2.89	−13.39	−11.02	21.88	23.89	2.415	0.414

To compare the results of bio-waste insulation panels, a sandwich panel with mineral wool was measured. The results of the sandwich panel with mineral wool are summarized in Table 6. The average thermal transmittance of the whole sandwich panel was 0.255 ± 0.016 W m^{-2} K^{-1} and the average thermal resistance was determined as 3.930 ± 0.250 $m^2 \cdot K \cdot W^{-1}$. The average thermal conductivity of mineral wool was 0.040 ± 0.003 W m^{-1} K^{-1}

Table 6. Thermal transmittance and thermal resistance values of sandwich panel filled with mineral wool insulation under steady-state conditions.

Test	Time (h)	φ_1 (W)	d (m)	A (m^2)	T_{ae} (°C)	T_{se} (°C)	T_{si} (°C)	T_{ai} (°C)	R (m^2 K W^{-1})	U (W m^{-2} K^{-1})
C1	4 h	25.85	0.174	2.89	−12.10	−10.40	22.37	23.53	3.983	0.251
C2	4 h	24.40	0.174	2.89	−11.92	−10.99	22.38	23.60	4.207	0.238
C3	4 h	28.58	0.174	2.89	−12.12	−11.58	22.28	23.49	3.601	0.278

3.2. Panel U-Value Calculation from Unsteady-State Conditions

The unsteady-state test was performed by changing the air temperature T_{ae} (°C) in the cold chamber after a certain amount of time, shown in Table 3. The air temperature was controlled by an electronic controller with a preset temperature program. The temperature in the hot chamber T_{ai} (°C) was continuously set at 24 °C and the temperature in the cold chamber T_{ae} (°C) was periodically changed between −13 °C and +6 °C for seven days. The preset temperature program was based on average real climate conditions recorded at the local meteorological station, Prague–Suchdol, Czech Republic, in winter 2017. The test panel was always conditioned for 24 h under the pre-set conditions with the air temperature in the hot chamber of 24 °C and air temperature in the cold chamber of 6 °C.

The following data were collected during the experimental tests: the air temperatures T_{ae} (°C) and T_{ai} (°C), the surface temperatures T_{se} (°C) and T_{si} (°C), and total energy transmitted through the panel (known as an $E_{xperimental}$ (Wh)) in Figure 7 at time interval of 1 min. The air temperature in the hot chamber T_{ai} (°C) was continuously recorded every 1 min as an indication of the test panel to thermal impulse. Figure 7 shows the automatic defrosting cycle of the cooling system after approximately 12 h. Total energy (W) transmitted through the test panel was recorded experimentally as a function of time. Furthermore, it was also calculated from Equation (5) and is shown as an $E_{searched}$ in Figure 7. Figure 7 shows results from two independent measurements, comparing the short-cut rapeseed, the short-cut woodchips and the mineral-wool panel reaction to dynamic thermal conditions as described in the previous paragraph. $U_{searched}$ was found after the experimental measurement using a solver function in MS Excel (version 2016, Microsoft, Redmond, WA, USA).

The steady-state measurements gave more optimistic thermal properties for the panel filled with mineral wool.

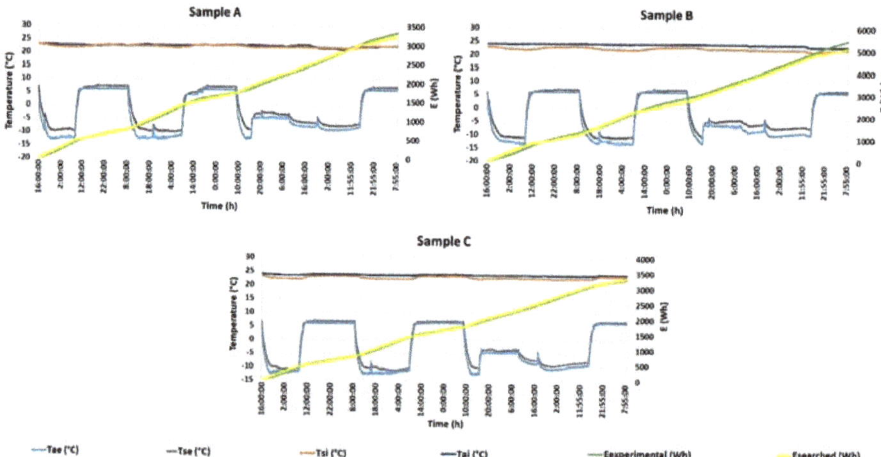

Figure 7. Total energy transmitted through the sample using searched U-value under unsteady-state conditions; (**A**) short-cut rapeseed; (**B**) short-cut woodchips; (**C**) mineral wool; T_{ae} = air temperature in the cold chamber, T_{ai} = air temperature in the hot chamber, T_{se} = surface temperature of the sample in the cold chamber, T_{si} = surface temperature of the sample in the hot chamber, Eexperimental = total energy transmitted through the sample (Wh), and Esearched = calculated total energy consumption using U-value given by Solver (Wh).

The total thermal transmittance of the entire sandwich panel filled with short-cut rapeseed insulation was 0.271 W m^{-2} K^{-1} and the total thermal resistance value was 3.690 m^2 K W^{-1}. The thermal conductivity of the short-cut rapeseed itself was 0.042 W m^{-1} K^{-1}. The total thermal transmittance of short-cut woodchips insulation panel was 0.404 W m^{-2} K^{-1} and the total thermal resistance value was 2.475 m^2 K W^{-1}. The thermal conductivity of the short-cut woodchips was 0.065 W m^{-1} K^{-1}.

To compare the results of the bio-waste insulation panels, a sandwich panel with mineral wool was measured. The total thermal transmittance of the entire sandwich panel filled with mineral wool insulation was 0.267 W m^{-2} K^{-1} and the total thermal resistance value was 3.745 m^2 K W^{-1}. The thermal conductivity of the mineral wool was 0.042 W m^{-1} K^{-1}.

Linear thermal transmittance ψ across the reinforcing stud (wood/HDF) was included in the resulting values of the thermal transmittance, thermal resistance, and thermal conductivity.

As can be seen, the U-values of the two tested insulation panels were very close from each other, i.e., panel with short-cut rapeseed filler and mineral wool filler. The short-cut woodchips showed inferior thermal properties compared to these panels. It could be caused due to the bulk density being double that of the short-cut rapeseed panel. It would be appropriate to test panels with identical material filler at various bulk densities.

Data from measurements conducted under unsteady-state conditions provided comparable values for two tested panels (rapeseed/mineral wool). Therefore, test conditions can strongly influence the performance of a panel in the experimental test, resulting in bio-waste thermal insulation being rejected in favor of conventional thermal insulation.

Thermal lag, the ability of the system to continuously store and dissipate heat energy when subjected to dynamic thermal loads, is a measure of the wall insulation panel efficiency to keep the indoor build environment at a constant temperature. Wooden sandwich panels with rapeseed short-cut shreddings core showed the same long-term thermal behaviour as wooden sandwich panels with mineral wool core, obviously at the benefit of reduced carbon footprint and lower environmental impact.

3.3. VOC Emissions

Within the frame of this work, to better understand the potential impact on indoor environment quality, short-cut rapeseed, short-cut woodchips and mineral wool insulation VOCs were analysed and compared. Fourteen key volatile organic compounds including aldehydes, alkanes, and terpenes were collected especially from bio-waste insulations using HS-SPME GC-MS. Target compounds for bio-based materials (Table 7) were selected based on a list published in ISO 16000-6 (Annex A) focused on building products' VOC emissions in indoor air [35] as the mineral wool proved to behave like an inert material with almost no VOC emissions. In general, short-cut woodchips' VOC detection proved to be the most abundant, resulting in a higher number of VOCs as well as higher detector responses observed in short-cut woodchips compared to short-cut rapeseed comparing the bio-based insulations—especially in the case of hexanal and alpha-pinene. An aldehyde pentanal together with even more important aldehyde–hexanal (described as "grassy" [36]) that is known as a product of unsaturated fatty acids oxidation [37] were emitted from a short-cut woodchips sample. Both compounds have been identified as causing unpleasant, irritating odors [38]. Compared to rapeseed and to mineral wool, a wide variety of terpenes (such as alpha-pinene, beta-pinene, camphene, 3-carene) were also observed emitting from woodchips' samples.

The chemical composition of raw material used for building insulation materials production represents only one of the factors affecting the quality of indoor air. In addition, the performance of building materials, and therefore VOCs' release, depends on prevailing thermal and moisture conditions, air pressure difference over the structure, and the quality of construction work [39] (these factors were not considered in presented study). However, it also depends on a structural design with a special focus on wall panel composition of all used materials. Therefore, it is necessary to identify the VOC emissions from insulation materials inside the sandwich panel, determine their environmental impact and select the appropriate enveloping material to prevent their emissions into the interior. It was also addressed in the studies of Little et al. [40], Yuan et al. [41], and Hodgson et al. [42].

Table 7. Detected VOCs from specific thermal insulation materials—intensities of monitored VOC emissions; intensities detector response for unique mass.

Insulation Sample	Short-Cut Rapeseed (Brassica Napus)	Short-Cut Woodchips	Mineral Wool
COMPOUND	–	–	–
Pentanal	–	30	–
Hexanal	63	180	–
Heptanal	–	11	–
Octanal	–	15	–
Nonanal	14	25	–
alpha-Pinene	9	155	–
beta-Pinene	–	69	–
3-Carene	–	9	–
Decane, 3,6-dimethyl-	–	14	–
Dodecane	12	–	–
Pentadecane	–	–	11
2-t-Butyl-4-methyl-5-oxo-[1,3]dioxolane-4-carboxylic acid	7	–	–
Hydrazine, (1,1-dimethylethyl)-	–	–	66
Nonane, 5-(2-methylpropyl)-	–	–	14
Note: selected mass areas were divided by 10^4		–	–

4. Conclusions

The present study demonstrates the applicability and viability of lignocellulosic bio-waste as a sustainable thermal insulation alternative to mineral-wool. After one-week of dynamic thermal loading, the mineral wool showed a 5% higher U-value compared to steady-state conditions, while short-cut rapeseed showed that a 12% lower U-value and short-cut woodchips reached 0.3% higher U-value. Compared to straw bales, the thermal conductivity and the heat capacity of the insulated wall remained homogeneous across the metering area, providing low thermal energy losses. Moreover, a modified guarded hot box method was successfully used for realistic simulations of thermal behavior of building envelopes and for the determination of insulation panels' thermal transmittance (U-values) under steady and unsteady state thermal conditions.

From the human wellbeing point of view, lignocellulosic bio-waste materials, especially rapeseed, emitted a similar amount of VOCs comparing to mineral wool. Therefore, it can be considered quite harmless for interiors' occupants.

Furthermore, the bio-based sandwich panel allows for complicated structural shapes (elements) engineering. It can be installed easily with loose-fill insulation, giving the architects more freedom in designing sustainable building envelopes, using only a 1.5% thicker layer of rapeseed than mineral wool, while reaching the same thermal resistance.

Author Contributions: The work presented in this paper is a collaborative development by all of the authors.

Funding: This work was supported by the Faculty of Forestry and Wood Sciences of the Czech University of Life Sciences Prague (Internal Grant Agency, Project No. A20/16: Thermal insulation properties of wood-based sandwich panel for use as structural insulated walls), and by grant "EVA 4.0", No. CZ.02.1.01/0.0/0.0/16_019/0000803 financed by the OP RDE (Operational Programme Research, Development and Education.

Conflicts of Interest: The authors declare no conflicts of interest.

Nomenclature

A	Area	m^2
c	Specific heat capacity	$J\,kg^{-1}\,K^{-1}$
d	Thickness	m
Δ	Difference	
E	Total energy transmitted through the sample	Wh
H	Heat loss	Wh
φ	Heat flow rate	W
λ	Thermal conductivity	$W\,m^{-1}\,K^{-1}$
m	Weight	kg
ϱ	Density	$kg\,m^{-3}$
ψ	Linear thermal transmittance	$W\,m^{-1}\,K^{-1}$
RH	Relative humidity	%
R	Thermal resistance	$m^2\,K\,W^{-1}$
T	Temperature	K, °C
t	Time	h
τ_S	Response time	h
U	Thermal transmittance	$W\,m^{-2}\,K^{-1}$

Subscripts

a	Air
e	External
i	Internal
I	Infiltration
s	Surface
t	Total
T	Transmission
V	Ventilation

References

1. Castro-Aguirre, E.; Iñiguez-Franco, F.; Samsudin, H.; Fang, X.; Auras, R. Poly (lactic acid)—Mass production, processing, industrial applications, and end of life. *Adv. Drug Deliv. Rev.* **2016**, *107*, 333–366. [CrossRef] [PubMed]
2. Azadi, P.; Malina, R.; Barrett, S.R.H.; Kraft, M. The evolution of the biofuel science. *Renew. Sustain. Energy Rev.* **2017**, *76*, 1479–1484. [CrossRef]
3. Wambua, P.; Ivens, J.; Verpoest, I. Natural fibres: Can they replace glass in fibre reinforced plastics? *Compos. Sci. Technol.* **2003**, *63*, 1259–1264. [CrossRef]
4. Oskouei, A.V.; Afzali, M.; Madadipour, M. Experimental investigation on mud bricks reinforced with natural additives under compressive and tensile tests. *Constr. Build. Mater.* **2017**, *142*, 137–147. [CrossRef]
5. Quoc-Bao, B.; Morel, J.-C.; Hans, S.; Walker, P. Effect of moisture content on the mechanical characteristics of rammed earth. *Constr. Build. Mater.* **2014**, *54*, 163–169.
6. Brischke, C.; Hanske, M. Durability of untreated and thermally modified reed (*Phragmites australis*) against brown, white and soft rot causing fungi. *Ind. Crop. Prod.* **2016**, *91*, 49–55. [CrossRef]
7. Asdrubali, F.; D'Alessandro, F.; Schiavoni, S. A review of unconventional sustainable building insulation materials. *Sustain. Mater. Technol.* **2015**, *4*, 1–17. [CrossRef]
8. Ramesh, T.; Prakash, R.; Shukla, K.K. Life cycle energy analysis of buildings: An overview. *Energy Build.* **2010**, *42*, 1592–1600. [CrossRef]
9. CSN 73 0540-1. *Thermal Protection of Buildings—Part 1: Terminology*; Czech Standards Institute: Prague, Czech Republic, 2005.
10. Vay, O.; De borst, K.; Hansmann, K.; Teischinger, A.; Müller, U. Thermal conductivity of wood at angles to the principal anatomical directions. *Wood Sci. Technol.* **2015**, *49*, 577–589. [CrossRef]

11. Huang, P.; Zeidler, A.; Chang, W.-S.; Ansell, M.P.; Chew, Y.M.J.; Shea, A. Specific heat capacity measurement of Phyllostachys edulis (*Moso bamboo*) by differential scanning calorimetry. *Constr. Build. Mater.* **2016**, *125*, 821–831. [CrossRef]
12. D'Alessandro, F.; Bianchi, F.; Baldinelli, G.; Rotili, A.; Schiavoni, S. Straw bale constructions: Laboratory, in field and numerical assessment of energy and environmental performance. *J. Build. Eng.* **2017**, *11*, 56–68. [CrossRef]
13. Binici, H.; Eken, M.; Dolaz, M.; Aksogan, O.; Kara, M. An environmentally friendly thermal insulation material from sunflower stalk, textile waste and stubble fibres. *Constr. Build. Mater.* **2014**, *51*, 24–33. [CrossRef]
14. Pinto, J.; Paiva, A.; Varum, H.; Costa, A.; Cruz, D.; Pereira, S.; Fernandes, L.; Tavares, P.; Agarwal, J. Corn's cob as a potential ecological thermal insulation material. *Energy Build.* **2011**, *43*, 1985–1990. [CrossRef]
15. Kymäläinen, H.-R.; Sjöberg, A.-M. Flax and hemp fibres as raw materials for thermal insulations. *Build. Environ.* **2008**, *43*, 1261–1269. [CrossRef]
16. Zhou, X.-Y.; Zheng, F.; Li, H.-G.; Lu, C.-L. An environment-friendly thermal insulation material from cotton stalk fibres. *Energy Build.* **2010**, *42*, 1070–1074. [CrossRef]
17. Charca, S.; Noel, J.; Andia, D.; Guzman, A.; Renteros, C.; Tumialan, J. Assessment of Ichu fibres as non-expensive thermal insulation system for the Andean regions. *Energy Build.* **2015**, *108*, 55–60. [CrossRef]
18. Costes, J.-P.; Evrard, A.; Biot, B.; Keutgen, K.; Daras, A.; Dubois, S.; Lebeau, F.; Courard, L. Thermal conductivity of straw bales: Full size measurements considering the direction of the heat flow. *Buildings* **2017**, *7*, 11. [CrossRef]
19. Chikhi, M.; Agoudjil, B.; Boudenn, A.; Gherabli, A. Experimental investigation of new biocomposite with low cost for thermal insulation. *Energy Build.* **2013**, *66*, 267–273. [CrossRef]
20. Alavez-Ramirez, A.; Chiñas-Castillo, F.; Morales-Dominguez, V.J.; Ortiz-Guzman, M. Thermal conductivity of coconut fibre filled ferrocement sandwich panels. *Constr. Build. Mater.* **2012**, *37*, 425–431. [CrossRef]
21. Panyakaew, S.; Fotios, S. New thermal insulation boards made from coconut husk and bagass. *Energy Build.* **2011**, *43*, 1732–1739. [CrossRef]
22. Wei, K.; Lv, C.; Chen, M.; Zhou, X.; Dai, Z.; Shen, D. Development and performance evaluation of a new thermal insulation material from rice straw using high frequency hot-pressing. *Energy Build.* **2015**, *87*, 116–122. [CrossRef]
23. Lagüela, P.; Bison, F.; Peron, P.; Romagnoni, P. Thermal conductivity measurements on wood materials with transient plane source technique. *Thermochim. Acta* **2015**, *600*, 45–51. [CrossRef]
24. Pavelek, M.; Prajer, M.; Trgala, K. Static and Dynamic Thermal Characterization of Timber Frame/Wheat (*Triticum aestivum*) Chaff Thermal Insulation Panel for Sustainable Building Construction. *Sustainability* **2018**, *10*, 2363. [CrossRef]
25. Tofanica, B.M.; Capelletto, E.; Gavrilescu, D.; Mueller, K. Properties of Rapeseed (*Brassica napus*) stalks fibers. *J. Nat. Fibres* **2011**, *8*, 241–262. [CrossRef]
26. Carré, P.; Pouzet, A. Rapeseed market, worldwide and in Europe. *Oilseeds Fats Crops Lipids* **2014**, *21*, D102. [CrossRef]
27. Modlinger, R.; Trgala, K. *Možné PŘÍČINY a důsledky Kůrovcové Kalamity v Lesích ČESKA s Ohledem na Specifika při Zpracování Kalamitního Dříví*; Odborná Studie; ČZU: Praha, Česká Republika, 2019; ISBN 978-80-213-2942-3.
28. Binici, H.; Aksogan, O. Eco-friendly insulation material production with waste olive seeds, ground PVC and wood chips. *J. Build. Eng.* **2016**, *5*, 260–266. [CrossRef]
29. FAO. Food and Agriculture Data. Available online: http://www.fao.org/faostat/en/#data/FO (accessed on 5 May 2019).
30. Xi, Z.; Zhiwei, L.; Wu, Y. Human physiological responses to wooden indoor environment. *Physiol. Behav.* **2017**, *174*, 27–34. [CrossRef]
31. *ISO 3310-1. Test. Sieves—Technical Requirements and Testing—Part 1: Test. Sieves of Metal. Wire Cloth*; International Organization for Standardization: Geneva, Switzerland, 2016.
32. *EN ISO 8990. Thermal Insulation—Determination of Steady State Thermal Transmission Properties—Calibrated and Guarded Hot Box*; European Committee for Standardization: Brussels, Belgium, 1996.
33. Buratti, C.; Belloni, E.; Lunghi, L.; Borri, A.; Castori, G.; Corradi, M. Mechanical characterization and thermal conductivity measurements using of a new'small hot-box'apparatus: Innovative insulating reinforced coatings analysis. *J. Build. Eng.* **2016**, *7*, 63–70. [CrossRef]

34. Trgala, K.; Pavelek, M.; Wimmer, R. Energy Performance of Five Different Building Envelope Structures Using a Modified Guarded Hot Box Apparatus—Comparative Analysis. *Energy Build.* **2019**, *195*, 116–125. [CrossRef]
35. *ISO 16000-6:2011. Indoor Air—Part 6: Determination of Volatile Organic Compounds in Indoor and Test Chambre Air by Active Sampling on Tenax TA®Sorbent, Thermal Desorption and Gas Chromatography Using MS or MS-FID*; Technical Committee ISO/TC 146; ISO: Geneva, Switzerland, 2011.
36. Risholm-Sundman, M.; Lundgren, M.; Vestin, E.; Herder, P. Emissions of acetic acid and other volatile organic compounds from different species of solid wood. *Holz Als Roh-Und Werkstolt* **1998**, *56*, 125–129. [CrossRef]
37. Stachowiak-Wencek, A.; Pradzyński, W. Emission of volatile organic compounds from wood of exotic species. *For. Wood Technol.* **2014**, *86*, 215–219.
38. NIOSH Pocket Guide to Chemical Hazards, Appendix: Aldehydes and NIOSH 2018: Aliphatic Aldehydes. Available online: http://www.cdc.gov/niosh/npg/npgd0652.html (accessed on 10 February 2019).
39. Summerscales, J.; Dissanayake, N.; Hall, W.; Virk, A.S. A review of bast fibres and their composites. Part 1: Fibres as reinforcements. *Compos. Part A: Appl. Sci. Manuf.* **2010**, *41*, 1329–1335. [CrossRef]
40. Little, J.C.; Kumar, D.; Cox, S.S.; Hodgson, A.T. Barrier materials to reduce contaminant emissions from structural insulated panels. In *Advances in Building Technology*; Elsevier: Amsterdam, The Netherlands, 2002; pp. 113–120.
41. Yuan, H. Modeling VOCs Emissions from Multi-Layered Structural Insulated Panels (SIPs). Ph.D. Thesis, Polytechnic Institute and State University Virginia, Blacksburg, VA, USA, 2005.
42. Hodgson, A.T. Volatile organic chemical emissions from structural insulated panel (SIP) materials and implications for indoor air quality. *Lawrence Berkeley Natl. Lab.* **2003**, *53768*, 1–30.

MDPI
St. Alban-Anlage 66
4052 Basel
Switzerland
Tel. +41 61 683 77 34
Fax +41 61 302 89 18
www.mdpi.com

Materials Editorial Office
E-mail: materials@mdpi.com
www.mdpi.com/journal/materials

www.ingramcontent.com/pod-product-compliance
Lightning Source LLC
LaVergne TN
LVHW071523170726
843515LV00008B/2051

* 9 7 8 3 0 3 6 5 6 4 0 7 4 *